Advances in sensor technology for sustainable crop production

It is widely recognised that agriculture is a significant contributor to global warming and climate change. Agriculture needs to reduce its environmental impact and adapt to current climate change whilst still feeding a growing population, i.e. become more 'climate-smart'. Burleigh Dodds Science Publishing is playing its part in achieving this by bringing together key research on making the production of the world's most important crops and livestock products more sustainable.

Based on extensive research, our publications specifically target the challenge of climate-smart agriculture. In this way we are using 'smart publishing' to help achieve climate-smart agriculture.

Burleigh Dodds Science Publishing is an independent and innovative publisher delivering high quality customer-focused agricultural science content in both print and online formats for the academic and research communities. Our aim is to build a foundation of knowledge on which researchers can build to meet the challenge of climate-smart agriculture.

For more information about Burleigh Dodds Science Publishing simply call us on +44 (0) 1223 839365, email info@bdspublishing.com or alternatively please visit our website at www.bdspublishing.com.

Related titles:

Managing soil health for sustainable agriculture Volume 2: Monitoring and management
Print (ISBN 978-1-78676-192-7); Online (ISBN 978-1-78676-195-8, 978-1-78676-194-1)

Precision agriculture for sustainability
Print (ISBN 978-1-78676-204-7); Online (ISBN 978-1-78676-206-1, 978-1-78676-207-8)

Robotics and automation for improving agriculture
Print (ISBN 978-1-78676-272-6); Online (ISBN 978-1-78676-274-0, 978-1-78676-275-7)

Advances in measuring soil health
Print (ISBN 978-1-78676-426-3); Online (ISBN 978-1-78676-326-6, 978-1-78676-327-3)

Chapters are available individually from our online bookshop:
https://shop.bdspublishing.com

climate•SMART•publishing
IN AGRICULTURAL SCIENCE

BURLEIGH DODDS SERIES IN AGRICULTURAL SCIENCE
NUMBER 122

Advances in sensor technology for sustainable crop production

Edited by Dr Craig Lobsey, University of Southern Queensland, Australia and Professor Asim Biswas, University of Guelph, Canada

burleigh dodds
SCIENCE PUBLISHING

Published by Burleigh Dodds Science Publishing Limited
82 High Street, Sawston, Cambridge CB22 3HJ, UK
www.bdspublishing.com

Burleigh Dodds Science Publishing, 1518 Walnut Street, Suite 900, Philadelphia, PA 19102-3406, USA

First published 2023 by Burleigh Dodds Science Publishing Limited

Library of Congress Control Number: 2022949278

British Library Cataloguing in Publication Data
A catalogue record for this book is available from the British Library

ISBN 978-1-78676-977-0 (print)
ISBN 978-1-78676-980-0 (PDF)
ISBN 978-1-78676-979-4 (ePub)
ISSN 2059-6936 (print)
ISSN 2059-6944 (online)

DOI 10.19103/AS.2022.0107

Typeset by Deanta Global Publishing Services, Dublin, Ireland

Contents

Part 3 Advances in sensor data analytics

Series list

Title	Series number
Achieving sustainable cultivation of maize - Vol 1 From improved varieties to local applications Edited by: Dr Dave Watson, CGIAR Maize Research Program Manager, CIMMYT, Mexico	001
Achieving sustainable cultivation of maize - Vol 2 Cultivation techniques, pest and disease control Edited by: Dr Dave Watson, CGIAR Maize Research Program Manager, CIMMYT, Mexico	002
Achieving sustainable cultivation of rice - Vol 1 Breeding for higher yield and quality Edited by: Prof. Takuji Sasaki, Tokyo University of Agriculture, Japan	003
Achieving sustainable cultivation of rice - Vol 2 Cultivation, pest and disease management Edited by: Prof. Takuji Sasaki, Tokyo University of Agriculture, Japan	004
Achieving sustainable cultivation of wheat - Vol 1 Breeding, quality traits, pests and diseases Edited by: Prof. Peter Langridge, The University of Adelaide, Australia	005
Achieving sustainable cultivation of wheat - Vol 2 Cultivation techniques Edited by: Prof. Peter Langridge, The University of Adelaide, Australia	006
Achieving sustainable cultivation of tomatoes Edited by: Dr Autar Mattoo, USDA-ARS, USA and Prof. Avtar Handa, Purdue University, USA	007
Achieving sustainable production of milk - Vol 1 Milk composition, genetics and breeding Edited by: Dr Nico van Belzen, International Dairy Federation (IDF), Belgium	008
Achieving sustainable production of milk - Vol 2 Safety, quality and sustainability Edited by: Dr Nico van Belzen, International Dairy Federation (IDF), Belgium	009
Achieving sustainable production of milk - Vol 3 Dairy herd management and welfare Edited by: Prof. John Webster, University of Bristol, UK	010
Ensuring safety and quality in the production of beef - Vol 1 Safety Edited by: Prof. Gary Acuff, Texas A&M University, USA and Prof. James Dickson, Iowa State University, USA	011
Ensuring safety and quality in the production of beef - Vol 2 Quality Edited by: Prof. Michael Dikeman, Kansas State University, USA	012
Achieving sustainable production of poultry meat - Vol 1 Safety, quality and sustainability Edited by: Prof. Steven C. Ricke, University of Arkansas, USA	013
Achieving sustainable production of poultry meat - Vol 2 Breeding and nutrition Edited by: Prof. Todd Applegate, University of Georgia, USA	014

Introduction

The global agri-food sector is under increasing pressure to produce more food from less inputs, while at the same time preserving important environmental resources. Greater emphasis is being placed on the sustainable management of food production systems, which are a function of numerous complex interactions between many different components, such as soil health and biodiversity.

Soil and crop information is critical to improving both our understanding, and the management of these systems. Better soil and crop information will enable new data-driven solutions that support more productive, resilient and sustainable agri-food systems. However, acquiring the necessary data is a significant challenge. This data must reflect the many key variables driving these systems, which are highly variable across different scales in space and time.

Sensors offer the opportunity to measure crop and soil health at unparalleled scales and resolution. The development of sensor technology will help improve our current understanding and optimisation of complex agri-food systems and support emerging data-driven solutions that improve the productivity and sustainability of crop production.

This volume provides a comprehensive review of key developments in sensor technology to improve monitoring and management of crop health, soil health, invasive plants and diseases. The volume is divided into three parts: In Part 1 each chapter will focus on advances in remote sensing technologies and techniques, that assess crop water status, crop health and soil health. Chapters in Part 2 highlight various proximal sensing technologies to assess soil health and water status. Part 3 draws attention to advances in sensor data analytics, such as machine vision technologies, the use of machine learning and proximal sensor fusion/multi-sensor platforms.

Part 1 Advances in remote sensing technologies

The first chapter of the book examines advances in remote and aerial sensing of crop water status. Chapter 1 begins by discussing the quantification of plant water status and the various methods used to assess plant water stress. It then moves on to examine the use of electromagnetic radiation and how it can interact with matter. A section on optical remote sensing of plant water status is also provided, followed by sections on thermal infrared remote sensing and microwave remote sensing of plant water status. The chapter also offers potential areas for future research development and other sources for further information.

The next chapter explores how current advances in remote sensing can contribute to improve the monitoring of plant health in response to stresses such as disease. Chapter 2 includes a case study that demonstrates how very high-resolution remote sensing data from drones, combined with deep learning models, can be used to evaluate stripe rust in the field. The chapter also discusses current limitations of remote sensing to provide accurate and specific data on plant health and disease and how these might be addressed.

The final chapter of Part 1 looks at advances in remote and aerial sensing techniques for monitoring soil health. Chapter 3 first discusses how active microwave remote sensing has been used to map soil moisture, providing theoretical, empirical and semi-empirical approaches that have been developed as a result of this form of sensor. A section on passive microwave sensing is also included, focusing on the impact of vertical soil moisture and temperature profiles, as well as the impact of surface roughness and the vegetation canopy layer. This section is then followed by an overview of how remote sensing can be used for the analysis of soil properties, such as soil moisture, roughness and salinity. A case study on the Murrumbidgee River Catchment in Australia is also provided to support the chapter's main discussion.

Part 2 Advances in proximal sensing technologies

Part 2 opens with a chapter that highlights advances in using proximal spectroscopic sensors to assess soil health. Chapter 4 introduces the concept of soil health and the need for sensor-based soil health measurements. It then reviews methods of soil spectroscopy, including instrumentation and modeling methods for both laboratory and in-situ sensing. The ability of spectroscopy to estimate key soil health properties is covered along with the potential advantages of merging other sensor data with spectral data. The chapter concludes with a case-study example, an exploration of future trends, and suggested sources of additional information.

Chapter 5 looks at advances in using proximal ground penetrating radar (GPR) sensors to assess soil health. The chapter summarises the GPR background needed to apply this technique to agriculture, including a review of basic principles, data acquisition, and data processing methods. Recent advances in each of these areas are described. Applications to soil mapping, soil water content characterization, compaction, and root mass detection are discussed. A case study using GPR groundwaves to map the soil water content at two depths is presented. The chapter concludes with a summary of current capabilities and suggestions for future work.

The subject of Chapter 6 is using proximal electromagnetic/electrical resistivity (ER)/electrical sensors to assess soil health. After a short summary of the soil health concern, the chapter recalls the definitions of the three relevant properties (conductivity, permittivity, magnetic susceptibility), and details the different electrical and electromagnetic techniques used in the soil domain. Two case studies in temperate and arid climates illustrate what can be obtained when using these techniques. A short discussion underlines the perspectives offered by a holistic approach to evaluate soil health characteristics from geophysical measurements.

Moving on from Chapter 6, Chapter 7 looks at using GPR to map agricultural subsurface drainage systems for economic and environmental benefit. The chapter first describes the evaluation of GPR against other proximal soil sensing methods. It then considers the factors potentially impacting GPR drainage pipe detection, goes on to examine GPR assessment of agricultural drainage pipe conditions and associated functionality implications, the effects of GPR antenna orientation relative to drain line directional trends and the integration of GPR with Real-Time Kinematic (RTK) Global Navigation Satellite System (GNSS) technology. A section on drainage mapping with a multichannel, stepped-frequency, continuous wave 3D-GPR system is also provided which is then followed by a review of complementary employment of GPR and unmanned aerial vehicle (UAV) imagery for drainage system characterization. The chapter concludes with a summary and recommendations for future research.

Part 3 Advances in sensor data analysis

The first chapter of Part 3 examines advances in machine vision technologies for the measurement of soil texture, structure and topography. Chapter 8 begins by providing an overview of the basic principles of machine vision technologies, focusing on areas such as 3D surface modelling and various methods of soil thin section microscopy. Two case studies are also provided in the chapter to support the main text discussion.

Chapter 9 draws attention to using machine learning to identify and diagnose crop disease. The chapter introduces how deep learning for image analysis and classification works and explain the requirements for collecting a dataset of plant disease images for use with deep learning networks. The chapter then discusses the results and successes of various previous studies and highlight pitfalls with individual methods. It is clear that deep learning is capable of handling complex disease classification problems where one disease is present. There is plenty of room for growth to work with the presence of multiple diseases in a single image or to quantify the amount of disease present.

The subject of Chapter 10 is advances in proximal sensor fusion and multi-sensor platforms for improved crop management. The chapter begins by first describing how a plant's height can determine how healthy it is and the importance of using proximal and remote sensing to assess this. The chapter moves on to examine remote sensing and weather data analysis and then goes on to describe multi-sensor approaches. A section on the statistical tools that can be used for fusing multi-sensor data is also included focusing on emerging machine learning tools.

The final chapter of the book reviews key issues in using sensor data in precision agriculture and, in particular, their mode of deployment (proximal or remote). Chapter 11 assesses relative strengths and weaknesses of proximal sensing techniques, compared with imaging data typically acquired from remote sensing platforms, before assessing trade-offs in sensor data resolution, as well as sources of error in the way data is processed. The chapter concludes by looking at ways of integrating remote and proximal sensor data, to utilise the beneficial characteristics of each type of data to improve the impact precision agriculture in improving efficiency and sustainability.

Part 1

Advances in remote sensing technologies

Chapter 1

Advances in remote/aerial sensing of crop water status

Wenxuan Guo, Texas Tech University and Texas A&M AgriLife Research, USA; and Haibin Gu, Bishnu Ghimire and Oluwatola Adedeji, Texas Tech University, USA

1 Introduction

Water is the most critical input for crop production, and agriculture is the top water user, especially in arid and semiarid regions. Irrigated agriculture accounts for approximately 70% of worldwide freshwater withdrawals, making agricultural water use one of the leading drivers of global water shortages (Tshwene and Oladele, 2016). As the world population increases, agriculture is under pressure to produce more in limited arable land while consuming less water per unit of output (Zwart and Bastiaanssen, 2004; Tshwene and Oladele, 2016; Guo et al., 2015). The extensive ranges in water use efficiency (WUE) indicate agricultural production can be sustained with 20–40% less water use if improved water management strategies are implemented (Zwart and Bastiaanssen, 2004). Various technologies are applied to improve WUE and water conservation. For example, precision irrigation technologies incorporate spatial and temporal plant water needs into irrigation scheduling for optimizing water management. Site-specific and real-time crop water status is critical for decision support in irrigation scheduling and precision water management.

http://dx.doi.org/10.19103/AS.2022.0107.01

Water typically accounts for more than 70% of the weight of non-woody plants (Hopkins and Huner, 2009). Most water in leaves resides in mesophyll cells (Fig. 1). Sufficient water supply is critical for the plant to maintain physiological processes and healthy growth and development (Gardner, 1984), such as transpiration, photosynthesis, phloem transport, respiration, and other metabolic activities. Water changes in the leaves affect internal conditions of the plant, such as tension in the cell walls, exchange of water and CO_2 across cell membranes, cell-to-cell contact and transport of water, and cell and tissue turgor (Govender et al., 2009).

The water status of a given tissue in a plant is determined by three factors, including soil water potential, transport resistance, and transpiration rate (Buckley, 2019). When water supply is insufficient, plants will suffer water stress with decreased leaf water potential and cell turgor that inhibits normal plant functions (Hsiao, 1973; Kim et al., 2018). Turgor pressure, also referred to as hydrostatic pressure, is the pressure exerted by the cell fluid against the cell wall. The stomata will close due to lack of water to maintain pressure in the guard cells. Stomatal closure prevents water loss, the movement of CO_2 into the plant, and photosynthetic rates of the leaves (Loka et al., 2011). Significant correlations between leaf water potential and stomatal conductance under water deficit stress have been reported (Loka et al., 2011). At the plant or canopy scale, plant water stress can result in decreased physiological activities and inhibition of growth, development, and survival (Govender et al., 2009). The effects of water stress on different physiological processes are complicated and interrelated (Loka et al., 2011). They are dependent on the duration and severity of water stress, plant genetics, and growth and developmental stages. Cell elongation is the most affected by water stress during early plant growth and development (Hsiao et al., 1976). Moderate water stress with relative water content (RWC) greater than 70% may cause structural variations, such as tissue thickness reductions, which in turn can affect foliar optical properties and

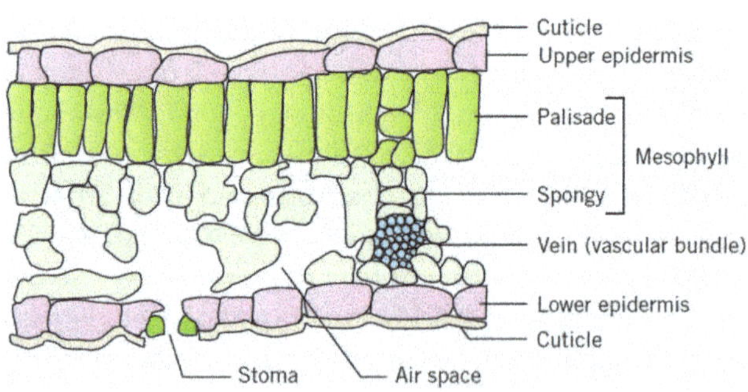

Figure 1 Diagrammatic cross-section representation of a typical mesomorphic leaf with extensive intercellular spaces with access to the ambient air through the open stomata. Source: Hopkins and Huner (2009).

spectral signatures. Severe water stress results in irreversible plant damage, accentuated drop in pigment content, reduced leaf area, dwarfed plant stature, or plant death (Baranoski and Van Leeuwen, 2017). In addition, plant water stress causes reduced transpiration rates and increased leaf temperature as less latent heat associated with transpiration is lost from the leaf. Water stress ultimately affects crop yield. The effects of water stress on yield depend on stress severity in relation to crop type, genotype, growth stage, management, and environmental conditions. Therefore, effective and efficient determination of plant water status can help minimize the impact of water stress on crop growth, maximize WUE, enhance water conservation, and improve crop production.

2 Quantification of plant water status

Accurate and timely information about plant water stress in relation to soil moisture and weather conditions at different crop growth stages is required for the evaluation of plant physiological conditions and decision support in precision water management, especially irrigation scheduling. Plant water status, soil moisture conditions, and crop evapotranspiration (ET_c) can be used as decision support in irrigation scheduling. The widely applied soil water balance approach calculates water inputs and losses to determine irrigation needs. A potential issue with this method is that many plant physiological features respond directly to changes in water status in the plant tissues rather than to changes in the bulk soil water content. In addition, the plant response to soil water content varies as a complex function of evaporative demand. Therefore, greater precision in the application of irrigation can be obtained through determining or sensing plant water stress conditions (Jones, 1990, 2004). Plant-based irrigation scheduling is complementary to soil- or ET_c-based methods as plant water status provides information on when to irrigate in response to plant water stress.

Assessment of plant water stress is usually based on the level of selected physiological parameters, such as water potential, RWC, stomatal reactions, photosynthesis rate, or osmotic adjustment (Bolat et al., 2014). The determination of leaf water content is widely used to estimate the water status of the plants. Plant water content is commonly described by the leaf RWC and equivalent water thickness (EWT). RWC is the ratio of the actual leaf water content to the maximum water content at full turgor pressure. It is calculated as:

$$\mathrm{RWC} = \frac{M_{\mathrm{fresh}} - M_{\mathrm{dry}}}{M_{\mathrm{turgor}} - M_{\mathrm{dry}}} \times 100\%,$$

where M_{fresh} is the fresh leaf mass immediately after sampling and M_{turgor} is the leaf mass at full turgor pressure obtained after saturating the leaves in water, and dry mass (M_{dry}) is obtained after oven drying samples at 85°C for 24 h (Yamasaki and Dillenburg, 1999; Turner, 1981). RWC is a measure of water deficit in the plant leaf. It is an important indicator of water status in the plant as it reflects the balance between water supply to the leaf tissue and transpiration rate (Lugojan and Ciulca, 2011; Schonfeld et al., 1988).

The leaf water content per unit leaf area or EWT is defined as the quantity of water per unit leaf area (Danson et al., 1992). It is calculated as:

$$\text{EWT} = \frac{M_{\text{fresh}} - M_{\text{dry}}}{A_{\text{leaf}}}\left(\text{g}/\text{cm}^2\right)$$

where A_{leaf} denotes the leaf area. Therefore, EWT represents a hypothetical thickness of a single layer of water averaged over the whole leaf area (Danson et al., 1992). EWT is an area-weighted indicator of leaf water content. It is related to a range of physiological and ecosystem processes, including leaf-level tolerance to dehydration (Wright et al., 2004). Therefore, EWT reflects crop water content relative to plant growth status (Yao et al., 2014).

Another plant water measurement is the fuel moisture content (FMC), expressed as the ratio between the quantity of water in vegetation and the dry weight of vegetation. It is calculated as:

$$\text{FMC} = \frac{\text{FM} - \text{DW}}{\text{DW}} \times 100\%$$

where FW is the fresh leaf weight of vegetation measured in the field and DW is the sample weight after it has been oven-dried. FMC is an optimum indicator of vegetation water status, especially for fire risk assessment in forestry or ecology (Maki et al., 2004). In recent years, it has been applied to monitor the water status of agricultural crops, such as corn, soybean, and wheat (Wu et al., 2009; Zhang et al., 2010; Shu et al., 2022).

The direct measurement of water content using RWC, EWT, FMC, or other methods is relatively challenging due to the complexity of determining the parameters in these calculations. Therefore, measures of water status based on the energy status of water in the plant have advantages over volumetric or absolute mass-based measures (Jones, 2007). A prevalent method of energy-based measurement is to determine the leaf water potential using a pressure chamber. A pressure chamber measures the leaf water potential by applying air pressure to a leaf and forcing the water out of it. In this process, the major part of a leaf is placed inside the chamber, with a small part of the leaf stem exposed to the outside of the chamber through a seal for observation. The pressure applied to the chamber that forces the water and sap out measures the water status of the leaf. A high-pressure value indicates a relatively low amount of

RWC and a high degree of water stress. The pressure that measures tension is commonly reported as negative values in bars.

The manual methods of determining crop water status, including the pressure approach, are time consuming and labor intensive. Besides, these methods only provide point measurements at sampling locations and thus are limited in accurately monitoring the plant water status on a field or regional basis. Furthermore, it is challenging to manually collect time-series data for quantifying changes in plant water status across different growth stages. Remote sensing offers various ways to estimate the crop water status or stress conditions at targeted spatial and temporal scales without the need for destructive sampling. In addition, remote sensing provides information for areas where ground measurements are inconvenient or impossible.

3 Electromagnetic radiation and interaction with matter

In agricultural remote sensing, information about a target is derived mainly based on the reflected or emitted electromagnetic (EM) radiation from the target surface. The amount and nature of the radiation detected by a sensor depend on the properties of the radiation, the properties of the object, the properties of the medium it travels through, and how the radiation interacts with the target object (Jones and Vaugham, 2010).

EM radiation is a form of energy propagated through space or other media in the form of electric and magnetic fields perpendicular to each other and the propagation direction. EM radiation carries energy, and it is the energy that determines how EM radiation interacts with matter. EM radiation exhibits the nature of particle-wave duality. Waves of EM radiation vary in energy levels according to their frequency. In the particle model, a wave contains discrete packets of energy, called quanta or photons. Each photon has energy following $E = hv = hc/\lambda$, where E is the energy of the photon, h is the Planck's constant (6.626×10^{-34} J·s), c is the light speed, v is the frequency, and λ is the wavelength of the wave. Thus, a photon with a higher frequency or lower wavelength has a higher energy level. The variation in energy of these EM waves plays a fundamental role in remote sensing as it affects the sensitivity of the sensors and the interaction of different wavelengths with the target and the atmosphere (Jones and Vaugham, 2010).

EM waves of all wavelengths form the EM spectrum, which is commonly classified into seven categories, in order of increasing wavelength, consisting of gamma rays, x-rays, ultraviolet (UV), visible, infrared (IR), microwave, and radio waves (Fig. 2). UV radiation has the shortest wavelengths for practical aerial or satellite remote sensing in agriculture. Shortwaves below UV and longwave radio waves are rarely applied in terrestrial remote sensing because

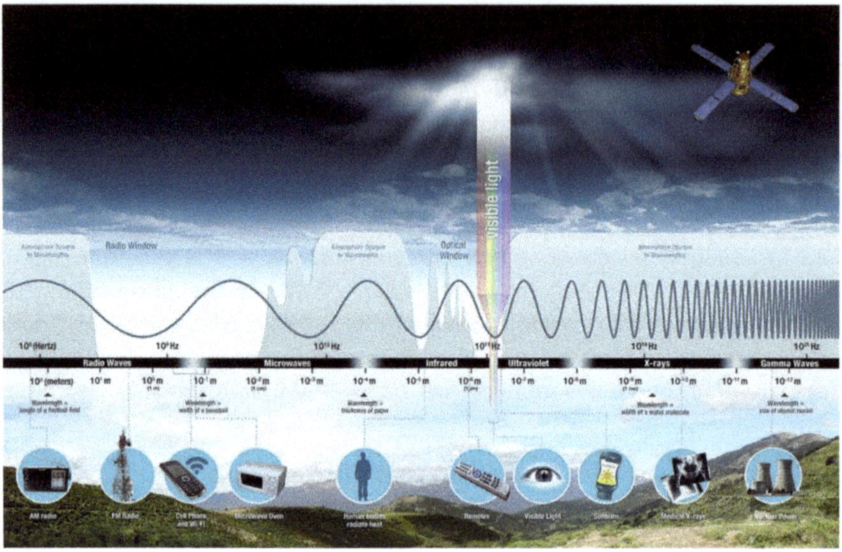

Figure 2 Electromagnetic radiation spectrum. Source: NASA: https://science.nasa.gov/ems/01_intro.

these waves are largely absorbed by the atmosphere (Merry, 2010). The IR region is often divided into the reflected IR and the emitted (or thermal) IR (TIR). The reflected IR covers wavelengths from approximately 0.7 μm to 3.0 μm. TIR energy is the radiation that is emitted from a surface in the form of heat. In terms of wavelength sensitivity to sensors, there are optical, thermal, and microwave remote sensing.

Understanding the interaction of EM radiation with water and plants, especially water in leaves and plant canopy, as affected by other factors, is a basis for remote sensing of plant water status. Water interacts with EM radiation mainly through absorption and scattering. Water absorbs over a wide range of EM radiation with electronic transitions, vibrational transitions, and rotational transitions (Bozhynov et al., 2020). The absorption of UV and visible radiation causes transitions between electronic energy levels, accompanied by simultaneous vibrational and rotational transitions; the absorption of IR generally causes transitions between vibrational levels and is accompanied by rotational transitions; the absorption of microwave radiation generally causes transitions between rotational energy levels (McQuarrie and Simon, 1997).

The magnitude of absorption depends on the wavelength of EM radiation and the state and temperature of the water. Liquid water has no rotational absorption features, so the main absorption in the visible and near-infrared (NIR) regions is much weaker than in the mid- and far-IR regions (Bozhynov et al., 2020; Fig. 3). The water molecule vibrates in several fundamental ways, each corresponding to a specific absorption wavelength, including symmetric stretch

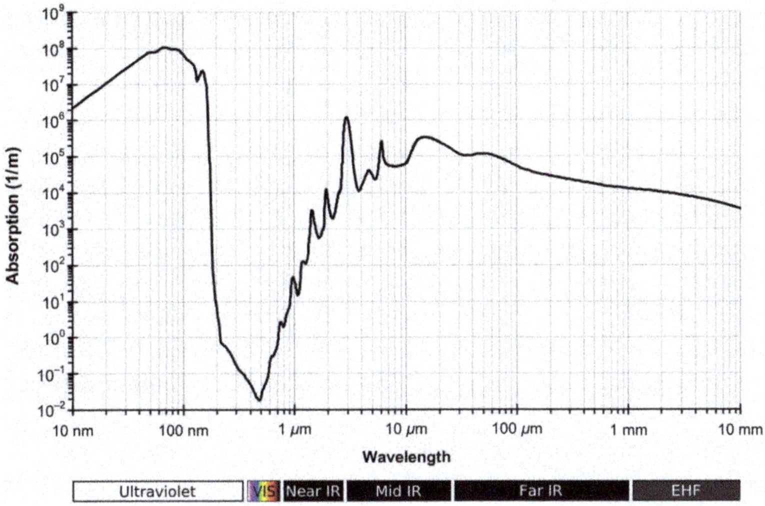

Figure 3 Liquid water absorption spectrum across a wide wavelength range. Source: Kebes (2008).

(v_1, 3050 nm), bending (v_2, 6080 nm), and asymmetric stretch (v_3, 2870 nm) of the OH bonds. Water absorption features due to O–H bonding in liquid water of plant canopy are at approximately 970 nm, 1200 nm, 1450 nm, and 1950 nm, with most pronounced features at 1450 nm and 1950 nm (Curran, 1989; Ustin and Jacquemoud, 2020). These four features are attributed to combinations at $2v_1+v_3$, $v_1+v_2+v_3$, v_1+v_3, and v_2+v_3, respectively (Jackquemond and Ustin, 2019; Bozhynov et al., 2020). Vibrational modes for liquid water are combinational at short wavelengths (400–1900 nm). In addition, vibrational modes are restricted in ice and liquid water by hydrogen bonds (Ustin and Jacquemoud, 2020).

When using remotely sensed observations, attention should be paid to water vapor in the atmosphere, which results in several absorption bands in the IR region, especially around 1450 nm and 1940 nm. These bands cause very noisy measurements and are typically not used for remote sensing. Imaging spectral data are often shown with these wavelengths removed. Areas of the EM spectrum where the atmosphere has little or no absorption due to water vapor and other gases are the atmospheric windows for remote sensing (Fig. 2). For example, spectral bands in the shortwave infrared (SWIR) region are suited for the remote sensing of canopy water content (Tucker, 1980; Ustin and Jackquemond, 2020).

4 Optical remote sensing of plant water status

The spectral range of 0.4–2.5 μm is the primary basis for optical remote sensing of vegetation that uses reflected EM radiation. The use of wavelengths beyond 2.5

μm is restricted due to the low energy output of the reflected radiation in these spectra (Tucker, 1980). In general, the spectral characteristics of vegetation in the optical wavelengths are governed primarily by scattering and absorption of the surface and internal leaf structure and biochemical constituents, including water (Govender et al., 2009). Plant leaves reflect (scatter), absorb, or transmit EM radiation of all wavelengths. Remote sensing of plant water status is mainly derived from reflectance characteristics of the plant surface. Specifically, the characteristic shape of the reflectance spectrum of green leaves is primarily determined by three factors: (1) absorption of EM radiation by leaf pigments in the visible spectrum; (2) scattering of EM radiation mainly by the internal physical structure related to air space of the leaf in the NIR; and (3) absorption of EM radiation by water, cellulose, and biochemicals in the leaf (Fig. 4). In general, the curves for green leaves of different plants have similar shapes but are different in magnitude. The internal absorption and scattering mechanisms account for the similarity in the shape of the reflectance spectra (Jacquemoud and Ustin, 2019).

The optical properties of actively growing healthy plants represent a contrast between high absorption and low reflectance in the visible spectrum and high reflectance in the NIR region. The surface condition, internal structure, biological, and optical properties of plant leaves largely determine plant canopy reflectance (Tucker, 1980; Gausman et al., 1969). Pigments are the main determinants controlling the spectral responses of leaves in the visible wavelengths (Gates et al., 1965; Govender et al., 2009). A substantial amount

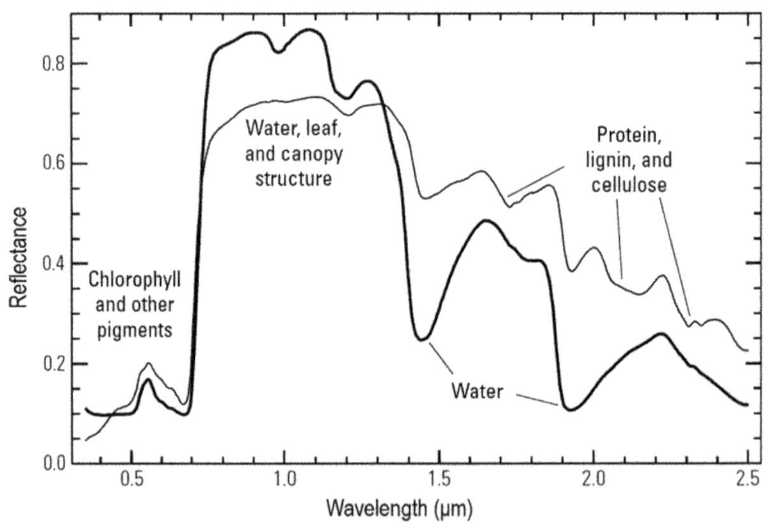

Figure 4 Laboratory reflectance spectra of an oak leaf in fresh (thick line) and dry (thin line) states. The causes of major plant absorption features are indicated. Source: Kokaly et al. (2007).

of EM radiation is absorbed by pigments, primarily chlorophylls a and b, in the visible spectrum of 400–700 nm. This high absorption, especially in the blue and red spectral regions, causes low reflectance and transmittance in the visible spectrum. This is evidenced by a high reflectance in the visible wavelengths as in NIR in leaf tissues with little pigments (Knipling, 1970; Woolley, 1971). The absorption rate is typically lower in the green wavelengths compared to the blue and red bands, leading to higher green reflectance and a green appearance for most higher plants.

A large fraction of the IR radiation is reflected or scattered at the leaf surface or the interfaces between cells and intercellular spaces. Plants reflect or transmit more than 90% of the incoming radiation in the NIR region (Jacquemoud and Ustin, 2008). Woolley (1971) conducted multiple experiments to evaluate the reflectance spectra of plant leaves with various water content levels for several plant species. This study revealed that, in the IR wavelengths, the shape of the reflectance curve for the corn leaf was similar to that of the mixture of small glass beads and water. Therefore, the leaf optical properties in the NIR and mid infrared are the combination of the absorption of EM radiation by water and the scattering of EM radiation by the leaf structures, especially cell walls. Reflectance over the 1.4–2.5 μm range is mainly due to scattering and absorption by leaf structure and water, as the absorption spectra of leaves are not significantly different from that of the liquid water (Allen et al., 1969, 1970; Knipling, 1970). The centers of major water absorption bands are 970 nm, 1200 nm, 1450 nm, 1780 nm, 1930 nm, and 2500 nm (Curran, 1989). Therefore, the spectral reflectance of NIR and SWIR bands can offer good indications of plant water status and overall crop health. For example, Fourty and Baret (1997) found that 1530 nm and 1720 nm are the most appropriate spectral wavelengths to assess plant water content. Reflectance typically increases with decreasing leaf or plant water content, with various responses in the visible and IR regions (Fig. 5).

The absorptions from different plant materials are similar and overlapping, so a single absorption band cannot be directly related to the amount of a plant constituent, including water (Kokaly and Clark, 1999). Various methods based on the absorption features have been developed to estimate plant canopy water content. Many remote sensing methods for assessing plant water status are based on the analysis of spectral data (as in Fig. 4). The derivative spectra method uses first or higher derivatives of absorbance with respect to wavelength for qualitative or quantitative analysis of plant constituents, including water content. The first-order derivative is the rate of change of absorbance with respect to wavelength, and it passes through zero at maximum absorption. This method also identifies inflection points, the wavelengths at local maxima of change in absorption in the spectrum. For example, it can determine the red edge position, the point of maximum slope on the reflectance spectrum of a plant leaf between red and NIR wavelengths.

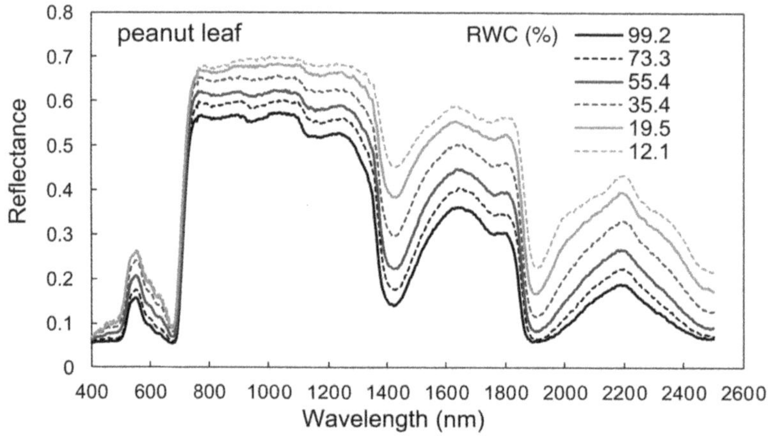

Figure 5 Relative water content and reflectance pattern for peanut. Source: Jacquemoud and Ustin (2019).

Clevers et al. (2008) applied the derivative spectra method to estimate canopy water content using absorption features at 970 nm and 1200 nm, with the best result in the NIR region at the left slope of the 970 nm absorption feature. A major advantage of derivative spectra is their insensitivity to variations in soil background, atmospheric effects, and the presence of yellow and dry vegetation.

Reflectance spectra of vegetation are affected by many factors, including atmospheric absorptions, soil exposed by incomplete vegetation coverage, or other absorbers. Band depth normalization from continuum-removed reflectance spectra is often utilized to minimize these influences (Clevers et al., 2008). The continuum is an estimate of the absorptions present in the spectrum, excluding the one of interest (Clark and Roush, 1984; Clevers et al., 2008). Linear segments are often used to approximate the continuum in practice (Clevers et al., 2008). The continuum-removed spectra are then calculated by the corresponding values of the continuum line. The band depth of each point in the absorption in the absorption feature is computed by $D = 1 - R'$, where R' is the continuum-removed reflectance value. The normalized depth D_n is calculated by dividing the band depth of each channel by the band depth at the band center D_c, as $D_n = D/D_c$ (Kokaly and Clark, 1999). This normalization allows the comparison of spectra acquired by different instruments or under different illumination conditions.

Vegetation indices (VIs) derived from a combination of specific reflectance of wavelengths ranging from visible, NIR, and SWIR are commonly used for estimating the plant water status (Table 1). A VI is a transformation of two or more bands of an image to maximize sensitivity to the vegetation characteristics of interest and minimize confounding factors such as atmospheric effects and

Table 1 Common vegetation indices for assessing plant water status or water stress using optical remote sensing

Index name	Abbreviation	Formula	Reference
Water index	WI	ρ900/ρ970	Penuelas et al. (1997)
	WI	P860/ρ1240	Zarco-Tejada and Ustin (2001)
Normalized water index	NWI	(ρ970 - ρ900) (ρ970 + ρ900)	Babar et al. (2006)
	NWI	(ρ970 - ρ850)/(ρ970 + ρ850)	Babar et al. (2006)
	NWI	(ρ970 - ρ920)/(ρ970 + ρ920)	Prasad et al. (2007)
	NWI	(ρ970 - ρ880)/(ρ970 + ρ880)	Prasad et al. (2007)
Normalized difference water index	$NDWI_{1240}$	(ρ860 - ρ1240)/(ρ860 + ρ1240)	Gao (1996)
	$NDWI_{1640}$	(ρ858 - ρ1640)/(ρ858 + ρ1640)	Chen et al. (2005)
	$NDWI_{2130}$	(ρ858 - ρ2130)/(ρ858 + ρ2130)	Chen et al. (2005)
Leaf water index	LWI	ρ1300/ρ1450	Seelig et al. (2008)
Moisture stress index	MSI	ρ1599/ρ819	Hunt and Rock (1989)
Shortwave infrared water stress index	SIWSI	(ρ858 – ρ1640)/(ρ858 + ρ1640)	Fensholt and Sandholt (2003)
Normalized difference infrared index	NDII	(ρ819 - ρ1649) / (ρ819 + ρ1649)	Hardisky et al. (1983)
Depth water index	DWI	2.044 × ρ1080 - 0.044 × ρ850 - ρ970 - 1200	Pasqualotto et al. (2018)
Water absorption area index	WAAI	$200 \times (1.857 \times \rho 858 + 0.097) - \int_{800}^{1200} \rho(\lambda) d\lambda$	Pasqualotto et al. (2018)

soil background. The VIs using red and NIR wavelengths, such as the normalized difference vegetation index (NDVI), are closely correlated to many physical properties of plant canopies, including leaf area index, ground cover, biomass, and plant health conditions. These indices can be used to infer plant water stress and the subsequent reduction of plant productivity. However, they are related to the total amount of water per unit of ground area but not to RWC or leaf water potential (Hunt and Rock, 1989). On the other hand, leaf characteristics, including the internal structure and dry matter, are also responsible for leaf reflectance variations in the SWIR. A combination of information from both the NIR (only influenced by the internal structure and the dry matter) and SWIR wavelength ranges is necessary to provide a better estimation of vegetation water content (Ceccato et al., 2001).

Various VIs have been developed to estimate the relative plant or leaf water content based on water absorption features. Penuelas et al. (1993) developed the water band index (WI) based on the ratio between the reflectance at the absorption feature of 970 nm and the one at 900 nm as a reference wavelength. This index was developed based on the measured water content of trees, shrubs, and a grass species. Caturegli et al. (2020) evaluated bermudagrass water status in a greenhouse and found that WI (900/970) and WI/NDVI, among the indices studied, were the more effective indicators of water stress. Specifically, lower values of WI indicated higher water stress, while higher values of WI/NDVI indicated higher water stress levels. Babar et al. (2006) proposed two normalized water indices (NWIs) for screening spring wheat genotypes under irrigated and water-stressed conditions. Prasad et al. (2007) proposed another two similar indices by incorporating absorption features at 880 nm and 920 nm for screening winter wheat genotypes under rainfed conditions. Zarco-Tejada and Ustin (2001) proposed the simple ratio water index (R860/R1240) through modeling to estimate the vegetation water content in relation to leaf thickness, leaf area index (LAI), and biomass.

The normalized difference water index (NDWI) is one of the most commonly adopted water indices for estimating plant water content. It is based on the absorption feature at 1200 nm and 860 nm as a reference wavelength (Gao, 1996). Jackson et al. (2004) compared the use of NDVI and NDWI from Landsat images in predicting corn and soybean water content and showed that the NDVI saturated at a lower water content than did the NDWI. This indicates that the NDWI is a better index for estimating crop water content over a full crop growing cycle. The moisture stress index (MSI, Hunt and Rock, 1989) is commonly used to estimate leaf water content. Hardisky et al. (1983) developed the normalized difference IR index (NDII) using ratios of different values of NIR and SWIR. NDII is sensitive to EWT, and it can be used to estimate the root zone water content and leaf water content (Yilmaz et al., 2008). Because leaf water amount and the suction pressure in the root zone are connected, the NDII can become an integrated, depth-independent estimation of how much water is available for vegetation in the subsurface for use by the plants. In addition to determining the water content of vegetation, the NDII can be effectively used to detect plant water stress according to the property of SWIR reflectance. However, these relationships appeared to be vegetation and crop type dependent (Sriwongsitanon et al., 2016).

Optical sensors, the most commonly adopted sensors in agricultural remote sensing, use visible, NIR and SWIR sensors to form images of the target surface by detecting the reflected radiation. These sensors can produce panchromatic, multispectral, or hyperspectral images. A panchromatic sensor is sensitive to a wide range of wavelengths and produces a single-band grayscale image. A multispectral sensor is often a multichannel radiometer that measures the intensity of EM radiation in multiple wavelengths, each capturing radiation within a wavelength range and forming an image with brightness and spectral information. One example of satellite optical sensors is the Landsat 8 Operational Land Imager that acquires multispectral images in nine spectral bands from visible to SWIR, including a panchromatic band (0.50–0.68 μm). Another example is the Sentinel-2 Multispectral Imager that delivers 13 spectral bands ranging from visible to SWIR. A hyperspectral sensor collects data, typically more than 100 spectral bands, each narrow spectral band forming an image.

One of the latest trends in remote sensors focuses on optimization and miniaturization, or the so-called SWaP - reducing size, weight, and power for unmanned aerial vehicle (UAV) remote sensing applications. Many small and lightweight optical sensors have been developed for UAV platforms in agricultural applications. Digital color cameras are the simplest UAV sensors that acquire a snapshot of a target surface in the visible wavelengths. A digital camera captures EM energy and converts it into electrical charges using a charge-coupled device or complementary metal-oxide-semiconductor technology and automatically combines the spectral bands of blue, green, and red into an image. Example color cameras include the DJI Zenmuse X7 (DJI, Shenzhen, China) and Autel EVO II Camera (Autel Robotics, Shenzhen, China). Multispectral sensors typically capture images in the visible to NIR regions (400–1000 nm), with more focus on monitoring plant health using NDVI, normalized difference red edge index, or other similar VIs. Examples of multispectral sensors include Micasense RedEdge (Micasense, Seattle, WA, USA), Parrot Sequoia (Parrot SA, Paris, France), DJI P4 (DJI, Shenzhen, China), and Sentera 6X Sensor (Sentera, St Paul, MN, USA), among others. Sensors covering SWIR wavelengths are rarely available for UAV-based applications in crop monitoring, especially water status assessment. Jenal et al. (2019) developed a compact and lightweight SWIR sensor sensitive to the 400–1700 nm wavelength range. Hyperspectral sensors are also available, such as the Bayspec OCI-F hyperspectral sensor (Bayspec, San Jose, CA, USA) covering 400–1000 nm with 60 bands, and the Headwall Hyperspec Co-Aligned visible and near-infrared (VNIR)/SWIR Sensor (Headwall Photonics, Bolton, MA, USA) that covers the VNIR (400–1000 nm) and SWIR (900–2500 nm).

Toth and Józków (2016) provided a survey on remote sensing platforms and sensors, including satellites, airborne platforms, and UAVs, and provided a short summary of imaging sensors. Qian (2021) provided a review of hyperspectral satellites and sensors, including their evolution and development history. Zhu et al. (2018) provided a review on remote sensing platforms and sensors with more focus on current sensors.

5 Remote sensing of plant water status using thermal infrared

The principle of thermal sensing of plant water status is mainly based on plant response to levels of water content in relation to the environment. Water is absorbed from the soil and transported through the plant, and, eventually, the majority of the absorbed water is lost via transpiration. The transpiration rate is controlled by stomatal pores on leaf surfaces that open or close in response to the turgidity of the surrounding guard cells (Gupta et al., 2020). During transpiration, energy is absorbed from the plant's environment to convert liquid water to vapor, facilitating a pronounced cooling effect at the leaf surface. As a result, a well-watered crop canopy is typically cooler than the ambient air in a dry and hot environment. If the plant water content decreases, leaf transpiration is reduced as a result of stomatal closure. The inhibited evaporative cooling effect causes higher canopy temperatures. Therefore, canopy temperature has been recognized as an indicator of plant water status, and quantifying leaf or canopy temperature has been established as an effective method to assess plant water status and detect water stress. The measurement of the temperature of an object is based on the Stefan-Boltzmann law that relates the amount of energy (E) emitted per unit time per unit area of a black body to its absolute temperature (T), as $E = \sigma T^4$, where σ is the Stefan-Boltzmann constant (5.67×10^{-8} W m^{-2} K^{-4}). For an object with an emissivity less than 1, Kirchhoff's law, $E = \varepsilon\sigma T^4$, describes the efficiency of emission in relation to temperature, where ε is the emissivity of an emitting body.

The use of canopy temperature alone cannot directly determine crop water status, because leaf temperature measured in field conditions is sensitive to many environmental factors, such as air temperature, wind speed, humidity, vapor pressure deficit (VPD), and incident radiation (Gerhards et al., 2019). The crop water stress index (CWSI, Idso et al., 1981; Jackson et al., 1981) provides a quantitative measure of plant water status based on the foliage-air temperature differential (T_c - T_a) as a function of air VPD. In other words, CWSI expresses water stress conditions using two baselines that relate canopy temperature under maximum stress and non-water stress conditions with VPD. Determination of CWSI requires the measurement of three environmental variables, canopy temperature (T_c), air temperature (T_a), and VPD. T_c can be obtained from a thermal image. The CWSI based on canopy temperature and meteorological parameters is calculated as:

$$\mathrm{CWSI} = \frac{(T_c - T_a) - (T_c - T_a)_{LL}}{(T_c - T_a)_{UL} - (T_c - T_a)_{LL}}$$

where (T_c - T_a) is the difference between canopy temperature and air temperature; $(T_c - T_a)_{LL}$ is the lower baseline of temperature difference between canopy and air for fully watered plants, where plants are transpiring at their potential rate with negligible canopy resistance; $(T_c - T_a)_{UL}$ is the upper baseline of temperature difference between canopy and air for water-stressed plants, with no transpiration flux. $(T_c - T_a)_{LL}$ and $(T_c - T_a)_{UL}$ are computed from the atmospheric VPD and vapor pressure gradient, respectively (Veysi et al., 2017; Virnodkar et al., 2020). Also, they must be calculated for the same environmental conditions at the time of measurement (Jackson et al., 1988). The crop water status or stress levels fall in between these two extremes, with larger values indicative of increasing water stress. It is challenging to determine the non-water stress baseline, which varies from crop to crop and local climatic factors (Berni et al., 2009; Clarke, 1997; Alchanatis et al., 2010). An alternative CWSI is defined by Jones (2014) as:

$$\text{CWSI} = \frac{\left(T_{\text{canopy}} - T_{\text{wet}}\right)}{\left(T_{\text{dry}} - T_{\text{wet}}\right)}$$

where T_{canopy} is the canopy temperature, which can be derived from the thermal image, T_{wet} gives the fully transpiring canopy temperature, and T_{dry} represents the water-stressed canopy temperature. Thus, a non-stressed crop has a CWSI value of 0, while a fully stressed non-transpiring crop has a CWSI value of 1.0.

Determining the non-stressed and water-stressed baselines is critical in the CWSI method. Artificial reference surfaces with known reflectances and aerodynamic attributes can be constructed and placed in the camera field of view when thermal images are acquired (Meron et al., 2003). Natural reference surfaces, such as well-watered crop sections, are also suggested for CWSI normalization, but they are difficult to maintain (Clawson et al., 1989). The artificial reference surface method is suitable for large-scale applications, as it provides reproducible surfaces and flexible deployment. It was found that the upper base level $(T_c - T_a)$ generally falls in the range of 4.6–5°C, when transpiration has ceased (Ehrler et al., 1978; Irmak et al., 2000). Therefore, some studies estimate T_{dry} by adding 5°C to the dry bulb temperature of air: $T_{dry} = T_a + 5°C$ (Cohen et al., 2005; Möller et al., 2007). The T_{wet}, as measured with the aid of the wet reference artificial surfaces, was found to provide the best representation of the lower baseline, while the use of air temperature plus 5.0°C as the upper baseline was better than the theoretical estimate (Alchanatis et al., 2010).

There are several limitations in applying the CWSI in assessing plant water status. Under changing atmospheric conditions, normalization of CWSI is more complicated than using VPD alone. The CWSI is limited to plant canopy only, and it is challenging to separate plant canopy temperature from that of the soil surface in images of low resolutions. Although high-resolution images can be applied to exclude soil pixels from the image scene, these procedures are time consuming and may be subjective. To address the limitations of CWSI under partial vegetation ground cover, Moran et al. (1994) developed the water deficit index (WDI) using a vegetation index/temperature (VIT) model, combining spectral VIs and the difference between surface temperature and air temperature (T_s - T_a). The concept is based on the trapezoidal shape formed by the relationship between the difference between the surface and air temperature (T_s - T_a) and a VI that represents the crop cover fraction (Fig. 6). The vertices of the trapezoid represent four extreme water stress conditions: (1) well-watered full-cover vegetation, (2) water-stressed full-cover vegetation, (3) saturated bare soil, and (4) dry bare soil. For each point inside the trapezoid, the WDI is calculated as the ratio of the distance to the left and right boundaries, which are considered to be wet and dry references, respectively. Thus, the VIT trapezoid becomes simply a relation between remotely sensed measurements of surface temperature and a VI derived from the surface reflectance factors in the red and NIR spectrum (Moran et al., 1994). For instance, WDI at Point X in Fig. 6 is equivalent to AX/AB, representing intermediate water stress under partial ground cover, as indicated by an intermediate VI value.

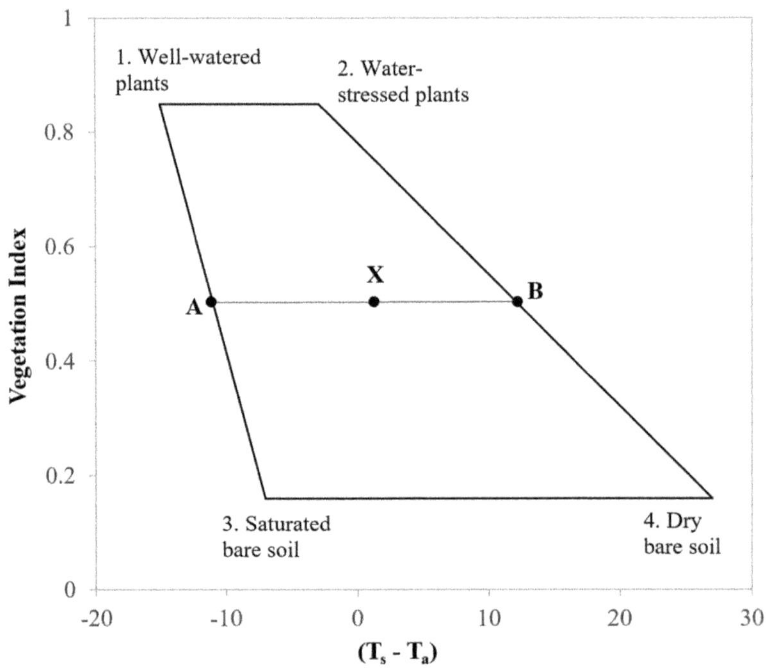

Figure 6 Water deficit index developed by Moran et al. (1994) based on the hypothetical vegetation index/temperature trapezoidal shape relating surface temperature (T_s) minus air temperature (T_a) and a vegetation index to represent ground cover.

Similar methods have been developed to determine the plant canopy water status using the combination of VIs and land surface temperature. A plot of T_s and a VI often results in a triangular shape or a trapezoid shape. The T_s/NDVI slope was found to be related to the evapotranspiration rate and the surface (Goetz, 1997). The temperature vegetation dryness index (TVDI) was developed on the basis of the triangular space (Sandholt et al., 2002). This index can be derived from satellite data only. Other indices originally designed for estimating soil moisture may have the potential to predict vegetation water content, such as the texture temperature vegetation dryness index (Gu et al., 2021), which incorporates soil texture into TVDI.

Temperature-based methods in determining crop water status typically use one broad TIR band under the assumption that emissivity is constant. However, the emissivity of vegetation is variable in nature. Thus, the accuracy of temperature retrieval is limited by this assumption (Huo et al., 2019). Hyperspectral TIR imaging devices can measure the emitted radiance of an object in narrow bands to perform stable temperature emissivity separation and temperature retrieval. For example, Huo et al. (2019) applied a hyperspectral TIR sensor to retrieve temperatures and emissivities of wheat plants and found that emissivity was sensitive to water variations and emissivity increased with water deficit levels. The TIR spectral features originate from primary absorption bands of biochemicals of leaf compounds and should exhibit higher spectral contrast. Therefore, changes in the compositions of leaf constituents induced by water stress should be accompanied by changes in the emissivity spectra (Gerhards et al., 2019; Ullah et al., 2013). Gerhards et al. (2019) reviewed the challenges and future perspectives of multi/hyperspectral TIR remote sensing for crop water-stressed detection.

Many studies have applied thermal remote sensing, especially CWSI in assessing crop water stress or irrigation scheduling. For example, Cohen et al. (2005) reported the CWSI and temperature measured from thermal images can be used to predict cotton leaf water potential and to evaluate cotton water status. Möller et al. (2007) used thermal and visible imagery and CWSI to estimate grapevine water status. Zhang et al. (2019) mapped corn water stress using UAV multispectral and thermal images. Gonzalez-Dugo et al. (2005) used airborne scanners to estimate canopy temperature and CWSI of cotton with fine spatial resolution for irrigation scheduling. Lacerda et al. (2022) applied UAV-based thermal imagery to detect cotton leaf water potential using the CWSI with good accuracy. Alchanatis et al. (2010) applied a high-resolution thermal imaging system to map cotton water status and found that midday is the optimal time for thermal image acquisition for estimating leaf water potential. Crusiol et al. (2020) presented a study to evaluate the water status of soybean plants under different water conditions via thermal images obtained using a UAV TIR camera.

Thermal sensors typically operate in two atmospheric windows, 3–5 μm range and 8–14 μm range; the former is not used in the daytime because of substantial reflected solar radiation in this range, while daytime data acquisition requires longwave imagers with the region having a negligible interference by sunlight. The 8–14 μm also includes the peak of blackbody emission at normal temperatures (Jones and Vaugham, 2010; Jacquemoud and Ustin, 2019). Commonly used thermal imaging sensors include IR imaging radiometers, imaging spectroradiometers, and IR imaging cameras. Example IR sensors in use include the Landsat Thermal Infrared Sensor instrument, the Moderate Resolution Imaging Spectroradiometer (MODIS), and the Advanced Spaceborne Thermal Emission and Reflection Radiometer (ASTER). There are two major thermal sensor technologies, cooled and uncooled sensors. A cooled thermal sensor has an integrated cooling system to lower the sensor temperature in order to achieve a higher signal-to-noise ratio, thereby allowing higher thermal sensitivity, higher spatial resolution, and higher frame rates. However, these sensors are larger in size and weight and consume a high amount of power. Uncooled sensors are smaller in both size and weight at the cost of inferior performance (Nguyen et al., 2021). Uncooled thermal sensors are normally mounted on UAVs because they are smaller and lighter and consume less energy.

6 Microwave remote sensing of plant water status

The microwave portion of the spectrum covers the range from approximately 1 mm to 1 m in wavelength. These wavelengths can be further divided into several bands, each with special characteristics and applications (Table 2). Compared with visible and IR, microwave radiation can penetrate through cloud, haze, and dust and is not easily affected by meteorological conditions and solar radiation levels (Steele-Dunne et al., 2017). This property allows microwave remote sensing under almost all weather and environmental conditions. Microwaves can also penetrate plant canopies and soils to some degree, providing some volumetric information that is not available in the visible and IR wavelengths (Jones and Vaugham, 2010).

Microwave sensing can leverage active or passive sensors. Passive sensors detect natural energy (radiation) that is emitted or reflected by the object or scene being observed, which is similar in concept to thermal remote sensing. Passive microwave sensors are typically radiometers or scanners with antennas for detecting and recording microwave energy. An active sensor emits radiation in the direction of the target and measures microwaves reflected or backscattered from the target. Active microwave sensors are generally divided into two categories: imaging and non-imaging sensors. The most common form of imaging active microwave sensors is RADAR (radio detection and ranging).

Table 2 IEEE microwave frequencies and application in agriculture

IEEE band	Frequency (GHz)	Wavelength (cm)	Application	Reference
P	0.3-1	30-100	Forest biomass estimation	Minh et al. (2014)
L	1-2	30-15	Canopy height and biomass estimation	Monsivais-Huertero et al. (2015);
S	2-4	15-7.5	Biomass estimation	Jia et al. (2013)
C	4-8	7.5-3.8	Detection and classifications of different crops; LAI estimation	Lee et al. (2001); McNairn et al. (2000); Brown et al. (1992)
X	8-12	3.8-2.5	Soil moisture retrieval, crop identification, and monitoring	Brown et al. (1992); Toan et al. (1989)
Ku	12-18	2.5-1.7	Estimation of LAI; retrieval of soil moisture; grain yield estimation	Frolking et al. (2006); Moran et al. (1998)
K	18-27	1.7-1.1	Canopy height estimation; soil moisture retrieval	Calvet et al. (2010); Rosenthal et al. (1985)
Ka	27-40	1.1-0.75	Grain yield estimation	Inoue et al. (2002)
V	40-75	0.75-0.40	Estimating vegetation water content	Calvet et al. (2010)
W	75-110	0.40-0.27	Estimating vegetation water content	Calvet et al. (2010)

Most crop monitoring activities rely on synthetic aperture radar (SAR) data due to their finer spatial resolution (Steele-Dunne et al., 2017). Non-imaging microwave sensors include altimeters and scatterometers.

Passive microwave sensing of a crop surface effectively measures its brightness temperature, which mainly depends on the microwave frequency, plant canopy properties, and observation angles. The plant canopy properties include the dielectric constant, emissivity, volume fraction, and arrangement of the plant constituents (Ulaby and Jedlicka, 1984; Ulaby et al., 2014; Konings et al., 2019). The dielectric constant (ε_r) is the relative electric permeability of a material when imposed by an electric field. It is expressed as a ratio of the electric permeability to that of a vacuum. Water has a high dielectric constant of 80 compared to 1.0 for air at 20°C. The microwave dielectric constant of water is much greater than that of dry vegetative matter. Thus, the dielectric properties of plant constituents are predominantly determined by the fractional volume composed of water content. Therefore, small changes in water content lead to significant changes in the canopy dielectric constant (Burke et al., 2005). The spectral and polarization properties of microwave emission from agricultural fields provide critical information about soil and plant biophysical parameters. At low frequencies, especially the L-band, soil dominates the emission from the surface with little influence from the crop canopy. At frequencies of the X-band or higher, emission from the soil is strongly attenuated by plants, and the measurement of microwave emission in these bands provides useful information on soil cover and plant water status. The microwave radiation from a surface at an angle away from the zenith is partially polarized, and the degree of polarization depends on the surface dielectric constant and its roughness. Radiation from a plant canopy is much less polarized than that from the bare soil. Therefore, vegetation attenuates the polarized emission from the soil and emits radiation that is either unpolarized or has a slight degree of H polarization, the degree of which is a function of crop type and frequency (Paloscia and Pampaloni, 1988; Ulaby et al., 1982). The density and structure of the plant canopy also influence the dielectric constant (Kraszevvski and Nelson, 2003; El-Rayes and Ulaby, 1987; Steele-Dunne et al., 2017). An incomplete canopy also complicates the interpretation of vegetation water content due to the partial emission from the soil surface (Fig. 7).

Approaches and models have been applied to retrieve plant water content using microwave radiometry. Similar to VIs in optical wavelengths, microwave band combinations have also been developed to quantify surface water conditions. For example, observations at 10 GHz and 36 GHz, especially if combined with TIR measurements, are useful in identifying moisture conditions of the surfaces (Paloscia and Pampaloni, 1992). Another example is the polarization index defined as the difference between the two linear polarizations (T_{bv} - T_{bh}) normalized to their average value [(T_{bv} + T_{bh})/2] (Paloscia and Pampaloni, 1988). Polarization indices are mostly related to plant water content without being significantly influenced by plant structure and surface temperature (Paloscia et al., 2018). A more widely adopted approach to retrieve plant and soil water content is based on radiative transfer modeling. For instance, the tau–omega (τ–ω) model represents a zeroth-order solution of the radiative transfer equation for modeling microwave interaction with vegetation-covered surfaces. This model estimates plant water status by simplifying the vegetation as a uniform layer with a constant temperature over a moist soil emitting polarized microwave radiation (Mo et al., 1982). Two parameters are mainly involved in this model: the vegetation optical depth (τ) and the single-scattering albedo (ω). For a certain vegetation type and frequency, the effective τ is the degree to which vegetation attenuates microwave radiation. It is proportional to the amount of water

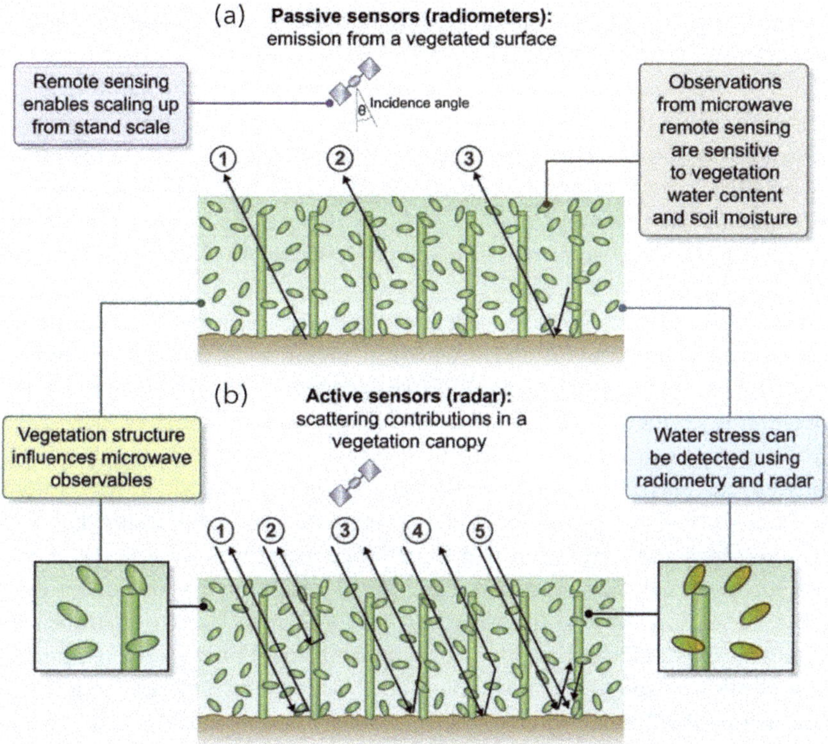

Figure 7 A model illustrating the cross-section of microwave interaction with vegetation and soil. (a) contributions to emission from a vegetated canopy: (1) emission from the soil attenuated by the overlying vegetation, (2) emission from the vegetation itself, and (3) emission from the vegetation that is reflected from the soil and attenuated by the vegetation. (b) Scattering contributions from a vegetation canopy: (1) direct backscattering from soil, including two-way attenuation by canopy, (2) direct backscattering from plants, (3) plant/ground scattering, (4) ground/plant scattering, and (5) ground/plant/ground scattering (Ulaby et al., 2014; Konings et al., 2019).

present in the plant canopy. The effective single-scattering albedo depends on the type and structure of vegetation. For example, Pampaloni and Paloscia (1986) estimated the plant water content of corn and alfalfa using the tau-omega model with different τ and ω values for each crop. They found that the τ value is correlated with plant water content by means of a logarithmic function. Togliatti et al. (2019) found that L-band τ is directly proportional to crop water in the US corn belt. To reduce the uncertainty due to temporal variations in the plant canopy, a time-varying parameterization of ω helps improve the model performance (Park et al., 2020). Chan and Bindlish (2019) explored a time-series approach using the tau-omega model to retrieve vegetation water content. This study used multiple days of brightness temperature to obtain effective vegetation optical depth that combined radiometric and polarization effects.

Active microwave systems, such as radar instruments, are applied to measure the backscattering coefficient of a surface. The backscattering coefficient is proportional to the ratio of the backscattered to the emitted energy by the sensor. For a soil-vegetation system, the backscattering coefficient consists of components from plant canopy, soil surface, and the interactions of the two (Fig. 7b). Their relative contributions detected by the sensor depend on the vegetation dielectric properties and geometry and sensor characteristics, including frequency, polarization, and incidence angle (Steele-Dunne et al., 2017; Knonings et al., 2019).

Two types of microwave models are applied to describe the propagation, scattering, and attenuation of EM waves in the vegetation layer: (1) models assuming a set of randomly distributed lossy scatters representing the different constituents of the vegetation, such as leaves, trunks, branches, and stalks; and (2) models representing a continuous layer with a randomly distributed dielectric constant (Chukhlantsev et al., 2003). The water cloud model (Attema and Ulaby, 1978) is widely applied to estimate the plant water status by characterizing radar backscattering from the vegetation canopy. In this model, water droplets are assumed to be randomly distributed and held in place by the vegetative matter. The expression of the model was derived for the backscattering coefficient as a function of three target parameters: volumetric water content of the vegetation, volumetric moisture content of the soil, and plant height. Additional parameters, such as leaf angle, may be incorporated into the model to improve the performance (Kweon and Oh, 2015). Another model (Eom and Fung, 1984) represents the vegetation layer as a layer of leaves above an irregular soil surface, assuming that the scattering from the vegetation is dominated by the leaves and a single leaf is modeled by a thin dielectric disk.

One of the active microwave indices is the radar vegetation index (RVI) (Kim and van Zyl, 2000), defined as:

$$\mathrm{RVI} = \frac{8\sigma_{HV}}{\sigma_{HV} + \sigma_{VV} + 2\sigma_{HV}}$$

where σ_{HV} is the cross-polarization backscattering cross-section and σ_{HH} and σ_{VV} are the copolarization backscattering cross-sections represented in power units. The RVI is a measure of volume scattering, usually caused by complex structural elements of vegetation, such as leaves and stems. Therefore, RVI has been used to monitor the level of vegetation growth as RVI is near zero for a smooth bare surface and increases with the amount of vegetation. The RVI was also applied for predicting plant water content in rice, soybean, wheat, corn, cotton, and pasture (Kim et al., 2012, 2014; Huang et al., 2015; Srivastava et al., 2015; Haldar et al., 2018). Szigarski et al. (2018) proposed two VIs, RVII and

RVIII, to improve the RVI by reducing dependence on soil roughness and soil moisture.

There are several active SAR missions in space, including RADARSAT-2 (C-band) of Canada, TerraSAR-X and TanDEM-X (X-band) of Germany, COSMO-SkyMed (X-band) of Italy, and Sentinel-1 of the ESA. Airborne SAR is a valuable tool in various scientific studies and also provides a test bed for new imaging techniques. An example of airborne systems is the uninhabited aerial vehicle synthetic aperture radar (UAVSAR), a polarimetric L-band SAR designed to acquire SAR data for differential interferometric measurements. Since 2018, the UAVSAR facility instrument suite has been enhanced with two additional bands: P-band and Ka-band. The P-band supports the EVS-1 AirMOSS mission to observe sub-canopy and subsurface root zone soil moisture. The Ka-band single-pass interferometric SAR generates high-precision, high-resolution, large-swath digital surface models for ice surface topography mapping. The instrument is designed to fly aboard a NASA aircraft and eventually on uninhabited aerial vehicles (NASA, 2021). Small and lightweight radar sensors onboard UAVs have been developed in the last several years. For example, Svedin et al. (2021) reported a frequency-modulated continuous-wave radar using separate transmit and receive antennas for a UAV system. The radar transmitted up to +23 dBm output power in the C-band (5.4–6.0 GHz), and a range compensation filter and a 10-bit analog-to-digital converter were used at the receiver side. SAR images generated from the system showed 3 dB lobe-widths, close to the theoretical resolution. Lort et al. (2018) developed a fully polarimetric X-band SAR system integrated into a UAV, overcoming restrictions of weight, space, robustness, and power consumption. Houtz et al. (2020) designed a portable L-band radiometer for UAV and ground-based applications. At an altitude of 10 m, the sensor provided 6 m spatial resolution. Li and Ling (2015) demonstrated radar sensors emitting a frequency range of 3.1–5.3 GHz for surface feature imaging.

7 Conclusion and future trends in research

The number of platforms and sensors and the associated remote sensing data have increased exponentially in the past few decades. The optical sensing capabilities have improved in quality and volume, especially with satellite and airborne hyperspectral sensors. In addition, the number of alternative modes of measurement has also grown, including airborne LiDAR and terrestrial laser scanning (TLS) and SAR sensors (Ghamisi et al., 2019). The open data policy adopted by the US Landsat and ESA Sentinel missions offers an unprecedented opportunity for large-area crop monitoring. Landsat 8, Landsat 9, Sentinel-2, and Sentinel-1 collectively represent the best available moderate-resolution satellite data at a global scale at 10–30 m spatial resolution, with weekly to biweekly data acquisition and covering visible-to-thermal and microwave wavelengths (Masek et al., 2020; Song et al., 2021). Commercial small satellites in constellations, such as those by Planet Labs, provide the ability to capture daily imagery of Earth at very high spatial resolutions for crop water monitoring. For example, Aragon et al. (2021) applied the CubeSats to monitor crop water use at daily and 3-m resolutions.

The availability of big data from various sources and formats also creates challenges to effectively and efficiently apply them in agricultural applications, including crop water status monitoring. Data fusion has the potential to address the challenge of continuous crop water status monitoring by combining multiple image sources with different spatial and temporal scales. Data fusion in remote sensing refers to the integration of data and information from different spatial, spectral, and temporal resolutions from sensors on satellites, aircraft, and other platforms to produce fused data with more detailed information than each of the sources. For example, Gao et al. (2006) proposed the Spatial and Temporal Adaptive Reflectance Fusion Model (STARFM) to blend the MODIS daily 500-m surface reflectance and the 16-day repeat cycle Landsat Enhanced Thematic Mapper Plus (ETM+) 30-m surface reflectance to produce a synthetic daily surface reflectance product at 30-m resolution. Xu et al. (2020) revised this model to monitor daily plant water content at 30-m spatial resolution with the fusion of MODIS and Landsat 8 NIR and SWIR reflectance to generate daily 30 m NDWI for the estimation of various scales of crop water content at both good spatial and temporal resolutions. Low-cost sensors onboard UAVs can be used to collect multispectral, thermal, and microwave images for calibrating canopy temperature or reflectance for satellite images with lower resolution. For instance, Sagan et al. (2019) applied data fusion approaches to combine UAV and satellite multiscale data for detecting crop early stress.

High computing power and advanced analytics tools are required to efficiently implement data fusion of big data for crop health and water status monitoring, which poses both challenges and opportunities. Cluster-based high-performance systems and cloud-based computation are dominant approaches for big data storage and analytics. Google Earth Engine (GEE) is a cloud-based platform that leverages high-performance computing resources for processing very large geospatial datasets and easily disseminating results. GEE substantially simplifies the work required to download and process satellite and other images. This platform hosts data from EOS and ESA satellites, such as Landsat, MODIS, ASTER, VIIRS, and Sentinel-2. The GEE is equipped with a library of more than 800 functions, which range in complexity from simple mathematical functions to powerful geostatistical, machine learning, and image processing operations (Gorelick et al., 2017). Studies have leveraged the GEE for crop water monitoring projects. For example, Yebra et al. (2019) developed an extensive global database of FMC measured from 161 717 records based on *in situ* destructive samples. This database is designed to calibrate and validate remote sensing algorithms for predicting live fuel moisture content. Zurqani et al. (2021) adopted GEE to map and quantify agricultural irrigation in heterogeneous landscapes of the South Carolina Coastal Plains. A variety of datasets, including climate variables and topographic attributes (elevation and slope), along with VIs derived from Sentinel-2 images, including NDWI, land surface water index, NDVI, etc., were included in the classification procedure using the Random Forest within GEE.

It is worth noting that remote sensing cannot directly determine the plant water status or assess plant water stress but the plant responses to water stress (Jones and Vaugham, 2010). Many different environmental factors can result in similar remote sensing signals or plant responses. For example, water stress, mineral deficiency, disease, and insect damage all lead to stomatal closure, and many stresses tend to reduce leaf area (Jones and Vaugham, 2010). The use of VIs that incorporate visible and NIR reflectance to assess plant water stress or other limiting factors only has validity in the absence of other factors that affect the reflectance of the plant in a similar manner (Pabuayon et al., 2019). Empirical relationships are often strong between a VI and plant water content for particular sites and plant communities, but they may not be applicable to other settings. Therefore, it is difficult to monitor the impact of water stress from only one observed response (Jones and Vaugham, 2010). Limited studies have developed methods to link RWC among different crops and environmental conditions. Studies are needed to assess different VIs or methods under various environmental conditions and crops. Hyperspectral sensors may have a good fit in these situations to measure reflectance at many narrow bands. Comprehensive methods that involve different data acquired using various sensors that incorporate physical measurements can be applied to eliminate confounding factors.

One of the challenges in determining plant canopy water content lies in the complexity of canopy structure. Each crop or cultivar may have a different canopy architecture, which may affect the EM radiation interaction with the plants. Wavelengths that are weakly absorbed and penetrate more deeply into canopies should detect a larger portion of the total water in canopies, whereas wavelengths that are strongly absorbed and thus penetrate only a short distance into canopies should be sensitive to just the water in the upper layers of the canopy. Therefore, VIs based on NIR and SWIR to estimate EWT cannot account for the total water content of plant canopies (Sims and Gamon, 2003). The direct application of LiDAR alone in crop water assessment is limited (Ahmad et al., 2021). However, LiDAR has the capability to quantify plant architecture and volumetric properties by providing a vertical profile representing the distribution of canopy elements (Damm et al., 2018). High-density 3D point clouds obtained with airborne laser scanning system and TLS also allow estimates of leaf area index (Kükenbrink et al., 2017; Moeser et al., 2014; Morsdorf et al., 2006). The incorporation of plant structure detected by LiDAR and other 3D point clouds can provide essential parameters for improving plant canopy water assessment using SAR data. There is generally a lack of research in the integration of microwave data and other types of images in assessing crop water status. Research in this direction has some potential to improve accuracy in quantifying crop water status.

The knowledge and information of plant water status enable precision irrigation management. Irrigation water and other crop inputs are managed at increasingly greater spatial and temporal scales (Guo et al., 2015). This requires data and information on crop water status at corresponding scales. Many remote sensing approaches, including proximal and UAV imaging, are applicable to leaf scales or plant canopy-only situations. However, common aerial or satellite systems provide images with resolutions of mixed pixels containing plant canopy and soil surface. Algorithms for plant water status based on pure plant canopy (without soil surface) will be biased if ground pixel sizes are greater or close to crop row spacings. Low-resolution images can be limited in the application of temperature-based methods for evaluating plant water stress due to the lack of full ground cover at water-critical crop stages. Soil background may bias the plant water status interpretation in thermal remote sensing (Torrion et al., 2014). There are at least two approaches to resolve this issue. One solution is to incorporate the soil background into the algorithm for assessing crop water status or stress conditions. For example, Torrion et al. (2014) evaluated a fusion approach that incorporated surface temperature, soil brightness, and ground cover to assess water stress. The combination of plant canopy temperature and the surrounding soil temperature is expressed as surface temperature. Soil brightness for an image scene varies with surface soil moisture. Results of this study showed that the fusion method was effective in assessing water stress without the use of additional energy balance calculation in partial ground-cover conditions. Another solution is to apply high-resolution images to extract pure plant canopy pixels for water status derivation. The UAV provides images at the centimeter resolution, enabling the extraction of plant canopy from full scenes, thus eliminating the soil effect.

Most aerial or satellite remote sensing products are discrete images acquired on a particular date or at a specific time during the day. On the other hand, plant water status or stress responses are dynamic processes even within the same day. Therefore, the water status derived from discrete remote sensing images may not represent the actual plant water content for irrigation management decision-making. The integration of remote sensing images into crop modeling appears to be a good solution to predict plant water status based on limited image data. Time-series remote sensing data can be applied to monitor the dynamics of plant water status. For example, Dennison et al.

(2003) developed a cumulative water balance index to incorporate time series of precipitation and reference evapotranspiration to estimate FMC and EWT derived from satellite images. They used this index to track and model temporal and spatial variation of regional drought stress. Limited research has addressed the within-day temporal variability of water status in relation to remote sensing images. Further research is needed to integrate image data from satellites, aerial platforms, and ground-based sensors for assessing plant water status.

Machine learning with remote sensing and other data sources is becoming a robust method for plant water status estimation with better accuracy than conventional remote sensing methods (Virnodkar et al., 2020). Machine learning is a type of artificial intelligence and a component of data science that emulates human intelligence by learning from the experience and the surrounding environment (El Naqa and Murphy, 2015; IBM, 2020; Liakos et al., 2018). Machine learning, along with computer vision, has been widely applied in agricultural applications, including yield prediction, weed detection, disease detection, soil and water management, plant phenotyping, etc. (Liakos et al., 2018; Lin and Guo, 2020, 2021). Deep learning can be defined as a computational model that learns representations of data with non-linear processing consisting of multiple layers (Kamarudin et al., 2021; Kamilaris and Prenafeta-Boldú, 2018; LeCun et al., 2015). Recent studies have applied various machine learning methods to predict crop water status or stress. Elsherbiny et al. (2021) applied an artificial neural network approach by integrating visible and thermal imagery for forecasting canopy water content in rice. The prediction variables included 14 VIs, normalized relative canopy temperature, and CWSI. The results of this study indicated that feature-based modeling from both visible and thermal images achieved better performance than features from the individual visible or thermal images. Ndlovu et al. (2021) used machine learning based on UAV and proximal imagery to estimate leaf water content in corn. The random forest regression method was found the most effective in predicting corn water content. Das et al. (2021) evaluated the water status of wheat genotypes using machine learning using CWSI, standardized canopy temperature index, stomatal conductance index, VPD, and crop stress index, derived from thermal imagery and on-site agrometeorological parameters. Kamilaris and Prenafeta-Boldú (2018) provided a survey on deep learning in agricultural applications. Kamarudin et al. (2021) conducted a comprehensive review on remote sensing and deep learning for crop water stress determination in various crops.

Precision irrigation is a direct application of crop water status monitoring. Variable rate irrigation can improve WUE and conserve water for sustainable agriculture with data and information for decision support (Neupane et al., 2021; Neupane and Guo, 2019). Remote sensing images along with field sensors and crop models can provide critical information for producers to manage irrigation site-specifically. For example, one may consider reducing irrigation in consistently low-yielding areas to optimize production and profitability (Guo, 2018). Many irrigation tools are available to assist in irrigation scheduling, but there is a lack of comprehensive, flexible, and efficient decision support systems to incorporate time series of remote sensing data and information into site-specific and time-specific irrigation management. The application of water stress indices in irrigation management needs evaluation of suitability and environmental conditions. For instance, the application of CWSI in irrigation scheduling is suitable in dry climate conditions but not in humid environments (Anda, 2009). Research and development are needed to integrate data, including satellite-based, aerial or UAV-based, and proximal datasets on crop conditions to fulfill high resolutions in space and time simultaneously to support daily decision-making in precision irrigation management.

8 Where to look for further information

8.1 Further reading

- A comprehensive introduction to principles of plant sensing is "Leaf Optical Properties" by Jacquemoud and Ustin, 2019.
- Good reviews on sciences and technologies on remote sensing of plant water status are Toth and Józków (2016), Qian (2021), and Zhu et al. (2018).
- Good review on remote sensing in water management, Calera et al. (2017).
- Excellent source for microwave SAR technology. Flores-Anderson AI, Herndon KE, Thapa RB & Cherrington E (2019) SAR Handbook: Comprehensive Methodologies for Forest Monitoring and Biomass Estimation. Available at: https://servirglobal.net/Global/Articles/Article/2674/sar-handbook-comprehensive-methodologies-for-forest-monitoring-and-biomass-estimation.

8.2 Key journals and conferences

- Two journals: Remote Sensing of Environment (Elsevier) and Remote Sensing (MDPI) are well attended by members of the remote sensing community.
- IEEE International Geoscience and Remote Sensing Symposium, IGARSS Annual Conference holds annual meetings.
- ASA, CSSA, and SSSA hold annual international meetings which cover remote sensing application in agriculture.
- ISPA (International Society of Precision Agriculture) conducts bi-annual international conferences.

9 References

Ahmad, U., Alvino, A. and Marino, S. (2021). A review of crop water stress assessment using remote sensing. *Remote Sensing* 13(20): 4155.

Alchanatis, V., Cohen, Y., Cohen, S., Sprinstin, M., Meron, M., Tsipris, J., Saranga, Y. and Sela, E. (2010). Evaluation of different approaches for estimating and mapping crop water status in cotton with thermal imaging. *Precision Agriculture* 11: 27–41.

Allen, W. A., Gansman, H. W., Rachardson, A. J. and Wiegand, C. L. (1970). Mean effective optical constants of thirteen kinds of plant leaves. *Applied Optics* 9(11): 2573–2577.

Allen, W. A., Gausman, H. W., Richardson, A. J. and Thomas, J. R. (1969). Interaction of isotropic light with a compact plant leaf. *Journal of the Optical Society of America* 59(10): 1376–1379.

Anda, A. (2009). Irrigation timing in maize by using the Crop Water Stress Index (CWSI). *Cereal Research Communications* 37(4): 603–610.

Aragon, B., Ziliani, M. G., Houborg, R., Fraz, T. E. and McCabe, M. F. (2021). CubeSats deliver new insights into agricultural water use at daily and 3 m resolutions. *Scientific Reports* 11(1): 12131.

Attema, E. P. W. and Ulaby, F. T. (1978). Vegetation modeled as a water cloud. *Radio Science* 13: 357–364.

Babar, M. A., Reynolds, M. P., van Ginkel, M., Klatt, A. R., Raun, W. R. and Stone, M. L. (2006). Spectral reflectance indices as a potential indirect selection criteria for wheat yield under irrigation. *Crop Science* 46(2): 578–588.

Baranoski, G. V. G. and Van Leeuwen, S. R. (2017). Detecting and monitoring water stress states in maize crops using spectral ratios obtained in the photosynthetic domain. *Journal of Applied Remote Sensing* 11(3): 036025.

Berni, J. A. J., Zarco-Tejada, P. J., Sepulcre-Cantó, G., Fereres, E. and Villalobos, F. (2009). Mapping canopy conductance and CWSI in olive orchards using high resolution thermal remote sensing imagery. *Remote Sensing of Environment* 113(11): 2380–2388.

Bolat, I., Dikilitas, M., Ercisli, S., Ikinci, A. and Tonkaz, T. (2014). The effect of water stress on some morphological, physiological, and biochemical characteristics and bud success on apple and quince rootstocks. *The Scientific World Journal* 2014: 769732.

Bozhynov, V., Mashchenko, O., Urbanová, P. and Kovacs, Z. (2020). Steps to visible aquaphotomics. In: *Proceedings of the Bioinformatics and Biomedical Engineering, 8th International Work-Conference, IWBBIO 2020*, May 6–8, 2020, Granada, Spain.

Brown, R. J., Manore, M. J. and Poirier, S. (1992). Correlations between X-, C- and L-band imagery within an agricultural environment. *International Journal of Remote Sensing* 13(9): 1645–1661.

Buckley, T. N. (2019). How do stomata respond to water status? *New Phytologist* 224(1): 21–36.

Burke, E. J., Harlow, R. C. and Ferré, T. P. A. (2005). Measuring the dielectric permittivity of a plant canopy and its response to changes in plant water status: An application of impulse time domain transmission. *Plant and Soil* 268(1): 123–133.

Calera, A., Campos, I., Osann, A., D'Urso, G. and Menenti, M. (2017). Remote sensing for crop water management: Trom ET modelling to services for the end users. *Sensors* 17(5): 1104.

Calvet, J. C., Wigneron, J. P., Walker, J., Karbou, F., Chanzy, A. and Albergel, C. (2010). Sensitivity of passive microwave observations to soil moisture and vegetation water content: L-band to W-band. *IEEE Transactions on Geoscience and Remote Sensing* 49(4): 1190–1199.

Caturegli, L., Matteoli, S., Gaetani, M., Grossi, N., Magni, S., Minelli, A., Corsini, G., Remorini, D. and Volterrani, M. (2020). Effects of water stress on spectral reflectance of Bermudagrass. *Scientific Reports* 10(1): 15055.

Ceccato, P., Flasse, S., Tarantola, S., Jacquemoud, S. and Gregoire, J. (2001). Detecting vegetation leaf water content using reflectance in the optical domain. *Remote Sensing of Environment* 77(1): 22–33.

Chan, S. K. and Bindlish, R. (2019). Retrieval of vegetation water content using brightness temperatures from the soil moisture active passive (SMAP) mission. *IGARSS* 2019: 5316–5319.

Chen, D., Huang, J. and Jackson, T. J. (2005). Vegetation water content estimation for corn and soybeans using spectral indices derived from MODIS near- and short-wave infrared bands. *Remote Sensing of Environment* 98(2–3): 225–236.

Chukhlantsev, A. A., Shutko, A. M. and Golovachev, S. P. (2003). Attenuation of electromagnetic waves by vegetation on canopies. *Journal of Communications Technology and Electronics* 48: 1177–1202.

Clark, R. N. and Roush, T. L. (1984). Reflectance spectroscopy quantitative analysis techniques for remote sensing applications. *Journal of Geophysical Research: Solid Earth* 89(B7): 6329–6340.

Clarke, T. R. (1997). An empirical approach for detecting crop water stress using multispectral airborne sensors. *Horttechnology* 7(1): 9–16.

Clawson, K. L., Jackson, R. D. and Pinter, P. J. (1989). Evaluating plant water stress with canopy temperature differences. *Agronomy Journal* 81(6): 858–863.

Clevers, J. G. P. W., Kooistra, L. and Schaepman, M. E. (2008). Using spectral information from the NIR water absorption features for the retrieval of canopy water content. *International Journal of Applied Earth Observation and Geoinformation* 10(3): 388–397.

Cohen, Y., Alchanatis, V., Meron, M., Saranga, Y. and Tsipris, J. (2005). Estimation of leaf water potential by thermal imagery and spatial analysis. *Journal of Experimental Botany* 56(417): 1843–1852.

Crusiol, L. G. T., Nanni, M. R., Furlanetto, R. H., Sibaldelli, R. N. R., Cezar, E., Mertz-Henning, L. M., Nepomuceno, A. L., Neumaier, N. and Farias, J. R. B. (2020). UAV-based thermal imaging in the assessment of water status of soybean plants. *International Journal of Remote Sensing* 41(9): 3243–3265.

Curran, P. J. (1989). Remote sensing of foliar chemistry. *Remote Sensing of Environment* 30(3): 271–278.

Damm, A., Paul-Limoges, E., Haghighi, E., Simmer, C., Morsdorf, F., Schneider, F. D., van der Tol, C., Migliavacca, M. and Rascher, U. (2018). Remote sensing of plant-water relations: An overview and future perspectives. *Journal of Plant Physiology* 227: 3–19.

Danson, F. M., Steven, M. D., Malthus, T. J. and Clark, J. A. (1992). High-spectral resolution data for determining leaf water content. *International Journal of Remote Sensing* 13(3): 461–470.

Das, S., Christopher, J., Apan, A., Choudhury, M. R., Chapman, S., Menzies, N. W. and Dang, Y. P. (2021). Evaluation of water status of wheat genotypes to aid prediction of yield on sodic soils using UAV-thermal imaging and machine learning. *Agricultural and Forest Meteorology* 307: 108477.

Dennison, P. E., Roberts, D. A., Thorgusen, S. R., Regelbrugge, J. C., Weise, D. and Lee, C. (2003). Modeling seasonal changes in live fuel moisture and equivalent water thickness using a cumulative water balance index. *Remote Sensing of Environment* 88(4): 442–452.

Ehrler, W. L., Idso, S. B., Jackson, R. D. and Reginato, R. J. (1978). Wheat canopy temperature relation to plant water potential. *Agronomy Journal* 70(2): 251–256.

El Naqa, I. and Murphy, M. J. (2015). What is machine learning? In: El Naqa, I., Li, R. and Murphy, M. (Eds.), *Machine Learning in Radiation Oncology*. Cham: Springer.

El-Rayes, M. A. and Ulaby, F. T. (1987). Microwave dielectric spectrum of vegetation-Part I: Experimental observations. *IEEE Transactions on Geoscience and Remote Sensing* GE-25(5): 541–549.

Elsherbiny, O., Zhou, L., Feng, L. and Qiu, Z. (2021). Integration of visible and thermal imagery with an artificial neural network approach for robust forecasting of canopy water content in rice. *Remote Sensing* 13(9): 1785.

Eom, H. J. and Fung, A. K. (1984). A scatter model for vegetation up to Ku-band. *Remote Sensing of Environment* 15(3): 185–200.

Fensholt, R. and Sandholt, I. (2003). Derivation of a shortwave infrared water stress index from MODIS near- and shortwave infrared data in a semiarid environment. *Remote Sensing of Environment* 87(1): 111–121.

Fourty, T. and Baret, F. (1997). Vegetation water and dry matter contents estimated from top-of-the-atmosphere reflectance data: A simulation study. *Remote Sensing of Environment* 61(1): 34–45.

Frolking, S., Milliman, T., McDonald, K., Kimball, J., Zhao, M. and Fahnestock, M. (2006). Evaluation of the SeaWinds scatterometer for regional monitoring of vegetation phenology. *Journal of Geophysical Research* 111(D17): D17.

Gao, B. C. (1996). NDWI–A normalized difference water index for remote sensing of vegetation liquid water from space. Remote sensing of Environment 58(3): 257–266.

Gao, F., Masek, J., Schwaller, M. and Hall, F. (2006). On the blending of the Landsat and MODIS surface reflectance: Predicting daily Landsat surface reflectance. *IEEE Transactions on Geoscience and Remote Sensing* 44(8): 2207–2218.

Gardner, M. L. (1984). Intestinal assimilation of intact peptides and proteins from the diet - A neglected field? *Biological Reviews of the Cambridge Philosophical Society* 59(3): 289–331.

Gates, D. M., Keegan, H. J., Schleter, J. C. and Weidner, V. R. (1965). Spectral properties of plants. *Applied Optics* 4(1): 11–20.

Gausman, H. W., Allen, W. A. and Cardenas, R. (1969). Reflectance of cotton leaves and their structure. *Remote Sensing of Environment* 1(1): 19–22.

Gerhards, M., Schlerf, M., Mallick, K. and Udelhoven, T. (2019). Challenges and future perspectives of multi-/hyperspectral thermal infrared remote sensing for crop water-stress detection: A review. *Remote Sensing* 11(10): 1240.

Ghamisi, P., Gloaguen, R., Atkinson, P. M., Benediktsson, J. A., Rasti, B., Yokoya, N., Wang, Q., Hofle, B., Bruzzone, L., Bovolo, F., Chi, M. and Anders, K. (2019). Multisource and multitemporal data fusion in remote sensing: A comprehensive review of the state of the art. *IEEE Geoscience and Remote Sensing Magazine* 7(1): 6–39.

Goetz, S. J. (1997). Multi-sensor analysis of NDVI, surface temperature and biophysical variables at a mixed grassland site. International Journal of remote sensing 18(1): 71-94.

Gonzalez-Dugo, M. P., Moran, M. S., Mateos, L. and Bryant, R. (2005). Canopy temperature variability as an indicator of crop water stress severity. *Irrigation Science* 24: 233–240.

Gorelick, N., Hancher, M., Dixon, M., Ilyushchenko, S., Thau, D. and Moore, R. (2017). Google Earth Engine: Planetary-scale geospatial analysis for everyone. *Remote Sensing of Environment* 202: 18–27.

Govender, M., Dye, P., Weiersbye, I., Witkowski, E. and Ahmed, F. (2009). Review of commonly used remote sensing and ground-based technologies to measure plant water stress. *Water SA* 35(5): 741–752.

Gu, H., Lin, Z., Guo, W. and Deb, S. (2021). Retrieving surface soil water content using a soil texture adjusted vegetation index and unmanned aerial system images. *Remote Sensing* 13(1): 145.

Guo, W. (2018). Spatial and temporal trends of irrigated cotton yield in the Southern High Plains. *Agronomy* 8(12): 298.

Guo, W., Cui, S., Torrion, J. and Rajan, N. (2015). Data-driven precision agriculture: Opportunities and challenges. In: Rattan, L. and Stewart, B. A. (Eds.), *Precision Agriculture: Soil-Specific Farming*. Boca Raton, FL: CRC Press, pp. 353–372.

Gupta, A., Rico-Medina, A. and Caño-Delgado, A. I. (2020). The physiology of plant responses to drought. *Science* 368(6488): 266–269.

Haldar, D., Tripathy, R., Dave, V., Dave, R., Bhattacharya, B. K. and Misra, A. (2018). Monitoring cotton crop condition through synergy of optical and radar remote sensing. *Geocarto International* 37(2): 377–395.

Hardisky, M. A., Klemas, V. and Smart, R. M. (1983). The influence of soil salinity, growth form, and leaf moisture on the spectral radiance of Spartina alterniflora canopies. *Photogrammetric Engineering and Remote Sensing* 49: 77–83.

Hopkins, W. G. and Huner, N. P. A. (2009). *Introduction to Plant Physiology* (4th edn.). Hoboken, NJ: John Wiley & Sons.

Houtz, D., Naderpour, R. and Schwank, M. (2020). A cost-effective portable L-band radiometer for drone and ground-based applications. In: *IEEE International Geoscience and Remote Sensing Symposium*, 2020, pp. 6531–6534.

Hsiao, T. C. (1973). Plant responses to water stress. *Annual Review of Plant Physiology* 24(1): 519–570.

Hsiao, T. C., Acevedo, E., Fereres, E. and Henderson, D. W. (1976). Water stress, growth and osmotic adjustment. Philosophical Transactions of the Royal Society of London B 273(927): 479–500.

Huang, Y., Walker, J. P., Gao, Y., Wu, X. and Monerris, A. (2015). Estimation of vegetation water content from the Radar Vegetation Index at L-band. *IEEE Transactions on Geoscience and Remote Sensing* 54(2): 981–989.

Hunt, E. R. and Rock, B. N. (1989). Detection of changes in leaf water content using near- and middle-infrared reflectances. *Remote Sensing of Environment* 30: 43–54.

Huo, H., Li, Z. and Xing, Z. (2019). Temperature/emissivity separation using hyperspectral thermal infrared imagery and its potential for detecting the water content of plants. *International Journal of Remote Sensing* 40(5–6): 1672–1692.

IBM. (2020). Deep learning. Available at: https://www.ibm.com/cloud/learn/deep-learning (last accessed 12 December 2022).

Idso, S. B., Jackson, R. D., Pinter, P. J., Reginato, R. J. and Hatfield, J. L. (1981). Normalizing the stress degree-day parameter for environmental variability. *Agricultural Meteorology* 24: 45–55.

Inoue, Y., Kurosu, T., Maeno, H., Uratsuka, S., Kozu, T., Dabrowska-Zielinska, K. and Qi, J. (2002). Season-long daily measurements of multifrequency (Ka, Ku, X, C, and L) and full-polarization backscatter signatures over paddy rice field and their relationship with biological variables. *Remote Sensing of Environment* 81(2-3): 194-204.

Irmak, S., Haman, D. Z. and Bastug, R. (2000). Determination of crop water stress index for irrigation timing and yield estimation of corn. *Agronomy Journal* 92(6): 1221-1227.

Jackson, R. D., Idso, S. B., Reginato, R. J. and Pinter, P. J. (1981). Canopy temperature as a crop water stress indicator. *Water Resources Research* 17(4): 1133-1138.

Jackson, R. D., Kustas, W. P. and Choudhury, B. J. (1988). A reexamination of the crop water stress index. *Irrigation Science* 9(4): 309-317.

Jackson, T. J., Chen, D., Cosha, M., Lia, F., Anderson, M., Walthalla, C., Doriaswamya, P. and Hunt, E. R. (2004). Vegetation water content mapping using Landsat data derived normalized difference water index for corn and soybeans. *Remote Sensing of Environment* 92(4): 475-482.

Jacquemoud, S. and Ustin, S. L. (2008). Modeling leaf optical properties. Photobiological Sciences Online. Environmental Photobiology. Available at: http://www.photobiology.info/#Environ (last accessed 12 December 2022).

Jacquemoud, S. and Ustin, S. L. (2019). *Leaf Optical Properties*. Cambridge, UK: Cambridge University Press.

Jenal, A., Bareth, G., Bolten, A., Kneer, C., Weber, I. and Bongartz, J. (2019). Development of a VNIR/SWIR multispectral imaging system for vegetation monitoring with unmanned aerial vehicles. *Sensors* 19(24): 5507.

Jia, M., Tong, L., Zhang, Y. and Chen, Y. (2013). Rice biomass estimation using radar backscattering data at S-band. *IEEE Journal of Selected Topics in Applied Earth Observations and Remote Sensing* 7(2): 469-479.

Jones, H. G. (1990). Plant water relations and implications for irrigation scheduling. *Acta Horticulturae* 278: 67-76.

Jones, H. G. (2004). Irrigation scheduling: Advantages and pitfalls of plant-based methods. *Journal of Experimental Botany* 55(407): 2427-2436.

Jones, H. G. (2007). Monitoring plant and soil water status: Established and novel methods revisited and their relevance to studies of drought tolerance. *Journal of Experimental Botany* 58(2): 119-130.

Jones, H. G. (2014). *Plants and Microclimate: A Quantitative Approach to Environmental Plant Physiology* (3rd edn.). Cambridge, UK: Cambridge University Press.

Jones, H. G. and Vaugham, R. A. (2010). *Remote Sensing of Vegetation: Principles, Techniques, and Applications*. New York: Oxford University Press.

Kamarudin, M. H., Ismail, Z. H. and Saidi, N. B. (2021). Deep learning sensor fusion in plant water stress assessment: A comprehensive review. *Applied Sciences* 11(4): 1403.

Kamilaris, A. and Prenafeta-Boldú, F. X. (2018). Deep learning in agriculture: A survey. *Computers and Electronics in Agriculture* 147: 70-90.

Kebes. (2008). Absorption spectrum of liquid water across a wide wavelength range. Available at: https://commonswikimediaorg/wiki/File:Absorption_spectrum_of_liquid_waterpng (last accessed 30 November 2021).

Kim, Y., Jackson, T., Bindlish, R., Lee, H. and Hong, S. (2012). Radar Vegetation Index for estimating the vegetation water content of rice and soybean. *IEEE Geoscience and Remote Sensing Letters* 2012(9): 564-568.

Kim, Y., Jackson, T., Bindlish, R., Lee, H., Hong, S., Jung, G. and Lee, K. (2014). Retrieval of wheat growth parameters with Radar Vegetation Indices. *IEEE Geoscience and Remote Sensing Letters* 11: 4259–4272.

Kim, Y. X., Stumpf, B., Sung, J. and Lee, S. J. (2018). The relationship between turgor pressure change and cell hydraulics of midrib parenchyma cells in the leaves of Zea mays. *Cells* 7(10): 180.

Kim, Y. and van Zyl, J. (2000). On the relationship between polarimetric parameters. In: *IEEE International Symposium on Geoscience and Remote Sensing*, Honolulu, HI, USA, July 24–28, 2000, Vol. 2000, pp. 1298–1300.

Knipling, E. B. (1970). Physical and physiological basis for the reflectance of visible and near-infrared radiation from vegetation. *Remote Sensing of Environment* 1(3): 155–159.

Kokaly, R. F. and Clark, R. N. (1999). Spectroscopic determination of leaf biochemistry using band-depth analysis of absorption features and stepwise multiple linear regression. *Remote Sensing of Environment* 67(3): 267–287.

Kokaly, R. F., Despain, D. G., Clark, R. N. and Livo, K. E. (2007). *Spectral Analysis of Absorption Features for Mapping Vegetation Cover and Microbial Communities in Yellowstone National Park Using AVIRIS Data*. Publications of the US Geological Survey 73. Available at: https://digitalcommonsunledu/usgspubs/73.

Konings, A. G., Rao, K. and Steele-Dunne, S. C. (2019). Macro to micro: Microwave remote sensing of plant water content for physiology and ecology. *New Phytologist* 223(3): 1166–1172.

Kraszevvski, A. W. and Nelson, S. O. (2003). Microwave techniques in agriculture. *Journal of Microwave Power and Electromagnetic Energy* 38(1): 13–35.

Kükenbrink, D., Schneider, F. D., Leiterer, R., Schaepman, M. E. and Morsdorf, F. (2017). Quantification of hidden canopy volume of airborne laser scanning data using a voxel traversal algorithm. Remote Sensing of Environment 194: 424–436.

Kweon, S. and Oh, Y. (2015). A modified water-cloud model with leaf angle parameters for microwave backscattering from agricultural fields. *IEEE Transactions on Geoscience and Remote Sensing* 53(5): 2802–2809.

Lacerda, L. N., Snider, J. L., Cohen, Y., Liakos, V., Gobbo, S. and Vellidis, G. (2022). Using UAV-based thermal imagery to detect crop water status variability in cotton. *Smart Agricultural Technology* 2: 100029.

LeCun, Y., Bengio, Y. and Hinton, G. (2015). Deep learning. *Nature* 521(7553): 436–444.

Lee, J. S., Grunes, M. R. and Pottier, E. (2001). Quantitative comparison of classification capability: Fully polarimetric versus dual and single-polarization SAR. *IEEE Transactions on Geoscience and Remote Sensing* 39(11): 2343–2351.

Li, C. J. and Ling, H. (2015). Synthetic aperture radar imaging using a small consumer drone. In: *IEEE International Symposium on Antennas and Propagation USNC/URSI National Radio Science Meeting*, July 2015, pp. 685–686.

Liakos, K. G., Busato, P., Moshou, D., Pearson, S. and Bochtis, D. (2018). Machine learning in agriculture: A review. *Sensors* 18(8): 2674.

Lin, Z. and Guo, W. (2021). Cotton stand counting from unmanned aerial system imagery using MobileNet and CenterNet deep learning models. *Remote Sensing* 13(14): 2822.

Lin, Z. and Guo, W. (2020). Sorghum panicle detection and counting using unmanned aerial system images and deep learning. *Frontiers in Plant Science* 11:534853.

Loka, D. A., Oosterhuis, D. M. and Ritchie, G. L. (2011). Water-deficit stress in cotton. In: Oosterhuis, D. M. (Ed.), *Stress Physiology in Cotton*. Cordova, TN: The Cotton Foundation, pp. 37–72.

Lort, M., Aguasca, A., López-Martínez, C. and Marín, T. M. (2018). Initial evaluation of SAR capabilities in UAV multicopter platforms. *IEEE Journal of Selected Topics in Applied Earth Observations and Remote Sensing* 11(1): 127–140.

Lugojan, C. and Ciulca, S. (2011). Evaluation of relative water content in winter wheat. *Journal of Horticulture, Forestry and Biotechnololgy* 15: 173–177.

Maki, M., Ishiahra, M. and Tamura, M. (2004). Estimation of leaf water status to monitor the risk of forest fires by using remotely sensed data. *Remote Sensing of Environment* 90(4): 441–450.

Masek, J. G., Wulder, M. A., Markham, B., McCorkel, J., Crawford, C. J., Storey, J. C. and Jenstrom, D. T. (2020). Landsat 9: Empowering open science and applications through continuity. *Remote Sensing of Environment* 248: 111968.

McNairn, H., Van der Sanden, J. J., Brown, R. J. and Ellis, J. (2000). The potential of RADARSAT-2 for crop mapping and assessing crop condition. In: *Proceedings of the Second International Conference on Geospatial Information in Agriculture and Forestry*, Vol. 2, pp. 81–88.

McQuarrie, D. A. and Simon, J. D. (1997). *Physical Chemistry: A Molecular Approach*. Sausalito, CA: University Science Books.

Meron, M., Tsipris, J. and Charitt, D. (2003). Remote mapping of crop water status to assess spatial variability of crop stress. In: Stafford, J. V. and Werner, A. (Eds.), *Precision Agriculture '03: Proceedings of the 4th European Conference on Precision Agriculture*. Wageningen, The Netherlands: Wageningen Academic Publishers, pp. 405–410.

Merry, C. J. (2010). Basic electromagnetic radiation. In: Bossler, J. D. (Ed.), *Manual of Geospatial Science and Technology* (2nd edn.). New York: CRC Press.

Minh, D. H. T., Toan, T. L., Rocca, F., Tebaldini, S., d'Alessandro, M. M. and Villard, L. (2014). Relating P-band synthetic aperture radar tomography to tropical forest biomass. *IEEE Transactions on Geoscience and Remote Sensing* 52(2): 967–979.

Mo, T., Choudhury, B. J., Schmugge, T. J., Wang, J. R. and Jackson, T. J. (1982). A model for the microwave emission of vegetation-covered fields. *Journal of Geophysical Research* 87(C13): 229–237.

Moeser, D., Roubinek, J., Schleppi, P., Morsdorf, F. and Jonas, T. (2014). Canopy closure, LAI and radiation transfer from airborne LiDAR synthetic images. Agricultural and Forest Meteorology 197*:* 158–168.

Möller, M., Alchanatis, V., Cohen, Y., Meron, M., Tsipris, J., Naor, A., Ostrovsky, V., Sprintsin, M. and Cohen, S. (2007). Use of thermal and visible imagery for estimating crop water status of irrigated grapevine. *Journal of Experimental Botany* 58(4): 827–838.

Monsivais-Huertero, A., Judge, J., Steele-Dunne, S. and Liu, P. W. (2015). Impact of bias correction methods on estimation of soil moisture when assimilating active and passive microwave observations. *IEEE Transactions on Geoscience and Remote Sensing* 54(1): 262–278.

Moran, M. S., Clarke, T. R., Inoue, Y. and Vidal, A. (1994). Estimating crop water deficit using the relation between surface-air temperature and spectral vegetation index. *Remote Sensing of Environment* 49(3): 246–263.

Moran, M. S., Vidal, A., Troufleau, D., Inoue, Y. and Mitchell, T. A. (1998). Ku-and C-band SAR for discriminating agricultural crop and soil conditions. *IEEE Transactions on Geoscience and Remote Sensing* 36(1): 265–272.

Morsdorf, F., Kötz, B., Meier, E., Itten, K. I. and Allgöwer, B. (2006). Estimation of LAI and fractional cover from small footprint airborne laser scanning data based on gap fraction. Remote Sensing of Environment 104(1): 50–61.

NASA. (2021). Uninhabited aerial vehicle synthetic aperture radar (UAVSAR) NASA airborne science program. Available at: https://airbornesciencenasagov/instrument/UAVSAR (Accessed November 20, 2021).

Ndlovu, H. S., Odindi, J., Sibanda, M., Mutanga, O., Clulow, A., Chimonyo, V. G. P. and Mabhaudhi, T. (2021). A comparative estimation of Maize leaf water content using machine learning techniques and unmanned aerial vehicle (UAV)-based proximal and remotely sensed data. *Remote Sensing* 13(20): 4091.

Neupane, J. and Guo, W. (2019). Agronomic basis, technology, and benefits of precision water management: A review. *Agronomy* 9(2): 87.

Neupane, J., Guo, W., West, C. P., Zhang, F. and Lin, Z. (2021). Effects of irrigation rates on cotton yield as affected by soil physical properties and topography in the Southern High Plains. *PLoS ONE* 16(10): e0258496.

Nguyen, T. X. B., Rosser, K. and Chahl, J. (2021). A review of modern thermal imaging sensor technology and applications for autonomous aerial navigation. *Journal of Imaging* 7(10): 217.

Pabuayon, I. L., Sun, Y., Guo, W. and Ritchie, G. (2019). High-throughput phenotyping in cotton: A review. *Journal of Cotton Research* 2(1): 2–18.

Paloscia, S. and Pampaloni, P. (1988). Microwave polarization index for monitoring vegetation growth. *IEEE Transactions on Geoscience and Remote Sensing* 26(5): 617–621.

Paloscia, S. and Pampaloni, P. (1992). Microwave vegetation indexes for detecting biomass and water conditions of agricultural crops. *Remote Sensing of Environment* 40(1): 15–26.

Paloscia, S., Pampaloni, P. and Santi, E. (2018). Radiometric microwave indices for remote sensing of land surfaces. *Remote Sensing* 10(12): 1859.

Park, C.-H., Jagdhuber, T., Colliander, A., Lee, J., Berg, A., Cosh, M., Kim, S., Kim, Y. and Wulfmeyer, V. (2020). Parameterization of vegetation scattering albedo in the Tau-Omega model for soil moisture retrieval on croplands. *Remote Sensing* 12(18): 2939.

Pasqualotto, N., Delegido, J., Van Wittenberghe, S., Verrelst, J., Rivera, J. P. and Moreno, J. (2018). Retrieval of canopy water content of different crop types with two new hyperspectral indices: water absorption area index and depth water index. International Journal of Applied Earth Observation and Geoinformation 67: 69–78.

Penuelas, J., Filella, I., Biel, C., Serrano, L. and Save, R. (1993). The reflectance at the 950–970 nm region as an indicator of plant water status. International Journal of Remote Sensing 14(10): 1887–1905.

Penuelas, J., Pinol, J., Ogaya, R. and Filella, I. (1997). Estimation of plant water concentration by the reflectance water index WI (R900/R970). *International Journal of Remote Sensing* 18(13): 2869–2875.

Prasad, B., Carver, B. F., Stone, M. L., Babar, M. A., Raun, W. R. and Klatt, A. R. (2007). Genetic analysis of indirect selection for winter wheat grain yield using spectral reflectance indices. *Crop Science* 47(4): 1416–1425.

Qian, S. (2021). Hyperspectral satellites, evolution, and development history. *IEEE Journal of Selected Topics in Applied Earth Observations and Remote Sensing* 14: 7032–7056.

Rosenthal, W. D., Blanchard, B. J. and Blanchard, A. J. (1985). Visible/infrared/microwave agriculture classification, biomass, and plant height algorithms. *IEEE Transactions on Geoscience and Remote Sensing* GE-23(2): 84–90.

Sagan, V., Maimaitijiang, M., Sidike, P., Maimaitiyiming, M., Erkbol, H., Hartling, S., Peterson, K. T., Peterson, J., Burken, J. and Fritschi, F. U. (2019). UAV/satellite multiscale data fusion for crop monitoring and early stress detection. In: *International Archives of the Photogrammetry, Remote Sensing and Spatial Information Sciences - ISPRS Archives 42 (XLII-2/W13)*, International Society for Photogrammetry and Remote Sensing, June 2019, pp. 715-722.

Sandholt, I., Rasmussen, K. and Andersen, J. (2002). A simple interpretation of the surface temperature/vegetation index space for assessment of surface moisture status. Remote Sensing of Environment 79(2–3): 213–224.

Schonfeld, M. A., Johnson, R. C., Carver, B. F. and Mornhigweg, D. W. (1988). Water relations in winter wheat as drought resistance indicators. *Crop Science* 28(3): 526–531.

Seelig, H.-D., Hoehn, A., Stodieck, L. S., Klaus, D. M., Adams, W. W. and Emery, W. J. (2008). Relations of remote sensing leaf water indices to leaf water thickness in cowpea, bean, and sugarbeet plants. *Remote Sensing of Environment* 112(2): 445–455.

Shu, M., Dong, Q., Fei, S., Yang, X., Zhu, J., Meng, L., Li, B. and Ma, Y. (2022). Improved estimation of canopy water status in maize using UAV-based digital and hyperspectral images. *Computers and Electronics in Agriculture* 197: 106982.

Sims, D. A. and Gamon, J. A. (2003). Estimation of vegetation water content and photosynthetic tissue area from spectral reflectance: a comparison of indices based on liquid water and chlorophyll absorption features. Remote sensing of Environment 84(4): 526–537.

Song, X., Huang, W., Hansen, M. C. and Potapov, P. (2021). An evaluation of Landsat, Sentinel-2, Sentinel-1 and MODIS data for crop type mapping. *Science of Remote Sensing* 3: 100018.

Srivastava, P. K., O'Neill, P., Cosh, M., Lang, R. and Joseph, A. (2015). Evaluation of radar vegetation indices for vegetation water content estimation using data from a ground-based SMAP simulator. In: *IGARSS*. Milan, Italy: IEEE Publications, pp. 1296–1299.

Sriwongsitanon, N., Gao, H., Savenije, H. H. G., Maekan, E., Saengsawang, S. and Thianpopirug, S. (2016). Comparing the Normalized Difference Infrared Index (NDII) with root zone storage in a lumped conceptual model. *Hydrology and Earth System Sciences* 20(8): 3361–3377.

Steele-Dunne, S. C., McNairn, H., Monsivais-Huertero, A., Judge, J., Liu, P. and Papathanassiou, K. (2017). Radar remote sensing of agricultural canopies: A review. *IEEE Journal of Selected Topics in Applied Earth Observations and Remote Sensing* 10(5): 2249–2273.

Svedin, J., Bernland, A., Gustafsson, A., Claar, E. and Luong, J. (2021). Small UAV-based SAR system using low-cost radar, position, and attitude sensors with onboard imaging capability. *International Journal of Microwave and Wireless Technologies* 13(6): 602–613.

Szigarski, C., Jagdhuber, T., Baur, M., Thiel, C., Parrens, M., Wigneron, J.-P., Piles, M. and Entekhabi, D. (2018). Analysis of the Radar Vegetation Index and potential improvements. *Remote Sensing* 10(11): 1776.

Toan, T. L., Lau, H., Mougin, E. and Lopes, A. (1989). Multitemporal and dual-polarization observations of agricultural vegetation covers by X-band SAR images. *IEEE Transactions on Geoscience and Remote Sensing* 27(6): 709–718.

Togliatti, K., Hartman, T., Arkebauer, T., Suyker, A., VanLoocke, A. and Hornbuckle, B. K. (2019). SMAP vegetation optical depth is directly proportional to crop water in the US corn belt. *IGARSS* 2019: 5445–5448.

Torrion, J. A., Maas, S. J., Guo, W., Bordovsky, J. P. and Cranmer, A. M. (2014). A three-dimensional index for characterizing crop water stress. *Remote Sensing* 6(5): 4025–4042.

Toth, C. and Jóźków, G. (2016). Remote sensing platforms and sensors: A survey. *ISPRS Journal of Photogrammetry and Remote Sensing* 115: 22–36.

Tshwene, C. and Oladele, I. (2016). Water use productivity and food security among smallholder homestead food gardening and irrigation crop farmers in North West Province, South Africa. *Journal of Agriculture and Environment for International Development* 110(1): 73–86.

Tucker, C. J. (1980). Remote sensing of leaf water content in the near infrared. *Remote Sensing of Environment* 10(1): 23–32.

Turner, N. C. (1981). Techniques and experimental approaches for the measurement of plant water status. *Plant and Soil* 58(1–3): 339–366.

Ulaby, F. T. and Jedlicka, R. P. (1984). Microwave dielectric properties of plant materials. *IEEE Transactions on Geoscience and Remote Sensing* GE-22(4): 406–415.

Ulaby, F. T., Long, D. G., Blackwell, W. J., Elachi, C., Fung, A. K., Ruf, C., Sarabandi, K., Zebker, H. A. and Van Zyl, J. (2014). *Microwave Radar and Radiometric Remote Sensing* (Vol. 4). Ann Arbor, MI: University of Michigan Press.

Ulaby, F. T., Moore, R. K. and Fung, A. K. (1982). *Microwave Remote Sensing: Active and Passive* (Vol. 11). Reading, MA: Addison-Wesley.

Ullah, S., Skidmore, A. K., Groen, T. A. and Schlerf, M. (2013). Evaluation of three proposed indices for the retrieval of leaf water content from the mid-wave infrared (2–6 μm) spectra. *Agricultural and Forest Meteorology* 171–172: 65–71.

Ustin, S. L. and Jacquemoud, S. (2020). How the optical properties of leaves modify the absorption and scattering of energy and enhance leaf functionality. In: Cavender-Bares, J., Gamon, J. and Townsend, P. (Eds.), *Remote Sensing of Plant Biodiversity*. Cham, Switzerland: Springer International Publishing, pp. 349–384.

Veysi, S., Naseri, A. A., Hamzeh, S. and Bartholomeus, H. (2017). A satellite based crop water stress index for irrigation scheduling in sugarcane fields. Agricultural Water Management 189*:* 70–86.

Virnodkar, S. S., Pachghare, V. K., Patil, V. C. and Jha, S. K. (2020). Remote sensing and machine learning for crop water stress determination in various crops: A critical review. *Precision Agriculture* 21(5): 1121–1155.

Woolley, J. T. (1971). Reflectance and transmittance of light by leaves. *Plant Physiology* 47(5): 656–662.

Wright, I. J., Reich, P. B., Westoby, M., Ackerly, D. D., Baruch, Z., Bongers, F., Cavender-Bares, J., Chapin, T., Cornelissen, J. H., Diemer, M., Flexas, J., Garnier, E., Groom, P. K., Gulias, J., Hikosaka, K., Lamont, B. B., Lee, T., Lee, W., Lusk, C., Midgley, J. J., Navas, M. L., Niinemets, U., Oleksyn, J., Osada, N., Poorter, H., Poot, P., Prior, L., Pyankov, V. I., Roumet, C., Thomas, S. C., Tjoelker, M. G., Veneklaas, E. J. and Villar, R. (2004). The worldwide leaf economics spectrum. *Nature* 428(6985): 821–827.

Wu, C., Niu, Z., Tang, Q. and Huang, W. (2009). Predicting vegetation water content in wheat using normalized difference water indices derived from ground measurements. *Journal of Plant Research* 122(3): 317-326.

Xu, C., Qu, J. J., Hao, X., Cosh, M. H., Zhu, Z. and Gutenberg, L. (2020). Monitoring crop water content for corn and soybean fields through data fusion of MODIS and Landsat measurements in Iowa. *Agricultural Water Management* 227: 105844.

Yamasaki, S. and Dillenburg, L. R. (1999). Measurements of leaf relative water content in Araucaria angustifolia. *Revista Brasileira de Fisiologia Vegetal* 11(2): 69-75.

Yao, X., Jia, W., Si, H., Guo, Z., Tian, Y., Liu, X., Cao, W. and Zhu, Y. (2014). Exploring novel bands and key index for evaluating leaf equivalent water thickness in wheat using hyperspectra influenced by nitrogen. *PLoS ONE* 9(6): e96352.

Yebra, M., Scortechini, G., Badi, A., Beget, M. E., Boer, M. M., Bradstock, R., Chuvieco, E., Danson, F. M., Dennison, P., Resco de Dios, V., Di Bella, C. M., Forsyth, G., Frost, P., Garcia, M., Hamdi, A., He, B., Jolly, M., Kraaij, T., Martín, M. P., Mouillot, F., Newnham, G., Nolan, R. H., Pellizzaro, G., Qi, Y., Quan, X., Riaño, D., Roberts, D., Sow, M. and Ustin, S. (2019). Globe-LFMC, a global plant water status database for vegetation ecophysiology and wildfire applications. *Scientific Data* 6(1): 155.

Yilmaz, M. T., Hunt, E. R. and Jackson, T. J. (2008) Remote sensing of vegetation water content from equivalent water thickness using satellite imagery. *Remote Sensing of Environment* 112: 2514–2522.

Zarco-Tejada, P. J. and Ustin, S. L. (2001). Modeling canopy water content for carbon estimates from MODIS data at land EOS validation sites. In: *Proceedings of IEEE*, 2001, pp. 342–344.

Zhang, J. H., Xu, Y., Yao, F. M., Wang, P. J., Guo, W. J., Li, L. and Yang, L. (2010). Advances in estimation methods of vegetation water content based on optical remote sensing techniques. *Science China Technological Sciences* 53(5): 1159–1167.

Zhang, L., Zhang, H., Niu, Y. and Han, W. (2019). Mapping maize water stress based on UAV multispectral remote sensing. *Remote Sensing* 11(6): 605.

Zhu, L., Suomalainen, J., Liu, J., Hyyppä, J., Kaartinen, H. and Haggren, H. (2018). A review: Remote sensing sensors. In: Rustamov, R. B., Hasanova, S. and Zeynalova, M. H. (Eds.), *Multi-Purposeful Application of Geospatial Data*. London, UK: IntelchOpen, pp. 19–42.

Zurqani, H. A., Allen, J. S., Post, C. J., Pellett, C. A. and Walker, T. C. (2021). Mapping and quantifying agricultural irrigation in heterogeneous landscapes using Google Earth Engine. *Remote Sensing Applications: Society and Environment* 23: 100590.

Zwart, S. J. and Bastiaanssen, W. G. M. (2004). Review of measured crop water productivity values for irrigated wheat, rice, cotton and maize. *Agricultural Water Management* 69(2): 115–133.

Chapter 2

Advances in remote sensing technologies for assessing crop health

Michael Schirrmann, Leibniz Institute for Agricultural Engineering and Bioeconomy, Germany

1 Introduction

Climate change and changes in social attitudes have led to a rethink in agricultural policy and practice. There has been a shift away from intensive, monocultural agriculture towards more diversified, species-rich and sustainable agricultural practices that meet the demands of citizens for less use of agrochemicals (such as pesticides), for being more resilient to extreme weather events and for causing less damage to the environment. The EU Biodiversity Strategy, e.g. commits to a 50% reduction in the use of synthetic pesticides by 2030 (European Commission, 2020). To meet these requirements, crop management decisions and methods need to be supported by more precise and timely information, in particular to assess the health status of the crop in fields. In this chapter, we explore the principles and advances in remote sensing of crop health to support more sustainable agricultural practices.

Crop health is not a clearly defined term. A common definition of crop health focuses on how far individual crop plants deviate from their full potential due to abiotic and biotic stresses. However, a broader definition of crop health may include factors that promote a healthier state. It may embrace the overall spectrum of well-being, including factors affecting crop vigour, abiotic and biotic stresses and the surrounding environment (Döring et al., 2012). An example is wheat plants growing on sandy and loamy soils next to each

http://dx.doi.org/10.19103/AS.2022.0107.03

other in a field. If rainfall is insufficient, plants on sandy soils will experience more stress and reduced vitality compared to counterparts on loamy soils given the water deficit caused by the dryer soil. Even without the presence of disease, such environmental factors result in differences in plant well-being. This broader and more holistic approach to the concept of crop health would help to better understand the underlying factors that ultimately influence crop performance. Promoting plant health following a holistic approach would also allow for better integration of alternative approaches to crop protection in line with agroecological practices (Vega et al., 2020).

Crops are not uniform in space and time. They vary significantly within the same field and between different fields. They vary also from season to season and within seasons due to a wide range of factors. Some factors affecting crop health may be less visible than others (such as soil water and nutrient status). The sheer complexity of the manifold factors affecting crop health and their interrelationships makes it difficult to map the spatiotemporal pattern of crop health in fields which can vary substantially. However, timely and comprehensive monitoring of plant diseases or the factors that facilitate disease development is key. This is particularly important before visual symptoms are observed by the farmer, in order to start crop protection measures at a sufficiently early stage for optimal impact before the infection spreads over the field or infects neighbouring fields.

Crop or field scouting can assess crop health at specific locations in fields quite accurately using expert knowledge of farmers and others. Manual assessment of plants exploits all the advantages of human observation. Leaves can be investigated from multiple viewing angles in the upper and lower leaf layers of the canopy, while experienced observation can focus on specific leaf disorders. Crop scouting can be further enhanced by magnifying glasses or handheld sensors to assess non-visual plant properties such as chlorophyll content or photosynthetic activity. Assessment results can reveal detailed information about crop health. However, crop scouting is very laborious when done by the farmer or expensive when outsourced to agricultural service companies. There are obvious limitations to manual spatial and temporal sampling frequency, usually restricting assessments to a few sample locations randomly chosen over a field, if it is done at all.

The need for accurate and timely spatial monitoring of crop health can be seen in the example of yellow (stripe) rust or *Puccinia striiformis* f. sp. Tritici. Spores are deposited on plant leaves by wind and rain, infecting wheat crops in particular. Once it has infiltrated the leaf, a fungal mycelium develops in the mesophyll tissue (Chen et al., 2014). By feeding on plant nutrients, yellow rust directly influences the health of infected plants and may potentially lead to complete yield losses in wheat crops. The fungal disease initially develops in small localized areas (foci) where conditions are favourable for disease infection.

These infected areas are usually difficult to detect on plant leaves in the lower layers in the canopy which are infected first. After a few days, the characteristic visual symptoms will become more detectable. Stripe rust e.g. develops long streaks between the leaf veins: the fungus develops urediniospore pustules which have a characteristic yellow to orange 'rusty' colour (Chen et al., 2014). However, by then, the latent infection will have already spread across the entire field. This pattern is characteristic of many diseases, depending on factors such as microclimate, weather conditions, soil conditions, transmitting vectors (e.g. mites) or surrounding vegetation (Oerke, 2020; Singh et al., 2018).

Remote sensing has the potential to play an important role in collecting high-quality spatiotemporal information about crop health, for example by monitoring disease outbreaks in the latent phase before visual symptoms appear. This is because remote sensing has advanced significantly in recent years. In the case of satellite imagery, the main improvements are in timely accessibility of and, in many cases, free availability of satellite data, as well as improvements in temporal resolution. While Landsat 8 (launched in 2013) revisits a location on earth every 16 days, Sentinel 2 A/B (launched in 2015/2017) has a revisit time of 5 days. A key commercial development has been satellite swarm technology which uses low orbit constellations of hundreds of micro-satellites and achieves revisit times of multiple times a day (SkySat constellation) with high spatial ground resolution (about 1 m). Unmanned aerial vehicle (UAV) platforms can also play a vital role in collecting very high-resolution data to assess crop health on selected dates and with individually selected sensor technology.

In the following sections, we will first look at the basics of and advances in remote sensing for crop health assessment. Later in the chapter, we will explore a case study on the detection of stripe rust from very high-resolution imagery using machine learning. Finally, the chapter reviews future trends in this area.

2 Remote sensing of crop health

Plant leaf reflectance over the visible and near-infrared range contains useful information about crop growth and health. Within the visible range, leaf reflectance is lower due to light absorption by leaf pigments involved in photosynthesis, in particular chlorophylls which absorb light in the blue and red region of the spectrum. In the near-infrared region, however, these leaf pigments do not absorb light and leaf reflectance is many times higher in magnitude. Near-infrared leaf reflectance is mainly the result of leaf cellular structure, in particular due to light scattering inside and outside the cell walls and air gaps between the cells. Non-scattered light will be transmitted through the leaf. There is almost no absorption due to chemical constituents in the near-infrared range except for water with wavelengths greater than 1000 nm (Jacquemoud and Ustin, 2019).

Given this pattern of light scattering, a sharp increase between red and near-infrared regions can be observed in the reflectance of green parts of plants, named the red edge. Red edge reflectance is correlated with many different crop parameters. Remote sensing has exploited the red edge since the early days of Earth observation. For the first Landsat satellite system, the "Earth Resources Technology Satellite-1" (ERTS-1) with the "Multi-Spectral Scanner" (MSS), Rouse et al. (1973) established the well-known NDVI (normalized difference vegetation index). The NDVI is a spectral index that relates near-infrared with red wavelength reflectance in a normalized form. NDVI has been shown to correlate well with biomass and moisture content of grassland sites in the Great Plains of the USA based on Landsat MSS data (Rouse et al., 1973).

Subsequently, other spectral indices have been developed that utilize variation in the red edge region to assess different aspects of crop performance. In contrast to the NDVI, these indices relate to narrow wavelength ranges. One such index is the normalized difference red edge (NDRE) used by Barnes et al. (2000) with the wavelengths 790 nm and 720 nm. They have developed the canopy chlorophyll content index (CCCI) using empirical minimum and maximum NDRE values. The CCCI was e.g. found to correlate well with nitrogen content in cotton plants. Many other studies have corroborated the strong correlation of red edge-related indices with nitrogen content in plants (Fitzgerald et al., 2010; González-Piqueras et al., 2017). It should be noted that these indices respond mainly to changes in chlorophyll in plant leaves and not directly to nitrogen present in leaves (Fitzgerald et al., 2010).

Red edge-related indices also help to assess leaf senescence in crops due to chlorophyll breakdown in later growth stages. Anderegg et al. (2020) have shown, using the plant senescence reflectance index (PSRI), that the visible range can also be used to analyse senescence dynamics over a season. The PSRI focuses on wavelengths that discriminate between chlorophyll and carotenoid leaf pigments and may be more appropriate than pure red edge indices because their ratio changes very early during senescence. Many diseases also lead to a reduction of leaf chlorophyll content that reduces absorbance in red wavelengths and cause a shift of the red edge position in the reflectance spectrum (Martinelli et al., 2015).

Vegetation monitoring has improved with the addition of narrowband channels to Sentinel-2 imaging, which were missing in Landsat. Sentinel-2 includes three channels in the red edge region of the spectrum, which enables the use of more specific and complex vegetation indices. Empirical relationships with Sentinel-2 vegetation indices have been shown for various crop performance parameters, including nitrogen uptake, canopy chlorophyll content and leaf area index. Based on winter wheat cropping systems in the Netherlands, Delloye et al. (2018) found that red edge bands could be used to measure biophysical variables such as canopy chlorophyll content, green area index and leaf chlorophyll content. They found a high correlation between canopy chlorophyll content and nitrogen uptake in conventionally cultivated fields but almost no correlation for organic fields. The authors related this to the higher variability of chlorophyll content commonly observed in organic fields which made it harder to establish a spatial pattern and relationship. This suggests that Sentinel-2 imagery spatial resolution may be suitable for conventional farming but remains insufficient for monitoring sustainable agricultural practices such as organic farming.

With the advent of the UAV platform in precision agriculture, spatial resolution has been greatly improved for assessing crop health because images taken with drones can even distinguish single plants with the right flight settings. Multispectral cameras that include spectral channels within the red edge and near-infrared range are now readily available, affordable and with a design that matches the requirements of many UAVs. Using a five-band multispectral camera (Red Edge, MicaSense, Inc., Seattle, WA, USA), Walsh et al. (2018) observed high correlations between several red edge-based vegetation indices in UAV imagery and nitrogen concentration in wheat crops. Since UAV imagery is taken using overlapping images and processed photogrammetrically to produce orthophotos, three-dimensional (3D) point clouds are a by-product of the structure-from-motion processing. The point clouds enable the construction of crop surfaces. When related to surface models generated from the bare soil surface before seeding, crop height can be derived for all subsequent UAV flight campaigns in the season. Crop heights obtained from photogrammetric UAV imagery were shown to be a linear predictor for estimating biomass and leaf area index of winter wheat and barley crops. In combination with the use of simple red, green and blue (RGB) imagery, it was possible to obtain sufficient model accuracies with regression analysis to measure earlier growth stages (Bendig et al., 2015; Schirrmann et al., 2016). Photogrammetrically derived 3D surfaces are unique to UAVs and are not so far possible with satellite-based imagery at this level of detail. Combining crop height estimates with multispectral vegetation indices would also allow for more complex assessments of plant undernourishment because estimates that relate to both biophysical and leaf chemical parameter can be included in data fusion models.

3 Remote sensing of crop diseases

Pathogen-related crop diseases have a significant impact on food crop production worldwide. Pathogens include fungi, viruses, bacteria or nematodes. Infection changes the metabolism and structure of a plant in a way that decreases performance, resulting in heavy yield losses in many cases. In the case of wheat crops, fungal diseases such as leaf rust, fusarium head blight, tritici blotch and stripe rust cause major yield losses globally (Savary et al., 2019).

Pesticides can be used to suppress pathogens but have harmful impacts on the environment and potentially on food safety. They are often applied preventatively before actual infection, e.g. using predictive modelling (Jalli et al., 2020). Specific sensor systems are potentially better able to detect diseases earlier than is possible with human observation. Monitoring able to robustly detect and map crop diseases very early after infection could help reduce pesticide use through more limited and targeted application, as well as allowing greater use of integrated pest management techniques. Scalable remote sensing approaches could then track subsequent disease development over the field because they can map entire fields.

Pathogen attack may have measurable effects long before visible symptoms occur. When a pathogen enters the plant's system, the first reactions are usually defence and stress responses. As part of these defence mechanisms, plants may emit biogenic volatile organic compounds (VOCs) to signal the plant is under attack and upregulate defence pathways by priming plant cells (Ameye et al., 2015). As infection in the cell tissue develops, it results in changes such as nutrient depletion, transport blockages or destruction of healthy cells. This results in chlorophyll degradation and changes in optical properties (i.e. reflectance) of infected plant parts (Chen et al., 2015; Choudhury et al., 2019). Such changes in photosynthetic performance have been observed via chlorophyll fluorescence measurements (Pérez-Bueno et al., 2019). Table 1 summarizes a range of approaches to detect crop diseases remotely.

During the latent phase of the disease development with no apparent visible symptoms, special sensor technologies may be needed. Electronic noses are a potential technology for identifying specific VOCs, which could allow detecting plant diseases shortly after an infection has occurred. Electronic noses aim to mimic mammalian olfactory systems using sensor arrays: odour molecules induce reversible changes on a substrate which are then transformed into electric signals (Cui et al., 2018). The sensor array produces a molecular fingerprint which can be analysed using machine learning. Non-destructive, in situ measurements with an electronic nose on tomato, cucumber and pepper plants have achieved very high classification accuracies in distinguishing healthy and diseased plants (using support vector machines for data analysis) after 10 days of natural infection with powdery mildew by mites, though this study was undertaken in controlled conditions (Ghaffari et al., 2012). Oerke (2020) argues if high sensitivity under natural field conditions can be achieved, an electronic nose can be used to guide pesticide application to target the original source of the VOCs in the field by following the gradient of volatiles. However, more in situ research on field crops is still needed, and a practical remote sensing application has not yet been developed.

Table 1 Remote sensing technologies for assessing crop changes due to disease infections

Technology	Potential target	Potential detection time after infection	Ground resolution
Electronic nose	Volatile organic compounds	Very short, first stress symptoms	Unspecific
Sun-induced chlorophyll fluorescence	Changes in the photosynthesis (photosystem II) due to the detection of chlorophyll fluorescence	Short, reacts to early stresses	Coarse due to very small spectral bands to be used
Hyperspectral reflectance camera	Early (subtle) changes to the chlorophyll content, specific disease discolorations	Medium, reacts to stresses and changes in the metabolism of the plants	Medium
Multispectral reflectance camera	Changes to the chlorophyll content, specific disease discolorations	Medium, reacts to stress and changes in the metabolism of the plants	High
RGB reflectance camera	Changes to the chlorophyll content, specific individual disease lesions	Late, after visual symptoms occurred but can be highly specific to the disease type when spatial resolution is high enough	Highest

Chlorophyll fluorescence is relatively hard to obtain with remote sensing. Chlorophyll fluorescence is normally measured using an active light technique inducing several light flashes on a plant in darkened conditions and recording the characteristics of the fluorescence signal. This approach is not practical at field scale. Sunlight absorption for photosynthesis also triggers chlorophyll fluorescence from plants with emission peaks in the red and far-red at wavelength near 683 nm and 736 nm: this is known as sun-induced chlorophyll fluorescence (Péres-Bueno et al., 2019). However, the signal is weak and largely filtered out of plant reflectance. To measure sun-induced chlorophyll fluorescence, the signal is obtained in wavelength regions where the sunlight is attenuated in the atmosphere, for example, by oxygen absorption bands at 760 nm and 687 nm (O_2-A and O_2-B band) (Rascher et al., 2015).

Both satellite and UAV platforms have been used to measure top-of-the canopy sun-induced fluorescence (SIF) with non-imaging and imaging spectrometers. A number of satellite missions have been associated with SIF estimation from space, e.g. GOSAT (Global Greenhouse Gas Observation) or GOME-2 (Global Ozone Monitoring Experiment). Guanter et al. (2014) investigated SIF data from the GOME-2 mission, retrieved using the simplified radiative transfer model proposed by Joiner et al. (2013). They found that the

GOME-2 SIF data captured high photosynthetic signals observed from tower flux measurements and could improve estimation of gross primary production of crops compared to standard modelling approaches using vegetation indices. Other studies assessed global drought effects on crops with GOME-2 SIF data more effectively (Yoshida et al., 2015; Wang et al., 2016). Unfortunately, the spatial resolution of space-borne SIF data is currently very low (kilometre scale), with significant limitations for analysis at a local or even regional scale (Bandopadhyay et al., 2020).

In contrast, UAV platforms have the potential to provide SIF estimates with high spatial resolution. There have been few studies of the potential of SIF measurement using UAVs, with just one on estimating the onset of disease. Calderón et al. (2013) investigated the development of Verticillium wilt (VW) caused by the soil-borne fungus *Verticillium dahliae* Kleb in olive trees using an integrated approach with multispectral, thermal and hyperspectral sensor data. The hyperspectral data was obtained with a line-scanning system mounted on a UAV recording 260 bands at 1.85 nm/pixel with a full width at half maximum (FWHM) of 3.2 nm. From this data, they estimated SIF using the Fraunhofer Line Depth principle, involving the three wavelength bands 763 nm, 750 nm and 780 nm (FLD3). The SIF data was found to correlate with the physiological stress invoked by *Verticillium* wilt. This data fusion approach was found to be helpful in assessing the early onset of the disease. SIF estimation is still challenging because the signal is strongly influenced by the atmosphere even at lower altitudes where UAVs operate (Bandopadhyay et al., 2020).

Hyperspectral measurements in the VisNIR range also have the potential to detect diseases early. Recent advances of technical development made it possible that hyperspectral cameras can be integrated on UAV platforms. Guo et al. (2021) used a hyperspectral imaging system mounted on a UAV to investigate the onset of stripe rust in an inoculated field experiment. The camera system (UHD 185, Cubert GmbH, Germany) worked in snapshot mode and provided a data cube with a spectral range of 450–950 nm and 4 nm spectral resolution, while the drone operated in 30 m altitude to provide a ground resolution of 1.2 cm. Even at an early stage of infection (7–16 days after inoculation), they found significant correlations between infestation rates and hyperspectral images using partial least square regression. With the same hyperspectral camera system, Liu et al. (2020) obtained high classification rates (>90%) in differentiating parts of fields of winter wheat with slight and severe infestation by *Fusarium* head blight infection in winter wheat using imagery obtained from a height of 60 m. Hyperspectral camera systems designed for UAVs are still very expensive (>€50 000) which limits regular use on UAV platforms.

More affordable, multispectral camera systems for UAVs or multispectral data from satellite systems have also been tested for disease assessment, e.g. stripe rust monitoring in winter wheat. Su et al. (2018) used a five-band multispectral camera (Red Edge, MicaSense, Inc.) with a UAV operating in 16 m altitude to monitor a field experiment of plots inoculated with stripe rust. They were able to classify the severity of stripe rust infestation at three levels (healthy, moderate and severe) with high accuracy (89%) using multispectral vegetation indices. However, this was only possible at a late stage in the disease (45 days after inoculation).

Zheng et al. (2018) has proposed the red edge disease stress index (REDSI) based on Sentinel-2 satellite imagery to differentiate stripe rust. The REDSI was developed from hyperspectral measurements on diseased plants in inoculated plots and uses the three Sentinel-2 bands B4, B5 and B7. With the REDSI as input, they mapped stripe rust in the Chuzhou and Hefei regions (China) with an accuracy of 85%, demonstrating the ability of Sentinel-2 data to document stripe rust damage on a regional scale.

4 Case study: detecting stripe rust using very high-resolution imaging

The following case study is based on Schirrmann et al. (2021). Our objective was to detect stripe rust symptoms when they become visible on the wheat leaves in experimental plots. We used very high-resolution imagery collected from two different altitudes (2 m and 10 m). The lower altitude permitted leaf differentiation at a millimetre scale (leaf-level imagery), while the higher altitude permitted differentiation of single plants (plant-level imagery).

Stripe rust is a common fungal disease in wheat crops caused by the fungus *Puccinia striiformis* Westend. f. sp. *tritici* Eriks. Wheat yield losses from stripe rust have been ranked fourth highest among common pests in wheat crops by Savary et al. (2019), particularly in Northwest Europe and sub-Saharan Africa. Almost 88% of global wheat production is susceptible to stripe rust disease (Beddow et al., 2015). The disease has high genetic variability, with new strains of stripe rust emerging quite frequently, complicating the development of resistant cultivars. This makes foliar fungicide application an important component of stripe rust control (Carmona et al., 2020).

Stripe rust is the earliest rust fungal disease occurring in the season. The disease develops in the leaf tissue but also affects ears and stems in its later stages. Leaves first develop yellow stripes along the vertical axis between the leaf veins, which later develop into long and narrow stripes with a yellow to orange coloration caused by densely located uredospore pustules (Chen et al., 2014). Each pustule is highly infectious with the potential to spread the disease throughout a field. Despite its conspicuous appearance, early detection is difficult because only a few leaves initially display these symptoms.

A randomized field experiment was conducted at the Fieldlab for Digital Agriculture (Marquardt, Potsdam, Germany) in 2018. Winter wheat with high susceptibility to stripe rust (variety 'Matrix B') was sown within 12 plots of 9 × 9 m dimensions with a seed row distance of 0.12 m and a seed rate of 350 grains per m^2. Each plot was separated by 3 m. Six plots were randomly chosen and inoculated with a spore solution (2.5 g stripe rust spores mixed with 500 ml purified mineral oil) on 11 April 2018, during cold and calm weather (stripe rust plots). The other plots were treated with the fungicide Osiris® (BASF, Germany) on 3 May 2018, to guarantee that they were free of disease (control plots). Plots were regularly assessed by scoring the percentage of visible infestation symptoms for the first three top leaf layers of ten plants at six locations in each plot.

Leaf-level imagery was recorded with a DSLM camera ILCE-6000 (Sony, Japan) with an APS-C type sensor chip (23.5 × 15.6 mm) and a 50 mm lens attached (SEL50F18, Sony, Japan) from 2.25 m altitude using an equipment carrier on which the camera system was attached in a nadir position (pointing directly down). Images were recorded on four days after inoculation (DAI). Training and test images for a deep learning model were recorded at opposite sides to each other in the plots. Images were further split into 224 × 224 sub-images along a regular, non-overlapping grid. Images and sub-images were visually inspected on screen for stripe rust symptoms and, if a symptom was present, the image or sub-image was marked as infected with stripe rust.

A deep residual convolution neural network (ResNet) was used to distinguish between images taken from healthy or stripe rust-infected areas. Convolutional neural networks are inspired by the receptive field of the human visual cortex (Kim et al., 2016). They include convolutional and subsampling layers (strided convolution layers), which perform feature learning, feature extraction and dimension reduction from the unstructured image data. This is mapped to a dense layer which decides the class label. ResNets are convolutional neural networks that include shortcut connections in the network architecture based on residual functions that make it possible to skip specific layers in the network (He et al., 2016).

The lightweight ResNet-18 architecture was used for this study. The ResNet-18 was trained in a supervised approach by teaching the network with the annotated sub-images taken from the training set. The training set of sub-images was artificially augmented by rotating and mirroring in order to produce more variety. Training was conducted on a GTX 1080 Ti GPU with 11 GB of memory on a server with two Intel Xeon E5-2640 CPUs and 512 GB of main memory running Debian Linux 8.11. To obtain the final prediction for the full image, a 224 × 224 pixel large window was evaluated with the trained ResNet-18 model across the large image without overlap. Around 442 probability scores were produced to estimate the chance of stripe rust symptom occurrence (one for each sub-image). A full image was flagged as stripe rust infected if one of the scores exceeded 0.99.

Plant-level imagery was obtained with a UAV platform near the acquisition date of the leaf-level imagery. A quadrocopter (HP-X4-E1200, Hexapilots, Germany) was used with the same camera system attached in the nadir position using a 16 mm lens. Flight programmes were planned at an altitude of 10 m and with an image overlap of 80%. Images were photogrammetrically processed with Metashape Professional (Agisoft LLC, Russia, 2019) using a structure-from-motion approach to produce orthophotos, which were geometrically corrected by ground-truthing markers. All plots were cut out to the edge. For each plot, the triangular greenness index (TGI) was calculated based on Hunt et al. (2011):

$$\mathrm{TGI} = 0.5\left[(\lambda_{\mathrm{Red}}\lambda_{\mathrm{Blue}})(R_{\mathrm{Red}} - R_{\mathrm{Green}}) - (\lambda_{\mathrm{Red}}\lambda_{\mathrm{Green}})(R_{\mathrm{Red}} - R_{\mathrm{Blue}})\right]$$

In this equation, red, green and blue refer to the channels of the RGB camera, λ to the centre wavelengths and R to the pixel values of the respective RGB channels. The TGI essentially calculates the area of the triangle that is spanned over the red, green and blue centre wavelengths by the pixel values of the RGB camera.

Figure 1 shows the typical symptoms of stripe rust along with other features in leaf-level images taken during the field experiment. It is apparent that visual symptoms have high variability and show a strong similarity to

Figure 1 Typical symptoms of stripe rust (upper line) in comparison with similar features (lower line) as they appeared on the wheat leaves during the field experiment. Top four images are listed as (a) to (d) (from left to right). Bottom four images are listed (e) to (h) (from left to right).

other symptoms of stress, damage and other diseases appearing on plants. The first visible symptoms caused by stripe rust included chlorotic lesions in the infected plots (Fig. 1a). They appeared as subtle, mostly whitish discolorations between the leaf veins due to chlorophyll reduction caused by fungal development in the plant. They were hard to detect even for a trained eye. Typical uredospore pustules later developed along these lesions with a striking orange colour (Fig. 1b). Their appearance could change rapidly after heavy rain, with colours changing more to yellow due to the lack of spores (Fig. 1c). In later stages, the entire leaf was affected by the fungus and necrotic symptoms also appeared (Fig. 1d).

Because the season was characterized by unusually dry weather, wheat plants developed many symptoms related to water-deficit and sunburn stresses that needed to be differentiated from the diseased symptoms (Fig. 1e). Moreover, we observed damages from the cereal leaf beetle (Fig. 1f–h). They feed between the leaf veins, and the damage they cause is very similar to the early symptoms of stripe rust. However, one distinguishing feature of beetle damage is rough edges to damaged areas, whereas stripe rust symptoms show a smoother transition from diseased to healthy areas of leaf. Finally, leaf rust appeared later in the field experiment, often on the same leaves that were affected by stripe rust (Fig. 1g–h). Compared to stripe rust, leaf rust symptoms were more widely distributed across the leaf and were not oriented in stripes along the leaf veins.

To determine the subtle differences between stripe rust symptoms and other symptoms requires high-resolution imagery that can resolve internal features of leaves in the crop canopy. Although the leaf-level images taken for this case study easily met this criteria, there were cases in which even an expert could not conclusively identify the symptoms of stripe rust on screen. This means it is challenging for deep learning algorithms to clearly distinguish between the symptoms, especially at the onset of infection.

Figure 2 shows the results of deep learning compared to crop scouting scoring results from the field experiment. The coloured lines show the observed infected leaf areas for flag (final leaf to emerge), first and second leaf layers from crop scouting. The bars in the background show the accuracy of the image classifier, both for each observation and for the date when the images were taken in days after inoculation (DAI).

The first symptoms developed initially in the lower leaf layers but spread to the higher leaf layers within a few days. They were first observed during field visits on 28 DAI. The symptoms were mainly chlorotic spots with a stripe-like appearance that were sporadically observed in the lower leaf layers and in only some parts of the infected plots. At 33–34 DAI, these symptoms were also found to a lesser degree in the higher leaf layers. Around this date, the first set of images was obtained to train and test the deep learning image classifier. For this set of images, model achieved an overall accuracy of 57%, meaning that

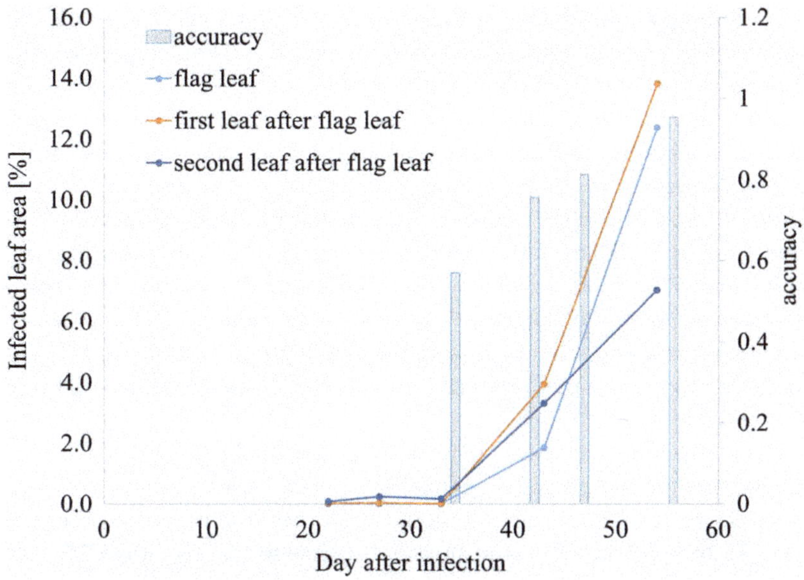

Figure 2 The connected points show the scoring of the infected leaf area of the flag, first and second wheat leaf in area percentages, while the bars in the background show the model accuracy of the image classifier based on the ResNet-18 deep learning model.

57% of all tested images were correctly classified as infected or not infected with stripe rust.

At 42–47 DAI, the infection rapidly spread towards the upper leaf layers covering 2% to 4% of the leaf areas. The symptoms showed the typical uredospore pustules that were now developing on some leaves. The stripe rust symptoms now had sufficiently distinctive characteristics to train the neural network better because the tested images obtained in this period had a much higher estimation accuracy of around 80% than the images collected the week before. In the later stages of the disease spreading after 56 DAI, an even higher estimation accuracy was reached by the model (95%), but during this time the crop canopy was already fully infested and the disease spread over the entire infected plots. It should be noted here that the estimation accuracy of the model tested on image patches, in which the symptoms are more centrally shown, was higher compared to the prediction on the full camera images for the image sets obtained during the earlier dates (86% to 93%). This can be explained by the fact that in the full images many regions are evaluated that are still 'unknown' to the model and where symptoms may not appear in a clean, clearly displayed position. Imagery was obtained in an unobserved manner to simulate a low flying drone. This is in contrast to using a smartphone, e.g. which is manually directed to the feature of interest and where the feature will be recorded with full exposure. This type of imagery can be evaluated much more effectively, of course.

Since the plant-level images captured by the drone at 10 m do not reach the spatial resolution to resolve individual features on leaves, we cannot rely on a form-based analysis of the leaf optical symptoms as with the deep learning approach using the leaf-level imagery taken at 2 m altitude. Figure 3 shows an image of a control plot adjacent to an image of an infected plot for the five UAV flight missions conducted during the spread of the disease in the field experiment. The RGB pixel values of the plot images were recalculated to TGI values. Hunt et al. (2011) determined a strong negative correlation for aerial imagery from RGB cameras, which means increasing TGI values corresponded to decreasing leaf chlorophyll content values. In general, TGI decreased as further plant development increased the integral photosynthetic activity of the canopy during the observed period. This was particularly pronounced from 36 DAI to 49 DAI.

The disease developed as diffuse hot spots in infected plots. Only at 49 DAI were the hot spots clearly identifiable within the plot image. For example, a well-defined stripe rust hot spot developed in the middle of the second row of the infected plot. In the image, this spot is represented by higher TGI values declining gradually into the surrounding area with lower TGI values. Other parts of the plot were also interspersed with discrete patterns indicative of localized

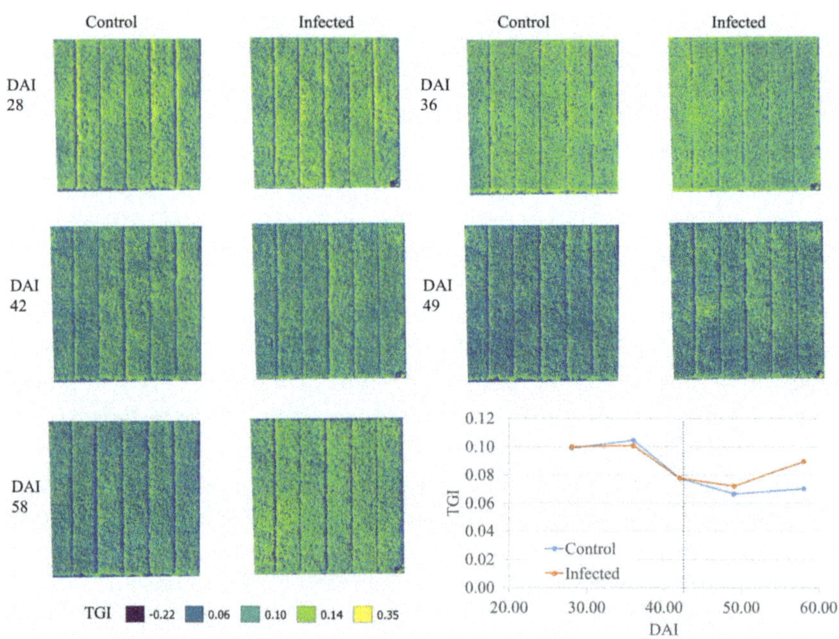

Figure 3 UAV images for a control and an infected plot calculated as TGI values shown for five dates along the disease outbreak. The diagram shows the average TGI for control and infected plot for each date.

infections with stripe rust. On the ground, these hot spots were scored with an infected leaf area of 20% to 60% of the flag leaves in the canopy.

The TGI images taken before and after 49 DAI were more problematic in identifying local stripe rust patterns. The average TGI was not distinguishable between infected and uninfected images until after 42 DAI (Fig. 3). Before 42 DAI, all TGI variability observed in the infected image was caused by natural background variability caused by other sources because they did not correspond to the infection hot spots observed in the later imagery. In the case of 42 DAI, the hot spots are only very vaguely visible in the image of the infected plot. It must be noted that, without knowing the exact location using the following image in the mission sequence, it is not possible to distinguish infected regions from those not affected by stripe rust. There is also no apparent difference in the overall spatial pattern between infected and uninfected images which may be caused by variability in stripe rust infection. For 58 DAI, the TGI image of the infected plot had noticeably higher values compared to the uninfected plot. However, by this point, the disease had dispersed over the entire plot and a distinct pattern was not recognizable. There was only a short time window in which the distinct pattern linked to the stripe rust infection was recognized from plant-level imagery.

In this study, we had the advantage of knowing the pattern of stripe rust in advance using ground-truthing information and imagery obtained after the outbreak. In the real setting, however, it is much harder to decide if a field is infested by stripe rust or not. In many cases, a distinct, hot spot-like pattern can suggest disease spreading in a field. Spatiotemporal profiles calculated over the course of the disease outbreak might be indicative of a specific disease.

However, such patterns may also be caused by factors such as low soil quality (e.g. sandy soils) or soil-borne diseases. Unless we have reliable ground-truth data or secondary information such as soil data, weather data or estimates of disease probability that can be integrated in a disease prediction model, it is difficult to detect the exact nature of plant disease purely from remote sensing data. This might change in coming decades if monitoring systems become available that can collect spatial canopy information with leaf-level resolution. Our study has shown that deep learning can be used to extract important information from unstructured image data, given leaf-level resolution, and can be used to create an image classifier capable of detecting stripe rust disease earlier and more effectively than current drone imagery because it is based on the evaluation of individual leaf symptoms.

5 Conclusion and future trends

As we have seen in the preceding sections, monitoring crop health, especially crop diseases, is not an easy task. For one thing, the time window in which a crop disease develops and spreads over the entire field is very short. Moreover, symptoms in the early stages of the disease are non-specific and can be attributed to many different causes unrelated to the crop disease in question. However, recent advances in remote sensing point in the right direction in better assessing crop health. In addition to technical advances in remote sensing, such as improvements in ground resolution and temporal availability, more remote sensing data and analysis are becoming more open and publicly available. A wider range of sensors and sensor platform designs, including UAV technology, has become more available to end users, and the wider community can play a greater role due to better collection and integration of remote sensing data, e.g. through smartphone technology. This will help end users, e.g. farmers and agricultural consultants, to adopt newly developed remote sensing tools and integrate them in their decision-making processes or decision-support systems.

The Sentinel-2 satellite system is a good example of an open and easily accessible remote sensing product that has recently improved crop monitoring. It provides better and more targeted spectral sampling for crop reflectance assessment. It also has higher spatial and temporal resolution compared to previous open earth observation systems. However, for crop health monitoring, especially for assessment of crop diseases, Sentinel-2 is still too limited in terms of spatial and temporal resolution. Satellite swarms operating with a constellation of many micro-satellites would overcome many of the current limitations because they allow mapping in sub-metre and intra-day degrees of resolution. If we could monitor the crops multiple times a day from space, temporal profiles of reflectance change could be analysed, which may give more specific information about the stress situation in a crop, early identification and possibly classification of crop disease. However, this is currently only possible to some extent for commercial satellite products such as Planet CubeSat data (www.planet.com). Research in an open satellite swarm system achieving this or an even better resolution would be a significant leap forward for crop health monitoring.

With the FLuorescence Explorer (FLEX, ESA) mission launching in 2022, intra-kilometre spatial resolution with 300 × 300 m^2 will be reached in assessing SIF from space (Drusch et al., 2017). Perhaps 1-day spatial resolutions assessing crops at field scale will become technically feasible from space. For crop health monitoring, research into a higher spatial resolution of such systems would be highly desirable as it would enable earlier assessment of crop stresses closely linked to many crop diseases. For UAV platforms, more research needs to be invested to make SIF estimation more practical and affordable. The same is true for hyperspectral systems for satellite programmes and UAV platforms, which are currently either too limited or too expensive for field-scale monitoring.

UAV platforms deliver unprecedented spatial resolution for agriculture. For now, these are the only available remote sensing tools that can distinguish single plants in arable crops. This is significant for agroecological farming because it allows differentiation between different crops used in mixed cropping systems such as intercropping or trap cropping. This may help to assess crop health over the season for each crop separately and may help to identify the benefits of the mixed cropping system. However, drone use is still challenging and needs experts to conduct field surveys although many agricultural service companies are now including drone services for the farmer. Research may lead to fully automated systems that coordinate drone swarms over the field. The drones could transmit data over 5G networks to servers on the ground, where data is fed into information systems that the farmer can use in management decisions. Currently, such systems are not yet feasible due to too many technical and legal issues. However, such automated monitoring systems could achieve a much more robust and timely early warning system for crop diseases than is currently possible.

6 Where to look for further information

This monograph provides a good overview of leaf optics and is a good read to understand the object of study par excellence in optical remote sensing for agriculture:

- Jacquemoud, S. and Ustin, S., 2019. *Leaf Optical Properties*. Cambridge University Press. https://doi.org/10.1017/9781108686457.

This review gives a nice overview about current remote sensing tools for assessing crop diseases:

- Oerke, E.-C. 2020. Remote sensing of diseases. *Annual Review of Phytopathology* 58, 225–252. https://doi.org/10.1146/annurev-phyto-010820-012832.

This conference is held every 2 years in Europe and is the standard conference for precision agriculture:

- ECPA – European Conference for Precision Agriculture.

7 References

Ameye, M., Audenaert, K., De Zutter, N., Steppe, K., Van Meulebroek, L., Vanhaecke, L., De Vleesschauwer, D., Haesaert, G. and Smagghe, G. 2015. Priming of wheat with the green leaf volatile *Z* -3-hexenyl acetate enhances defense against *Fusarium graminearum* but boosts deoxynivalenol production. *Plant Physiology* 167(4), 1671–1684. https://doi.org/10.1104/pp.15.00107.

Anderegg, J., Yu, K., Aasen, H., Walter, A., Liebisch, F. and Hund, A. 2020. Spectral vegetation indices to track senescence dynamics in diverse wheat germplasm. *Frontiers in Plant Science* 10, 1749. https://doi.org/10.3389/fpls.2019.01749.

Bandopadhyay, S., Rastogi, A. and Juszczak, R. 2020. Review of top-of-canopy sun-induced fluorescence (SIF) studies from ground, UAV, airborne to spaceborne observations. *Sensors* 20(4), 1144. https://doi.org/10.3390/s20041144.

Barnes, E. M., Clarke, T. R., Richards, S. E., Colaizzi, P. D., Haberland, J., Kostrzewski, M., Waller, P., Choi, C., Riley, E., Thompson, T., Lascano, R. J., Li, H. and Moran, M. S. 2000. Coincident detection of crop water stress, nitrogen status, and canopy density using ground based multispectral data. *Proceedings of the Fifth International Conference on Precision Agriculture*. Minnesota, USA, 1–15.

Beddow, J. M., Pardey, P. G., Chai, Y., Hurley, T. M., Kriticos, D. J., Braun, H. J., Park, R. F., Cuddy, W. S. and Yonow, T. 2015. Research investment implications of shifts in the global geography of wheat stripe rust. *Nature Plants* 1, 15132.

Bendig, J., Yu, K., Aasen, H., Bolten, A., Bennertz, S., Broscheit, J., Gnyp, M. L. and Bareth, G. 2015. Combining UAV-based plant height from crop surface models, visible, and near infrared vegetation indices for biomass monitoring in barley. *International Journal of Applied Earth Observation and Geoinformation* 39, 79–87. https://doi.org/10.1016/j.jag.2015.02.012.

Calderón, R., Navas-Cortés, J. A., Lucena, C. and Zarco-Tejada, P. J. 2013. High-resolution airborne hyperspectral and thermal imagery for early detection of Verticillium wilt of olive using fluorescence, temperature and narrow-band spectral indices. *Remote Sensing of Environment* 139, 231–245. https://doi.org/10.1016/j.rse.2013.07.031.

Carmona, M., Sautua, F., Pérez-Hérnandez, O. and Reis, E. M. 2020. Role of fungicide applications on the integrated management of wheat stripe rust. *Frontiers in Plant Science* 11, 733. https://doi.org/10.3389/fpls.2020.00733.

Chen, W., Wellings, C., Chen, X., Kang, Z. and Liu, T. 2014. Wheat stripe (yellow) rust caused by *Puccinia striiformis* f. sp. *tritici*. *Molecular Plant Pathology* 15(5), 433–446. https://doi.org/10.1111/mpp.12116.

Chen, Y.-E., Cui, J.-M., Su, Y.-Q., Yuan, S., Yuan, M. and Zhang, H.-Y. 2015. Influence of stripe rust infection on the photosynthetic characteristics and antioxidant system of susceptible and resistant wheat cultivars at the adult plant stage. *Frontiers in Plant Science* 6, 779. https://doi.org/10.3389/fpls.2015.00779.

Choudhury, S., Larkin, P., Meinke, H., Hasanuzzaman, M. D., Johnson, P. and Zhou, M. 2019. Barley yellow dwarf virus infection affects physiology, morphology, grain yield and flour pasting properties of wheat. *Crop and Pasture Science* 70(1), 16. https://doi.org/10.1071/CP18364.

Cui, S., Ling, P., Zhu, H. and Keener, H. M. 2018. Plant pest detection using an artificial nose system: a review. *Sensors* 18(2), 378. https://doi.org/10.3390/s18020378.

Delloye, C., Weiss, M. and Defourny, P. 2018. Retrieval of the canopy chlorophyll content from Sentinel-2 spectral bands to estimate nitrogen uptake in intensive winter wheat cropping systems. *Remote Sensing of Environment* 216, 245–261. https://doi.org/10.1016/j.rse.2018.06.037.

Döring, T. F., Pautasso, M., Finckh, M. R. and Wolfe, M. S. 2012. Concepts of plant health - reviewing and challenging the foundations of plant protection: concepts of plant health. *Plant Pathology* 61(1), 1–15. https://doi.org/10.1111/j.1365-3059.2011.02501.x.

Drusch, M., Moreno, J., Del Bello, U., Franco, R., Goulas, Y., Huth, A., Kraft, S., Middleton, E. M., Miglietta, F., Mohammed, G., Nedbal, L., Rascher, U., Schüttemeyer, D. and Verhoef, W. 2017. The FLuorescence Explorer mission concept–ESA's earth Explorer 8. *IEEE Transactions on Geoscience and Remote Sensing* 55(3), 1273–1284. https://doi.org/10.1109/TGRS.2016.2621820.

European Commission 2020. EU Biodiversity strategy for 2030: bringing nature back into our lives; European Commission: Brussels, Belgium, 2020; Document 52020DC0380, COM(2020) 380 Final.

Fitzgerald, G., Rodriguez, D. and O'Leary, G. 2010. Measuring and predicting canopy nitrogen nutrition in wheat using a spectral index–the canopy chlorophyll content index (CCCI). *Field Crops Research* 116(3), 318–324. https://doi.org/10.1016/j.fcr.2010.01.010.

Ghaffari, R., Laothawornkitkul, J., Iliescu, D., Hines, E., Leeson, M., Napier, R., Moore, J. P., Paul, N. D., Hewitt, C. N. and Taylor, J. E. 2012. Plant pest and disease diagnosis using electronic nose and support vector machine approach. *Journal of Plant Diseases and Protection* 119(5–6), 200–207. https://doi.org/10.1007/BF03356442.

González-Piqueras, J., Lopez-Corcoles, H., Sánchez, S., Villodre, J., Bodas, V., Campos, I., Osann, A. and Calera, A. 2017. Monitoring crop N status by using red edge-based indices. *Advances in Animal Biosciences* 8(2), 338–342. https://doi.org/10.1017/S2040470017000243.

Guanter, L., Zhang, Y., Jung, M., Joiner, J., Voigt, M., Berry, J. A., Frankenberg, C., Huete, A. R., Zarco-Tejada, P., Lee, J. E., Moran, M. S., Ponce-Campos, G., Beer, C., Camps-Valls, G., Buchmann, N., Gianelle, D., Klumpp, K., Cescatti, A., Baker, J. M. and Griffis, T. J. 2014. Global and time-resolved monitoring of crop photosynthesis with chlorophyll fluorescence. *Proceedings of the National Academy of Sciences of the United States of America* 111(14), E1327–E1333. https://doi.org/10.1073/pnas.1320008111.

Guo, A., Huang, W., Dong, Y., Ye, H., Ma, H., Liu, B., Wu, W., Ren, Y., Ruan, C. and Geng, Y. 2021. Wheat yellow rust detection using UAV-based hyperspectral technology. *Remote Sensing* 13(1), 123. https://doi.org/10.3390/rs13010123.

He, K., Zhang, X., Ren, S. and Sun, J. 2016. Deep residual learning for image recognition. *Proceedings of the 2016 IEEE Conference on Computer Vision and Pattern Recognition (CVPR)*. IEEE Publications: Las Vegas, NV, 770–778. https://doi.org/10.1109/CVPR.2016.90.

Hunt, E. R., Daughtry, C. S. T., Eitel, J. U. H. and Long, D. S. 2011. Remote sensing leaf chlorophyll content using a visible band index. *Agronomy Journal* 103(4), 1090–1099. https://doi.org/10.2134/agronj2010.0395.

Jalli, M., Kaseva, J., Andersson, B., Ficke, A., Nistrup-Jørgensen, L., Ronis, A., Kaukoranta, T., Ørum, J.-E. and Djurle, A. 2020. Yield increases due to fungicide control of leaf blotch diseases in wheat and barley as a basis for IPM decision-making in the Nordic-Baltic region. *European Journal of Plant Pathology* 158(2), 315–333. https://doi.org/10.1007/s10658-020-02075-w.

Jacquemoud, S. and Ustin, S. 2019. *Leaf Optical Properties*. Cambridge University Press. https://doi.org/10.1017/9781108686457.

Joiner, J., Guanter, L., Lindstrot, R., Voigt, M., Vasilkov, A. P., Middleton, E. M., Huemmrich, K. F., Yoshida, Y. and Frankenberg, C. 2013. Global monitoring of terrestrial chlorophyll fluorescence from moderate-spectral-resolution near-infrared satellite measurements: methodology, simulations, and application to GOME-2.

Atmospheric Measurement Techniques 6(10), 2803–2823. https://doi.org/10.5194/amt-6-2803-2013.

Kim, J., Sangjun, O., Kim, Y. and Lee, M. 2016. Convolutional neural network with biologically inspired retinal structure. *Procedia Computer Science* 88, 145–154. https://doi.org/10.1016/j.procs.2016.07.418.

Liu, L., Dong, Y., Huang, W., Du, X. and Ma, H. 2020. Monitoring wheat fusarium head blight using unmanned aerial vehicle hyperspectral imagery. *Remote Sensing* 12(22), 3811. https://doi.org/10.3390/rs12223811.

Martinelli, F., Scalenghe, R., Davino, S., Panno, S., Scuderi, G., Ruisi, P., Villa, P., Stroppiana, D., Boschetti, M., Goulart, L. R., Davis, C. E. and Dandekar, A. M. 2015. Advanced methods of plant disease detection. A review. *Agronomy for Sustainable Development* 35(1), 1–25. https://doi.org/10.1007/s13593-014-0246-1.

Oerke, E. C. 2020. Remote sensing of diseases. *Annual Review of Phytopathology* 58, 225–252. https://doi.org/10.1146/annurev-phyto-010820-012832.

Pérez-Bueno, M. L., Pineda, M. and Barón, M. 2019. Phenotyping plant responses to biotic stress by chlorophyll fluorescence imaging. *Frontiers in Plant Science* 10, 1135. https://doi.org/10.3389/fpls.2019.01135.

Rascher, U., Alonso, L., Burkart, A., Cilia, C., Cogliati, S., Colombo, R., Damm, A., Drusch, M., Guanter, L., Hanus, J., Hyvärinen, T., Julitta, T., Jussila, J., Kataja, K., Kokkalis, P., Kraft, S., Kraska, T., Matveeva, M., Moreno, J., Muller, O., Panigada, C., Pikl, M., Pinto, F., Prey, L., Pude, R., Rossini, M., Schickling, A., Schurr, U., Schüttemeyer, D., Verrelst, J. and Zemek, F. 2015. Sun-induced fluorescence - a new probe of photosynthesis: first maps from the imaging spectrometer HyPlant. *Global Change Biology* 21(12), 4673–4684. https://doi.org/10.1111/gcb.13017.

Rouse, W., Haas, R. H., Well, J. and Deering, D. W. 1973. *Monitoring Vegetation Systems in the Great Plains with ERTS. Remote Sensing Center, Texas A&M*. University College Station, Texas.

Savary, S., Willocquet, L., Pethybridge, S. J., Esker, P., McRoberts, N. and Nelson, A. 2019. The global burden of pathogens and pests on major food crops. *Nature Ecology and Evolution* 3(3), 430–439. https://doi.org/10.1038/s41559-018-0793-y.

Schirrmann, M., Giebel, A., Gleiniger, F., Pflanz, M., Lentschke, J. and Dammer, K.-H. 2016. Monitoring agronomic parameters of winter wheat crops with low-cost UAV imagery. *Remote Sensing* 8(9), 706. https://doi.org/10.3390/rs8090706.

Schirrmann, M., Landwehr, N., Giebel, A., Garz, A. and Dammer, K. H. 2021. Early detection of stripe rust in winter wheat using deep residual neural networks. *Frontiers in Plant Science* 12, 469689. https://doi.org/10.3389/fpls.2021.469689.

Singh, K., Wegulo, S. N., Skoracka, A. and Kundu, J. K. 2018. *Wheat streak mosaic virus*: a century old virus with rising importance worldwide: Wheat streak mosaic virus. *Molecular Plant Pathology* 19, 2193–2206. https://doi.org/10.1111/mpp.12683.

Su, J., Liu, C., Coombes, M., Hu, X., Wang, C., Xu, X., Li, Q., Guo, L. and Chen, W.-H. 2018. Wheat yellow rust monitoring by learning from multispectral UAV aerial imagery. *Computers and Electronics in Agriculture* 155, 157–166. https://doi.org/10.1016/j.compag.2018.10.017.

Vega, D., Gazzano Santos, M. I., Salas-Zapata, W. and Poggio, S. L. 2020. Revising the concept of crop health from an agroecological perspective. *Agroecology and Sustainable Food Systems* 44(2), 215–237. https://doi.org/10.1080/21683565.2019.1643436.

Walsh, O. S., Shafian, S., Marshall, J. M., Jackson, C., McClintick-Chess, J. R., Blanscet, S. M., Swoboda, K., Thompson, C., Belmont, K. M. and Walsh, W. L. 2018. Assessment of UAV based vegetation indices for nitrogen concentration estimation in spring wheat. *Advances in Remote Sensing*. Anaesthetics Research Society 07(2), 71–90. https://doi.org/10.4236/ars.2018.72006.

Wang, S., Huang, C., Zhang, L., Lin, Y., Cen, Y. and Wu, T. 2016. Monitoring and assessing the 2012 drought in the Great Plains: analyzing satellite-Retrieved Solar-Induced chlorophyll Fluorescence, Drought Indices, and Gross Primary Production. *Remote Sensing* 8(2), 61. https://doi.org/10.3390/rs8020061.

Yoshida, Y., Joiner, J., Tucker, C., Berry, J., Lee, J.-E., Walker, G., Reichle, R., Koster, R., Lyapustin, A. and Wang, Y. 2015. The 2010 Russian drought impact on satellite measurements of solar-induced chlorophyll fluorescence: insights from modeling and comparisons with parameters derived from satellite reflectances. *Remote Sensing of Environment* 166, 163–177. https://doi.org/10.1016/j.rse.2015.06.008.

Zheng, Q., Huang, W., Cui, X., Shi, Y. and Liu, L. 2018. New spectral index for detecting wheat yellow rust using Sentinel-2 multispectral imagery. *Sensors* 18(3), 868. https://doi.org/10.3390/s18030868.

Chapter 3

Advances in remote/aerial sensing techniques for monitoring soil health

Jeffrey P. Walker and Nan Ye, Monash University, Australia; and Liujun Zhu, Monash University, Australia and Yangtze Institute for Conservation and Development, Hohai University, China

1 Introduction

Due to its dependencies on weather, topography, land surface type, and texture, the status of soil varies significantly in time and space. Although conventional ground monitoring stations have been widely used to capture the temporal variation of environment parameters at point-based scales, their applications are limited by poor spatial representativeness, high cost of installation and maintenance, and interference with farming activities. Remote sensing provides a cost-effective alternative for soil health monitoring. In principle, sensors are mounted on a remote sensing platform, such as satellite, aircraft, drone or vehicle, and subsequently used to measure the energy from the soil surface. The properties of the received signals can be linked to soil-related parameters (e.g. soil moisture, roughness, salinity) within the field of view through physical and/or empirical models. Compared with in situ approaches, remote sensing techniques have the desired capability of providing spatially explicit maps of soil parameters, with up to global coverage. Several remote sensing techniques (e.g. optical, microwave, nuclear, and gravity) have been developed for monitoring the land surface conditions. However, due to its

http://dx.doi.org/10.19103/AS.2022.0107.05

all-weather capability and ability to penetrate the vegetation layer, even into the soil itself, microwave remote sensing has become widely acknowledged as a key technique for the agricultural industry.

Microwave remote sensing measures the electromagnetic radiation in the microwave spectrum, ranging from 0.3 GHz to 40 GHz. This spectrum region is further divided into eight bands as listed in Table 1. By measuring intensity, polarization, phase, and/or other properties of microwave radiation, the dielectric and/or geometric properties of the sensed target can be detected. For soil material, the dielectric constant varies with soil water content from approximately 3.5 for very dry soil to approximately 40 for saturated soil (Ulaby et al., 1986), forming the fundamental basis of microwave remote sensing for soil moisture content. The relationship between soil dielectric constant and soil moisture is shown in Fig. 1, with both real and imaginary parts of the soil dielectric constant increasing with soil water content, and the relationship influenced by soil particle size distribution (Ulaby et al., 1986).

A dielectric constant is a complex number ($\varepsilon' - j \cdot \varepsilon''$), where $j = \sqrt{-1}$, with the real (ε') and imaginary (ε'') parts determining the propagation speed of the electromagnetic wave through the soil medium and the loss of electromagnetic energy, respectively. Normally, it is expressed as a relative value which is the ratio of the dielectric constant of material to that of free space. Due to the three phases of soil, soil dielectric constant is a combination of the individual constituents including soil particles, water, and air components. The value of soil dielectric is also affected by other factors, including soil texture, temperature, salinity, and electromagnetic wavelength. To relate soil dielectric constant to volumetric soil moisture, several semi-empirical mixing models have been developed to estimate soil dielectric constant from the knowledge on electromagnetic wavelength, soil texture, bulk density, and salinity (Dobson et al., 1985; Hallikainen et al., 1985; Mironov et al., 2004; Wang and Schmugge, 1980). In these, the dielectric constant of moist soil was found to be slightly dependent on temperature, and so under most natural temperature conditions,

Table 1 Microwave band designations

Band designation	Wavelength (cm)	Frequency (GHz)
P	100.0-30.0	0.3-1.0
L	30.0-15.0	1.0-2.0
S	15.0-7.50	2.0-4.0
C	7.50-3.75	4.0-8.0
X	3.75-2.40	8.0-12.5
K_u	2.40-1.67	12.5-18.0
K	1.67-1.10	18.0-26.5
K_a	1.10-0.75	26.5-40.0

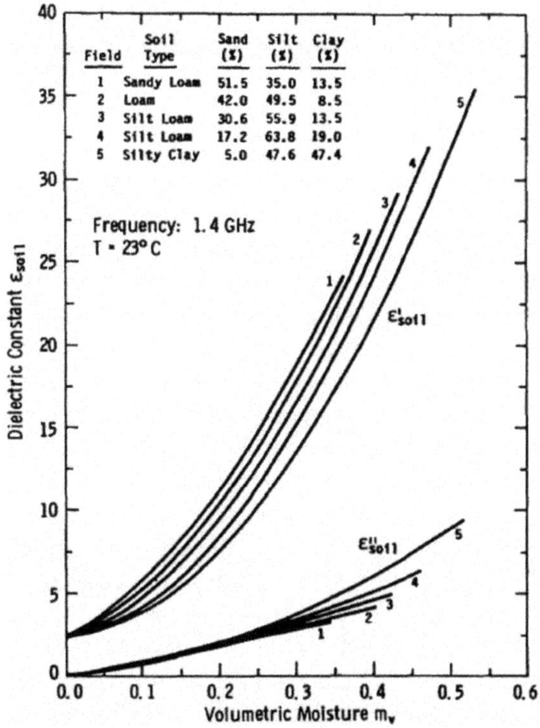

Figure 1 Soil dielectric constant as a function of volumetric soil moisture for five soils at 1.4 GHz. Smooth curves were drawn through measured data points. Source: Ulaby et al., (1986).

the effect of temperature on soil dielectric constant can be ignored. However, when soil gets frozen, its dielectric constant is reduced significantly since the dielectric constant of the water constituent changes from that of liquid water (approximately 80) to that of ice (approximately 3).

Based on the provision of electromagnetic radiation sources, microwave remote sensing techniques are divided into two categories: active and passive. Active microwave remote sensing instruments, known as radars, transmit a pulse of microwave radiation and measure the signal scattered back in the direction of the sensor. The coefficient between the power of the transmitted and received signal is dependent on the reflectivity of the target, which in soil material is related to the soil moisture content. In contrast, passive remote sensing instruments, referred to as radiometers, do not transmit any electromagnetic waves but only receive the self-emitted radiation from the land surface at a specific microwave frequency. The intensity of microwave emission of soil relies mainly on soil temperature and soil surface emissivity, which in turn correlates with soil moisture content through the soil dielectric constant.

The following sections provide a general description of the principles and features of active and passive microwave remote sensing of soil moisture.

2 Active microwave remote sensing

Active microwave remote sensing has been widely used to map soil moisture at regional to global scales. The radar system generally consists of a transmitter, receiver, antenna, and processer. An electromagnetic pulse in the microwave frequencies is generated in the transmitter and transmitted to the target through the radar antenna. Over land surfaces, a part of the transmitted electromagnetic wave is scattered by the vegetation canopy and/or soil surface, and returned back to the radar system. The backscatter signal is collected by the same antenna and its intensity is measured by the receiver. The coefficient between the power of the transmitted and backscattered signal is obtained in the processer, which can be related to the water content of the sensed soil target. The most common active microwave mapping configuration is the synthetic aperture radar (SAR), which can provide a spatial resolution in the order of tens of meters over a swath of 50–500 km. Currently, five space-borne SAR systems are operating at microwave frequencies for soil moisture observations: European Space Agency (ESA)'s ERS-1/2 C-band SAR, ESA's ENVISAT (ERS-3) C-band ASAR (Advanced SAR), the Canadian C-band RADARSAR-1/2, the Japanese L-band ALOS (Advanced Land Observing Satellite) PALSAR (Phased Array type L-band SAR), and the German X-band TerraSAR.

The electromagnetic waves sent and received by radars are normally polarized either horizontally (H) or vertically (V), and therefore there can have four polarization combinations: HH, VV, HV, and VH, where the first and second letters represent the polarization of transmitted and received signals, respectively. The backscattering coefficient σ_{pp} in decibels [dB] at the polarization of p is used to describe the intensity of the backscattered radiation (Ulaby et al., 1982). For a given target, the backscattering coefficient is dependent on wave polarization, frequency, and incidence angle (Ulaby et al., 1982). Under bare soil conditions, the backscattering coefficient obtained using a radar system operating at consistent polarization, frequency, and incidence angle is affected by the dielectric constant of the soil and surface roughness (Ulaby et al., 1986). Under vegetated soil conditions, the backscattering coefficient is dependent also on the attenuation effect of the vegetation layer which makes the backscattering response more complicated (Ulaby et al., 1982). In addition, since the soil moisture retrieval is normally based on a flat land surface assumption, the topographic relief causes a variation of local incidence angle and significantly affects the backscattered signal (Van Zyl et al., 1993).

The total co-polarized backscatter σ_{pp}^{T} from the land surface is the sum of three components, given as:

$$\sigma_{pp}^{T} = \sigma_{pp}^{vol} + \sigma_{pp}^{S} \cdot \exp(-2 \cdot \tau_C) + \sigma_{pp}^{int}, \quad (1)$$

where the first term is the backscatter from the vegetation volume σ_{pp}^{vol}, and the second term is the soil surface backscatter σ_{pp}^{S} attenuated by a vegetation layer with opacity of $\tau_C$during both transmitting toward and reflecting from the soil surface. The third term σ_{pp}^{int} is the interaction between the vegetation and soil surface (Ulaby et al., 1996). For bare or low vegetated soils, the total backscattering coefficients σ_{pp}^{T} is dominated by σ_{pp}^{S} and thus dependent mainly on the soil moisture and surface roughness, while for highly vegetated soil, the σ_{pp}^{T} is determined primarily on the volumetric scattering from the vegetation canopy σ_{pp}^{vol}. Numerous theoretical, empirical, and semi-empirical models have been developed to retrieve the soil electric constant and subsequent soil moisture content from radar backscattering data (Dubois et al., 1995; Fung et al., 1992; Oh et al., 1992; Shi et al., 1995) (Fig. 2).

2.1 Theoretical approaches

Theoretical approaches have been developed based on the diffraction theory of electromagnetic waves to describe the microwave backscattering from land surfaces with known roughness characteristics. Their applicability is limited to the frequency of electromagnetic waves and the range of surface roughness (D'urso and Minacapilli, 2006; Fung et al., 1992).

Most of the currently used surface scattering models have been derived from the small perturbation model (Rice, 1951) and Kirchhoff model (Beckmann and Spizzichino, 1963), which are restricted to slightly rough surfaces and very rough surfaces, respectively. The integral equation model (IEM) (Fung, 1994b; Fung et al., 1992) combines these two theories and is thus applicable to a wider

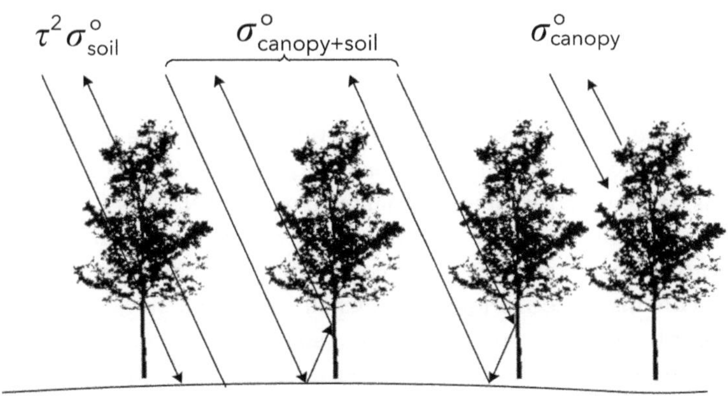

Figure 2 Backscattering mechanics of vegetated soil. Source: modified from Ulaby et al. (2014).

range of roughness conditions than conventional models, such as the physical optical model and geometric optical mode (Fung, 1994b; Shi et al., 2005).

Although the theoretical models can predict the general variation of backscattering coefficient in response to changes in roughness and soil moisture content (Dubois and Van Zyl, 1994), their complexity and the restrictive requirement for the parameterization of the vegetation and soil surface layer limits their effective applicability for the soil moisture retrieval (Ulaby et al., 1986).

2.2 Empirical approaches

Being limited by their validity regions, theoretical backscattering models are not valid for many natural land surface conditions. In addition, theoretical models fail to estimate backscatter in good agreement with experimental radar backscatter measurements (Oh et al., 1992; Walker et al., 2004). Therefore, many empirical models have been developed from experimental measurements to establish the relationship between soil moisture and backscattering observations (Walker et al., 2004).

The most commonly used empirical method is a linear assumption between soil moisture and radar backscattering polarization index. For example, Shoshany et al. (2000) proposed an empirical soil moisture retrieval method using the normalized backscatter moisture index (NBMI) which is defined as:

$$\mathrm{NBMI} = \frac{\sigma_{t1} - \sigma_{t2}}{\sigma_{t1} + \sigma_{t2}}, \tag{2}$$

where σ_{t1} and σ_{t2} are the backscatter coefficients at different time steps. Subsequently, the volumetric soil content is calculated through:

$$SM_v = a \cdot \mathrm{NBMI} + b, \tag{3}$$

where a and b are empirical parameters regressed from in situ soil moisture measurements. This approach estimates soil moisture through change detection rather than a direct relationship between microwave backscattering observations and soil moisture content (Engman, 1990; Kite and Pietroniro, 1996). It is based on an assumption that the change of NBMI is caused solely by the variation of soil moisture, and therefore the effects of other factors including soil texture, surface roughness, and vegetation which are relative temporally consistent are minimized (Engman and Chauhan, 1995).

Although empirical methods can result in an accurate soil moisture retrieval with less complexity and reduced calculation cost than theoretical methods, their applications are restricted in the calibration conditions (Chen et al., 1995; Dubois et al., 1995). To establish a widely applicable empirical relationship for soil moisture retrieval from radar backscattering observations, a large number of experimental measurements are required (Oh et al., 1992), while current empirical models are generally developed from a limited amount of field measurements and therefore valid only under the specific land surface conditions from which they were derived (Wang and Qu, 2009).

2.3 Semi-empirical approaches

By combining theoretical and empirical approaches, semi-empirical models of backscattering have been developed based on a theoretical foundation with model parameters derived from experimental data.

Oh et al. (1994) developed the first semi-empirical backscattering model, and found that the depolarization ratio (σ_{vh}/σ_{vh}) is very sensitive to soil moisture, and developed a semi-empirical model based on empirical fitting of scatterometer measurements over bare soil surfaces with different roughness conditions. In the Dubois et al. (1995) method, the co-polarized backscattering coefficients σ_{hh} and σ_{vv} are related to the surface dielectric constant, incidence angle, electromagnetic frequency, and root-mean-squared height of soil surface in a nonlinear way.

Compared with empirical models, semi-empirical backscattering models are not expected to have the same site-specific problems (Walker et al., 2004). Generally, these models are more suitable for bare soil surface conditions than vegetated soil conditions.

3 Passive microwave remote sensing

Due to the thermal motion of atoms, any objects at a physical temperature above absolute zero (~−273.15°C or 0 K) radiates electromagnetic energy. A radiometer is used to measure the intensity of this emission, which increases proportionally with the increase in temperature. To explain the relationship between physical temperature and microwave emission, the blackbody concept is used, as introduced by Planck in his quantum theory in 1901. A blackbody is defined as an ideal material that absorbs all incidence radiation and reflects none; it is also a perfect emitter since otherwise its temperature would infinitely increase. Therefore, for a thermodynamic equilibrated blackbody, it emits all absorbed energy outward. In addition, the intensity of electromagnetic emission can be quantified using the term brightness temperature which is defined as the physical temperature of the blackbody emitting the same amount of energy. In contrast to a blackbody, a white body is defined as a perfect reflector that reflects all incidence energy and therefore emits none. Actually, most materials behave between blackbody and white body (refereed as gray body), meaning that a part of incidence energy is reflected with the remaining absorbed and emitted when at thermodynamic equilibrium. Using the emissivity (e) to describe the ability of materials to emit electromagnetic energy, the brightness temperature (TB) of the material is expressed as:

$$TB_p = e_p \cdot T, \tag{4}$$

where T is the physical temperature of the material in Kelvin (K), and the subscript p indicates the polarization, either horizontal or vertical. This equation is derived from Plank's blackbody radiation law through the Rayleigh–Jeans approximation for microwave frequencies (Njoku and Entekhabi, 1996; Ulaby et al., 1981b). Therefore, the emissivity of a gray body varies from 0 for a white body to 1 for a blackbody.

For soil material, the emissivity varies from ~0.95 for dry soil (with moisture content of 0.05 m^3/m^3) to ~0.6 for wet soil (with moisture content of 0.4 m^3/m^3), depending on the electromagnetic wavelength, incidence angle, surface roughness, and soil properties (Jackson and Le Vine, 1996; Njoku and Entekhabi, 1996). Assuming soil at a physical temperature of 300 K, this variation in emissivity corresponds to a brightness temperature variation of 90 K (Njoku and Entekhabi, 1996), which is much larger than the typical radiometric sensitivity of microwave radiometers (approximately 1 K).

Following Kirchoff's reciprocity theorem, the microwave emissivity (e) of the target can be related to its microwave reflectivity through:

$$\Gamma_p = 1 - e_p. \tag{5}$$

The reflectivity is dependent mainly on the polarization, electromagnetic wavelength, surface roughness, and dielectric constant of materials. For flat specular surfaces, the reflectivity (Γ_p^*) is determined by the Fresnel equation as:

$$\Gamma_H^* = 1 - \left| \frac{\cos\theta - \sqrt{\varepsilon - \sin^2\theta}}{\cos\theta + \sqrt{\varepsilon - \sin^2\theta}} \right|^2 \tag{6}$$

$$\Gamma_V^* = 1 - \left| \frac{\varepsilon \cdot \cos\theta - \sqrt{\varepsilon - \sin^2\theta}}{\varepsilon \cdot \cos\theta + \sqrt{\varepsilon - \sin^2\theta}} \right|^2, \tag{7}$$

where ε is the relative dielectric constant of the material, and θ is the incidence angle of microwave radiation. The subscripts H and V represent horizontal and vertical polarizations, respectively.

The intensity of the emission at microwave frequencies measured by a radiometer, known as brightness temperature, can therefore be related to the dielectric constant through reflectivity. For bare soil with a smooth surface, the emissivity of soil at a given polarization and incidence angle can be determined using Eqns. (5)–(7) from volumetric soil moisture content and soil texture properties. However, for more general land surface conditions, the effects of temperature profiles in the soil, the roughness of the soil surface, and the vegetation coverage over the soil layer are significant to the relationship between brightness temperature observations and soil moisture (Choudhury et al., 1979; Jackson and Schmugge, 1991; Njoku and Entekhabi, 1996).

3.1 Impact of vertical soil moisture and temperature profiles

In natural land surfaces, soil moisture is not consistent in-depth, and passive microwave observations are only affected by water content in the top soil layer. The effective depth of estimated soil moisture from emitted radiation at microwave frequencies, known as the penetration depth γ_D, is defined as the depth above which soil contributes 63% (1−1/e) of the microwave emission (Ulaby et al., 1981b), and can be expressed as a function of electromagnetic wavelength (λ) and complex dielectric constant of soil ($\varepsilon'_S - j \cdot \varepsilon''_S$):

$$\gamma_D = \frac{\lambda\sqrt{\varepsilon'_S}}{2\pi\varepsilon''_S}. \tag{8}$$

The penetration depth is very sensitive to the soil moisture conditions, with γ_D varying for L-band from approximately 75 cm for dry soil with a dielectric constant of $5 - j \cdot 0.1$ to approximately 3.7 cm for wet soil with a dielectric constant of $30 - j \cdot 5$. Therefore, the penetration depth is a significant parameter to determine the thickness of the soil surface layer for which the variations in moisture content and temperature make a major contribution to the microwave emission.

The simple relationship of microwave emission in Eqn. (4) is based on an assumption that soil moisture and temperature are constant with depth. At low microwave frequencies, the top several centimeters of soil makes a major impact on the microwave emission (Njoku and Entekhabi, 1996), and in natural soil the vertical distributions of moisture content and temperature can be substantial over this layer, determined by solar radiation, precipitation, evapotranspiration, infiltration rate, and vegetation root distribution. Therefore the uniform soil moisture and temperature profile assumption is not satisfied to estimate soil brightness temperature and emissivity over most natural land surfaces. To account for this variability, an effective soil temperature (T_{eff}), the equivalent temperature in a uniform profile having the same microwave response to the nonuniform temperature profile can be calculated through radiative transfer theory (Choudhury et al., 1982) as:

$$T_{eff} = \int_0^\infty T_S(z) \cdot \alpha(z) \cdot \exp\left[-\int_0^z \alpha(z')\,dz'\right] dz, \tag{9}$$

where $T_S(z)$ is the soil temperature at depth z, and the attenuation coefficient α is dependent on the real and imaginary parts of complex soil dielectric constant as :

$$\alpha(z) = (4\pi/\lambda) \cdot \varepsilon''_S(z)/2 \cdot (\varepsilon'_S(z))^{1/2}, \tag{10}$$

where λ is the electromagnetic wavelength. Using this theoretical method, the effective soil temperature can be calculated from the measured soil temperature profile and moisture content, which in turn can be used to estimate the profile of soil dielectric constant. However, the required soil moisture and temperature profile data are only available in limited controlled experiments and so are difficult to obtain over large areas. Therefore, Choudhury et al. (1982) and Wigneron et al. (2001) developed simple linear parameterizations based on Eqn. (9) and experimental data collected at L-band, expressed as:

$$T_{eff} = T_{deep} + C_t \cdot (T_{surf} - T_{deep}), \tag{11}$$

where T_{deep} and T_{surf} are the deep soil temperature (approximately at 50 cm or 100 cm) and surface temperature (approximately corresponding to a depth interval of 0–5 cm). The parameter C_t is an empirical attenuation coefficient to determine the proportion of the contributions from the deep and surface soil layers to the effective soil temperature. Here the surface temperature can be estimated from thermal infrared observations, or near-surface air temperature derived from meteorological data, while the deep soil temperature can be modeled based on geographic location and season (Choudhury et al., 1982). The constant values of the C_t parameter were calibrated at several frequency bands, and C_t found to be equal to 0.246 at L-band (Choudhury et al., 1982). In reality, the C_t, similar to penetration depth, is also influenced by soil moisture. For very dry conditions, soil layers at depth (deeper than 1 m for dry sand) contribute significantly to the microwave emission from soil, with the C_t being lower than 0.5. In contrast for very wet conditions, the soil emission derives mainly from layers at the soil surface and $C_t \approx 1$. To take the dependence of $C_t$on soil moisture into account, Wigneron et al. (2001) proposed a slightly improved formula based on Eqn. (11) in which C_t is a function of soil moisture:

$$C_t = (SM_{surf}/w_0)^{b_{w0}}, \tag{12}$$

where SM_{surf} [m^3/m^3] is the volumetric water content in the top 0–3 cm soil. The w_0 [m^3/m^3] and b_{w0} are semi-empirical parameters depending on the soil properties. The long-term suitability of Eqn. (12) was tested over several sites at the seasonal to interannual temporal scales (De Rosnay et al., 2006). The value of w_0 was found to be close to 0.3 m^3/m^3 over two bare soil sites: INRA Avignon (Wigneron et al., 2001) and SMOSREX (De Rosnay et al., 2006). The value of b_{w0} was close to 0.3 over the INRA Avignon site and close to 0.65 over the SMOSREX site.

3.2 Impact of surface roughness

Generally, natural land surfaces are not flat and smooth like the assumption made in Eqns. (9) and (10). Newton and Rouse Jr (1980) and Wang (1983) found from field measurements that the rougher the soil surface, the higher the soil emissivity and the lower the sensitivity to soil moisture content (see Fig. 3). It was also found that the effects of surface roughness decreased with increased wavelength.

In order to take the effect of surface roughness into account, scattering of the radiation at the soil-air interface was introduced in Eqns. (9) and (10). The reflectivity (Γ_p) of a rough surface generally consists of two components: the noncoherent component (Γ_p^{non}) and the coherent component (Γ_p^{coh}) (Shi et al., 2002). Accordingly, Γ_p^{non} is calculated by integrating over the upper hemisphere bistatic scattering coefficient ($\sigma_{pp}(\vartheta,\vartheta_S,\varphi_S-\varphi)$), which characterizes the scattering of radiation from an incidence direction (ϑ,φ) to the scattered direction (ϑ_S,φ_S):

$$\Gamma_p^{non}(\theta) = 1/(4\pi\cdot\cos\theta)\cdot\int_0^{\pi/2}\int_0^{2\pi}\left(\sigma_{pp}(\theta,\theta_S,\phi_S-\phi)\right) + \sigma_{pq(\theta,\theta_S,\phi_S-\phi)}\sin\theta_S d\phi_S d\theta_S, \quad (13)$$

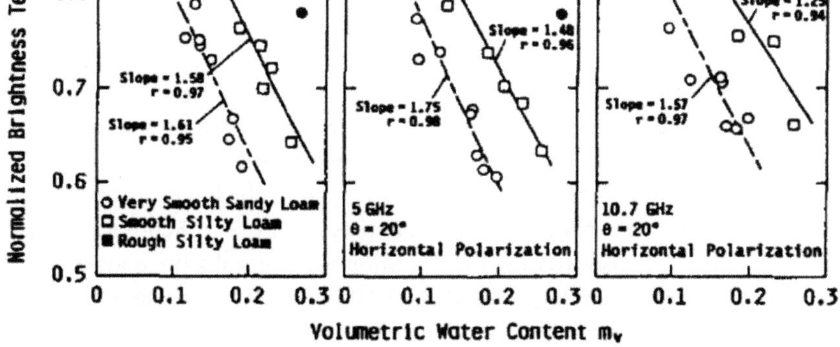

Figure 3 Variations in brightness temperature as a function of moisture content; for soils of different roughness at 1.4 GHz, 5 GHz and 10.7 GHz. Source: Wang et al. (1983).

where the subscripts p and q indicate horizontal and vertical polarizations, or vice versa. The subscript s indicates the direction of the scattered radiation. The bistatic scattering coefficients σ_{pp} and σ_{pq} can be calculated from complex modeling approaches such as the advanced IEM (Chen et al., 2003). The Γ_p^{coh} is expressed as a function of the Fresnel reflectivity (Γ_p^*) such that:

$$\Gamma_p^{coh}(\theta) = \Gamma_p^*(\theta) \cdot \exp\left[-(4\pi \cdot SD \cdot \cos\theta/\lambda)^2\right] \tag{14}$$

where SD is the standard deviation of the surface height and λ is the electromagnetic wavelength.

This approach is useful to understand the physics of the scattering effect of a rough soil surface. For instance, Shi et al. (2002) demonstrated a large difference in roughness effects at different incidence angles and polarizations. At large incidence angles ($\vartheta \approx 50°$), the soil emission was found to increase as the geometric surface roughness increases at horizontal polarization. This has a good agreement with earlier experimental observations (Choudhury et al., 1979; Wang, 1983; Wang et al., 1983). Conversely, soil emission at vertical polarization decreases as the geometric surface roughness increases. However, this approach does not account for the fine-scale roughness (Schwank and Mätzler, 2006), and only surface scattering effects were considered that ignored volume scattering effects which might substantially affect the soil emissivity. In addition, it is difficult to calculate the emissivity from the scattering coefficients obtained using complex theoretical models and performing a two-dimensional integral on the soil upper hemisphere.

A simple semi-empirical model of soil reflectivity for the rough surface was initially developed by Wang and Choudhury (1981) based on two best-fit parameters Q_R and H_R:

$$\Gamma_P(\theta) = \left[\left(1 - Q_R \cdot \theta_p^*(\theta)\right) + Q_R \cdot \theta_q^*\right] \cdot \exp\left(-H_R \cdot \cos^2(\theta)\right) \tag{15}$$

where the subscripts p and q indicate horizontal and vertical polarizations, or vice versa. The Q_R is a polarization mixing parameter, and the H_R is a surface height parameter that can be related to the SD of surface heights. Wang et al. (1983) considered in a more detailed study that the $\cos\theta$ dependence was much too strong. In addition, H_R in Eqn. (15) increases with surface roughness effects resulting in an increase in soil emissivity at both H and V polarizations, which is in contradiction with theoretical analysis (Mo and Schmugge, 1987; Shi et al., 2002). Consequently, the H_R parameter should be considered as dependent on incidence angle and polarization, and so a generalized semi-empirical equation of roughness effects has been proposed as (Wigneron et al., 2007):

$$\Gamma_P(\theta)=\left[\left(1-Q_R(\theta)\right)\cdot\Gamma_p^*(\theta)+Q_R(\theta)\cdot\Gamma_q^*\right]\cdot\exp\left(-H_{Rp}(\theta)\cdot\cos^{N_{Rp}}(\theta)\right) \quad (16)$$

In this generalized formulation, the dependence of Q_R and H_R on reflectivity and polarization is accounted for and the N_{Rp} exponent is inserted in the exponential term.

The Q_R was found to be dependent on the electromagnetic frequency and has very small values at L-band (from 0 to 0.12 for three soil types) (Wang et al., 1983). This is in agreement with most of the published studies based on a large experimental data set which considered that $Q_R = 0$ (Mo and Schmugge, 1987; Wegmuller and Mätzler, 1999; Wigneron et al., 2001). The dependence of the model roughness parameter $H_{Rp}(\theta)$ on the surface roughness characteristics, such as SD and autocorrelation length (LC), is not well known. Two studies (Mo and Schmugge, 1987; Wigneron et al., 2001) found that the best geophysical parameters to model H_R were the slope parameter (m = SD/LC) and the surface soil moisture *SM*. The dependence of H_R on soil moisture content could be explained by a volume scattering effect such that as the soil dries out deeper layers of soil contribute to the emission. Wigneron et al. (2007) suggested that the spatial fluctuations of the dielectric constant within the soil volume may be strong during drying out, having an important 'dielectric' roughness effect, and therefore H_R could be considered as an effective parameter that accounts for (1) 'geometric roughness' effects, corresponding to spatial variations of soil surface height, and (2) 'dielectric roughness' effects, corresponding to the variation of the dielectric constant at the soil surface. The results obtained by Escorihuela et al. (2007) over the SMOSREX (De Rosnay et al., 2006) bare soil confirmed the general soil moisture dependence of H_R and found that a linear dependence was preferable to the exponential one given by (Wigneron et al., 2001).

Wang et al. (1983) found that $N_R = 0$ was consistent with measurements at frequencies of 1.4, 5, and 10.7 GHz. This result was also found in the studies at L-band (Mo and Schmugge, 1987; Wigneron et al., 2001). Based on long-term measurements over a relatively smooth soil during the SMOSREX experiment, Escorihuela et al. (2007) found that $N_R \approx 1$ at horizontal polarization and $N_R \approx -1$ at vertical polarization.

3.3 Impact of vegetation canopy

Over vegetated soil, the microwave emission from the soil layer is affected by the vegetation canopy layer which attenuates (absorbs and scatters) the soil emission and adds its own contribution to the overall microwave emission. As vegetation density increases, the contribution of the vegetation layer increases and that of the soil layer decreases. When the density of the canopy is adequate, the radiation emitted from the soil layer is masked, and the observed microwave emission relies mainly on the vegetation. The magnitude of the attenuation effect of the canopy depends upon the wavelength and the vegetation water content (VWC).

To date, a number of models have been developed to estimate the microwave emission from the soil-vegetation layer (Jackson et al., 1982; Kirdiashev et al., 1979; Meesters et al., 2005; Mo et al., 1982; Ulaby and Wilson, 1985; Wigneron et al., 1995). In these models, the microwave emission from the vegetated soil surface is usually expressed as a zero-order solution of the radiative transfer equations since it assumes that the scattering phase matrix term can be neglected (Mätzler et al., 2006; Ulaby et al., 1981b, 1982, 1986). The τ - ω model (Mo et al., 1982) is therefore defined as:

$$TB_p = (1-\Gamma_p)\cdot\gamma_p\cdot T_s + (1-\omega_p)\cdot(1-\gamma_p)\cdot T_v + (1-\omega_p)\cdot(1-\gamma_p)\cdot\gamma_p\cdot\Gamma_p\cdot T_v \quad (17)$$

where T_v and T_s are the effective temperature [K] of the vegetation and soil layers. The ω_p and γ_p parameters are the single scattering albedo and transmissivity of the vegetation layer, respectively, and Γ_p is the reflectivity of a rough soil surface at p polarization (either horizontal or vertical). The microwave emission from a vegetated soil surface is considered as the sum of three parts corresponding to the three terms in Eqn. (17). The first term represents the upward radiation from the soil layer and attenuated by the overlying vegetation. The second term is the upward radiation directly from the vegetation layer. The third term denotes the downward radiation from the vegetation layer, reflected by the soil surface, and attenuated by the vegetation layer again.

The single scattering albedo ω_p indicates the scattering of the soil emissivity, and is a function of vegetation geometry. At microwave frequencies, the value of ω_p is almost zero, varying between 0.05 and 0.10 (Jackson and Schmugge, 1991; Wigneron et al., 2004, 2007). The transmissivity of the vegetation γ_p can be further defined as a function of the vegetation optical depth at nadir (τ_{NAD}) and the incidence angle (θ):

$$\gamma_p = \exp\left[-\tau_{NAD}\cdot\left(tt_p\cdot\sin^2(\theta)+\cos^2(\theta)\right)\cdot\cos^{-1}(\theta)\right] \quad (18)$$

where tt_p is an empirical parameter in relation to vegetation structure and polarization. The optical depth (τ_{NAD}) is dependent on the vegetation density and frequency, and can be linearly related to the VWC (kg/m^2) at L-band using an empirical parameter (b) (Van de Griend and Wigneron, 2004):

$$\tau_{NAD} = b\cdot \mathrm{VWC}. \quad (19)$$

Alternatively, the vegetation optical depth could also be linearly related to the log of the normal difference vegetation index (NDVI) (Burke et al., 2001) using two empirical factors (α and β):

$$\tau_{NAD} = \alpha + \beta \cdot (1 - \log(\text{NDVI})) \tag{20}$$

There is some experimental evidence indicating possible polarization and angle dependence of both τ and ω. However, this dependence was found mainly from experimental data collected over non-isotropic vegetation, such as vertical stalks in tall grasses, grains, and maize (Hornbuckle et al., 2003; Kirdiashev et al., 1979; Wigneron et al., 1995). The canopy and stem structure of most vegetation covers are randomly oriented, and the effects of any systematic orientation of vegetation would be mostly minimized at satellite scales (Martinez-Vazquez et al., 2009; Owe et al., 2001).

For soil moisture retrieval from passive microwave observations of Soil Moisture and Ocean Salinity (SMOS) and SMAP, the vertical temperature gradients within the soil and vegetation layers are assumed uniform, since it reaches equilibrium at around their 6 am/pm overpass times. Therefore, Eqn. (17) can be simplified assuming equaled soil and vegetation temperatures ($T_s = T_v$) (Hornbuckle and England, 2005), expresses as:

$$TB_p = \left[(1-\omega_p)\cdot(1-\gamma_p)\cdot(1+\gamma_p\Gamma_p)+(1-\Gamma_p)\cdot\gamma_p\right]\cdot T_s \tag{21}$$

The sensitivity of the microwave brightness temperature observation to the water content of the soil layer decreases with the increase of the vegetation opacity depth (Jackson and Schmugge, 1991). The brightness temperature variation reduced by the attenuation effect of the vegetation canopy is much larger than the noise sensitivity threshold of a microwave radiometer (typically <1 K). Therefore, the passive microwave technique can obtain a large signal-to-noise ratio (SNR) for accurate soil moisture remote sensing.

4 Remote sensing of soil properties

Remote sensing is a means to acquire information about an object by measuring its emitted, reflected, and/or scattered radiation from a distance without physical contact. Based on the provision of radiation sources, remote sensing techniques can be divided into active and passive categories. Active remote sensing measures the radiation emitted primarily from an illumination source, but also secondary reflection/scattering from surface of the sensed object. In contrast, passive remote sensing only measures the self-emitted radiation from the sensed object due to its inherent physical temperature, which normally has a lower SNR than active remote sensing. In most cases, the remotely sensed radiation is in the form of electromagnetic waves, but other forms (e.g. acoustic wave, neutron scattering, gamma attenuation, and gravity change) can also be used for ground, airborne, and satellite remote sensing. According to the electromagnetic frequency spectrum, remote sensing can be categorized into optical and microwave remote sensing. Optical remote sensing has a high spatial resolution and a high spectral resolution, with current optical remote sensing satellites reaching submeter (<1 m) spatial resolutions (e.g. WorldView, GeoEye, QuickBird, Ikonos, Pleiades,GF-2, and KF01) and ~10 nm spectral resolutions (e.g. MightSat-II, EO-1, PROBA, ENVISAT-1, MRO, HySI, TacSat-3, HICO, GOSAT, OCO-2, and EnMAP). However, optical imagery is adversely affected by the atmosphere and clouds, temporal repeat, and its inability to penetrate the land surface cover. With longer wavelength, microwave remote sensing has a high penetration capability and a sensitivity to target dielectric constant. But microwave sensors require a larger antenna and achieve a lower spatial resolution. Being limited by current antenna technology, the best spatial resolution that can be achieved by L-band (1–2 GHz) radiometers from the space is only ~40 km (e.g. SMOS and SMAP).

4.1 Soil moisture

Extensive research has been conducted on soil moisture retrieval from remote sensing techniques over the past four decades, with the main difference in techniques being the measured frequency within the electromagnetic spectrum and the source of the radiation. Optical methods exploit the relationship between soil moisture and spectral reflectance in the visible/near-infrared bands to empirically determine the soil moisture in the top millimeters or so of the soil surface, while thermal infrared-based methods start from capturing the thermal inertia of soil which is then related to the soil moisture through the thermal conductivity and/or the heat capacity (Petropoulos et al., 2015). In general, the absorption of visible bands increases as soil moisture is increased, resulting in a decrease of reflectance at visible bands (Gao et al., 2013; Liu et al., 2002). However, this tenuous relationship and the optimal spectrum bands vary significantly from site to site (Liu et al., 2009), because of the additional dependence of reflectance on organic matter content, roughness, texture, and observation geometry (Petropoulos et al., 2015). Moreover, the optical and thermal signal has limited capability to penetrate cloud and vegetation, requiring careful correction to eliminate atmospheric effects (Zhao and Li, 2013). These are the main reasons for limited use of visible/thermal bands in soil moisture retrieval, despite the multitude of optical sensors currently available.

The microwave technique has been widely acknowledged as the most promising for near-surface soil moisture mapping at regional to global scales, due to its penetration capability of the vegetation layer, independence of solar illumination, and direct relationship to soil moisture via the dielectric constant (Ulaby et al., 1981a). The microwave radiometers operating at higher frequency such as C-band (i.e. AMSR-E, Wind-Sat), or X-band (i.e. TMI) are also sensitive to soil moisture, however, microwave emission from the soil layer at these wavelengths is significantly affected by attenuation from the vegetation canopy and scattering by the atmosphere. Since the penetrability of microwave radiation is dependent mainly on the wavelength of the signal, microwave radiometers at these higher frequencies need to account for the atmospheric contribution, are only sensitive to water content in the top less than 1 cm layer of soil, and are limited to areas with low vegetation cover (VWC < 3 kg/m^2). In contrast, L-band radiometer observations are unaffected by the atmosphere, able to penetrate the top ~5 cm layer of soil, and sensitive to soil moisture with vegetation coverage of up to ~5 kg/m^2, corresponding to about 70% of the nonfrozen land regions on Earth, excluding dense forests. Consequently, passive microwave at L-band (~1.4 GHz) is considered the most promising remote sensing technique for near-surface soil moisture mapping at regional to global scales. Not only due to its penetration capability of cloud and vegetation, but also its independence of solar illumination, direct relationship with volumetric soil moisture through the soil dielectric constant, and reduced sensitivity to land surface roughness and vegetation cover (Jackson and Schmugge, 1989; Njoku et al., 2002; Schmugge et al., 1986; Shutko and Reutov, 1982; Ulaby et al., 1981b).

The first space-borne passive microwave observations at L-band were collected by the SkyLab mission within a limited duration in the 1970s, being the earliest demonstration of soil moisture retrieval from passive L-band observations from space (Jackson et al., 2004). However, the first satellite dedicated to soil moisture was not launched until November 2009, carrying a 2D interferometric radiometer to measure microwave emission from the Earth surface at L-band (1.413 GHz). Development of the SMOS mission was led by the ESA, aiming to map global soil moisture every 2–3 days with a target accuracy of 0.04 m^3/m^3 (Kerr et al., 2010). Additionally, the Soil Moisture Active and Passive (SMAP) mission developed the National Aeronautic and Space Administration was launched in January 2015, designed to combine L-band radiometer (1.41 GHz) and radar (1.26 GHz) instruments in order to achieve soil moisture retrieval with the same target accuracy as SMOS but at an enhanced special resolution (Entekhabi et al., 2010).

Based on the current level of antenna technology, the spatial resolution of space-borne radiometers is limited to around 40 km. For a higher spatial resolution, extensive research has been conducted using active microwave data, especially the data collected by SAR. A SAR is a coherent mostly side-looking radar system that utilizes signal processing and the movement of the platform to simulate an extremely large antenna or aperture, resulting in a high resolution in azimuth (up to half of the real aperture length) (Ulaby et al., 2014). Over land surfaces, a part of the transmitted electromagnetic wave is scattered by the vegetation canopy and/or soil surface, and returned back to the SAR system (known as backscatter). In general, backscatter increases as soil moisture (relative permittivity) is increased, and thus soil moisture can be inverted from scattering models describing this relationship. The commonly used scattering models include the IEM (Fung, 1994a) for bare soil and the water cloud model (Attema and Ulaby, 1978) for vegetated areas.

Apart from soil moisture, radar measurements are also sensitive to soil surface roughness and vegetation parameters, which results in high uncertainty in the soil moisture retrieval algorithms for SAR observations (Ulaby et al., 2014). To decouple the effect of soil roughness and vegetation, some studies decomposed the received signal into individual contributions of the soil surface, vegetation and multiple scattering between soil surface and vegetation, with soil moisture then being retrieved from the soil surface contribution (Hajnsek et al., 2009; Jagdhuber et al., 2013). Some other studies assume that roughness and vegetation do not change between repeat passes of radar observations, and thus the difference of backscatter can be empirically related to the variation of soil moisture (Wagner et al., 1999a,b). This assumption was also used to remove the unknowns representing the temporal variation of roughness and vegetation, resulting in a well-constrained inversion process (Kim et al., 2012; Zhu et al., 2019b). While space-borne SAR has shown promising results for soil moisture retrieval at high spatial resolution (0.1–1 km), it suffers from a narrow beamwidth that cannot satisfy the temporal and global coverage requirements of hydrology applications (~3 days). Currently, only the Sentinel-1 (Torres et al., 2012), SAOCOM, PALSAR-2 (Kankaku et al., 2014), Biomass mission (Le Toan et al., 2011), and NIRSAR (Rosen et al., 2017) can plan to provide regular global observations, however, their repeat coverages range from 6 days to 60 days. Algorithms that can merge data from multiple SAR missions was thus proposed recently to enhance the revisit (Zhu et al., 2019a).

4.2 Soil roughness

The collected radar signal is heavily dependent on surface roughness. When a wave impinges on a smooth soil surface, some of the energy is scattered in non-specular directions, with a small amount returning to the radar as backscatter. As surface roughness increases, the amount of reflection in the specular direction decreases, and consequently the surface scattering increases (Ulaby et al., 2014). Radar measurements are thus an effective venue to characterize surface roughness. In general, the soil surface is described either as a one-scale stationary random process or a random process superimposed on a periodic surface for periodic plowed surfaces (Ulaby et al., 2014). For random surfaces, two parameters describing the vertical (RMS height) and horizontal (correlation length) variation are commonly used, while a few extra parameters (e.g. the period, orientation, and the maximum height of the periodic surface) are required for periodic plowed surfaces. These parameters are well considered in various surface scattering models, e.g. the IEM (Fung, 1994a) and the multi-scale IEM (Monsivais-Huertero et al., 2018). Roughness can thus be estimated by inverting these scattering models with or without ground soil moisture measurements (Zhu et al., 2019a). However, the use of ground-measured roughness parameters commonly result in substantial inconsistency between the modeled and observed radar observation (Baghdadi et al., 2002; Lievens et al., 2011; Zhu et al., 2016), suggesting that the RMS height and correlation length cannot represent the roughness perceived by radar systems. Accordingly, the roughness values estimated from radar data are effective without a determined physical definition. Moreover, many studies have confirmed that the effective roughness varies at different radar frequency bands and incidence angles. For example, larger effective roughness values were observed at lower incidence angles over the same location (Baghdadi et al., 2006; Lievens et al., 2011; Martinez-Agirre et al., 2017). The use of radar-based roughness in other applications is still questionable.

4.3 Soil salinity

The presence of salt in the shallow Earth surface can be detected by remote sensing either directly on bare soil, or indirectly through the vegetation type and growth that is controlled or affected by the salinity. Salt mineralogy determines the presence or absence of absorption in optical bands (Metternicht and Zinck, 2008), while the microwave spectrum is mainly affected by the quantity of salt in the soil water (Lasne et al., 2008). Salts also cause variations of surface roughness. For example, the concentration of salt at the soil surface can result in smooth surfaces (e.g. puffy crusts), leading to higher reflectance in the optical bands but lower emissivity and backscatter in the microwave spectrum than nonsaline and cultivated surfaces (Metternicht and Zinck, 2008).

Optical remote sensing is mainly used to capture the spatial pattern of salt-affected surface features and their temporal evolutions, with some sophisticated remote sensing indexes (Khan et al., 2005; Sidike et al., 2014; Wang et al., 2019) proposed for aerial photos, high spatial resolution satellite imagery (e.g. WorldView series), and sensors with a wider spectrum up to 2500 nm (e.g. Landsat series and Sentinel-2). The end of dry season is widely recognized as the optimal time slot for salinity mapping as salts dissolve during the rainy season

(Metternicht and Zinck, 2003; Wang et al., 2019). In contrast, some soil moisture is required for microwave-based remote sensing, as the measurements are mainly dependent on dielectric constant determined by moisture and salt that dissolved in moisture (Ulaby et al., 2014). Previous studies using microwave data have mainly focused on (1) saline water detection in water logging areas by analyzing the dependence of brightness temperature on temperature and salinity (Singh and Srivastav, 1990); and (2) soil salinity mapping by relating the imaginary part of the dielectric constant with measured backscatter (Bell et al., 2001; Periasamy and Ravi, 2020; Taghadosi et al., 2019). The presence of halophytic salt-tolerant vegetation interferes with the reflectance, emissivity, or backscatter of salinity features, being more challenging for direct salinity mapping. The type of vegetation and its growth however provide alternative indicators of soil salinity. For example, the health status of cotton plants and cotton yield have strong correlation with electrical conductivity (Metternicht and Zinck, 2003).

Despite the great advance, precise estimation of salt quantities from remote sensing is rather challenging due to the multiple dependence of remote sensing data on surface parameters, e.g. soil moisture, soil texture, roughness, organic matter (Jackson and O'Neill, 1987; Petropoulos et al., 2015). Ground samples remain necessary to provide required field data (e.g. water table depth and salinity data) and is still the basis of empirical relationships between remote sensing data and salinity features.

5 Case study

5.1 Murrumbidgee River Catchment

The Murrumbidgee River Catchment is located in southeast of Australia, with an area of ~82 000 km^2, and an elevation ranging from ~40 m in the western plains to ~2000 m in the eastern mountainous areas as shown in Fig. 4. The annual rainfall varies from 300 mm in the west to 1900 mm in the high elevated ranges in the east (Australian Bureau of Rural Science, 2001). Accordingly, the land surface of the Murrumbidgee River Catchment varies from bare soil with sparse vegetation in the west, to irrigated fields mixed with grasslands in the middle, and to forest in the east. Due to the variability in topography, climate, land cover, and soil textures of the Murrumbidgee catchment, the long-term hydrological monitoring network: OzNet (Smith et al., 2012) has been operating for 15 years. At the same time, a series of airborne field experiments have been conducted for the development of soil moisture remote sensing techniques, such as NAFE'05 (Panciera et al., 2008), NAFE'06 (Merlin et al., 2008), AACES-1/-2 (Peischl et al., 2012), SMAPEx-1 to -3 (Panciera et al., 2014), and SMAPEx-4/-5 (Ye et al., 2021).

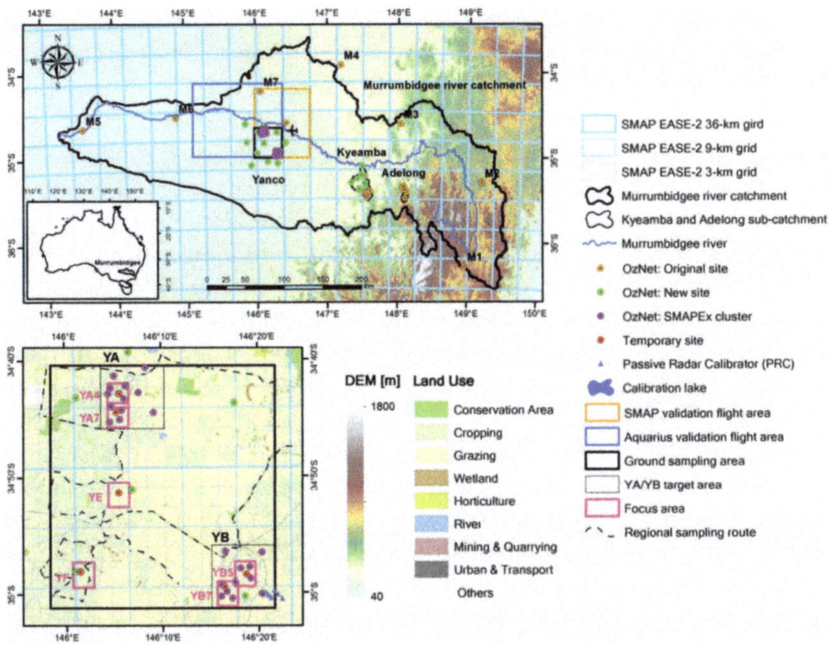

Figure 4 Location of the OzNet monitoring stations and the SMAPEx-4/5 study area in the Murrumbidgee River Catchment with the digital elevation model (DEM) and the SMAP EASE-2 grid at 36 km scale as backdrop (top). Layout of the ground sampling focus areas and monitoring stations with land use map and the SMAP EASE-2 grids at 9 km and 3 km scales as backdrop (bottom). Source: Adapted from Ye et al. (2021).

5.2 Soil moisture active passive experiments

The soil moisture active passive experiments (SMAPEx) comprised a series of five airborne field campaigns for calibration and validation of SMAP under Australian land surface conditions. The first three experiments (SMAPEx-1 to -3; Panciera et al., 2014) were conducted between July 2010 and September 2011, for the development of SMAP prelaunch soil moisture retrieval and downscaling algorithms. The SMAPEx-4 and -5 (Ye et al., 2021) were conducted at the beginning of the SMAP operational phase in May and September 2015 respectively, for in-orbit validation of the SMAP observations and soil moisture products in different seasons. The specific objectives were to: (1) evaluate SMAP active-passive downscaled 9 km radiometer observations; (2) inter-compare SMAP, SMOS, and Aquarius with airborne radiometer/radar observations; (3) validate SMAP radar-only, radiometer-only, and radar-radiometer soil moisture retrieval algorithms using airborne soil moisture retrieval results from field experiments and long-term soil moisture monitoring network measurements; and (4) further develop radar-only soil moisture retrieval algorithms.

A flight area of 71 km × 89 km was identified over the Yanco area (Fig. 4), and airborne active and passive microwave observations were collected coincident with SMAP coverage during 3 weeks in each experiment. The main airborne instruments used in the SMAPEx-4 and -5 were the Polarimetric L-Band Multi-beam Radiometer (PLMR) and the Polarimetric L-band Imaging Synthetic (PLIS) aperture radar, which operate at the same frequencies of the SMAP radiometer and radar, respectively. During the 3 weeks period of SMAPEx-4, eight flights were taken over two study areas according to the 3 dB footprints of the SMAP and Aquarius radiometers, respectively (Fig. 4). Airborne PLMR brightness temperature observations at 1 km resolution and airborne PLIS backscatter observations at ~30 m resolution were collected in each flight between 3 am and 9 am (local time), in alignment with the SMAP nominal local overpass time of 6 am. Figure 5 shows an example of PLMR brightness temperature normalized to 38.5°and PLIS backscatter normalized to the reference incidence angle of SMAP (40°), using the Ye et al. (2015) approach. Ground sampling of the top 5 cm soil moisture, VWC, and surface roughness were collected concurrently with airborne sampling over the six 3 km focus areas (Fig. 4), in order to provide spatial soil moisture data for validation and the ancillary data used in soil moisture retrieval. Prior to SMAPEx-4 and -5, heavy rainfall events occurred and were followed by a drying out period, providing ideal opportunities to validate SMAP products under a variety of soil moisture, land surface types, and topography conditions. A total of 16 flights were conducted during the SMAPEx-4 and -5.

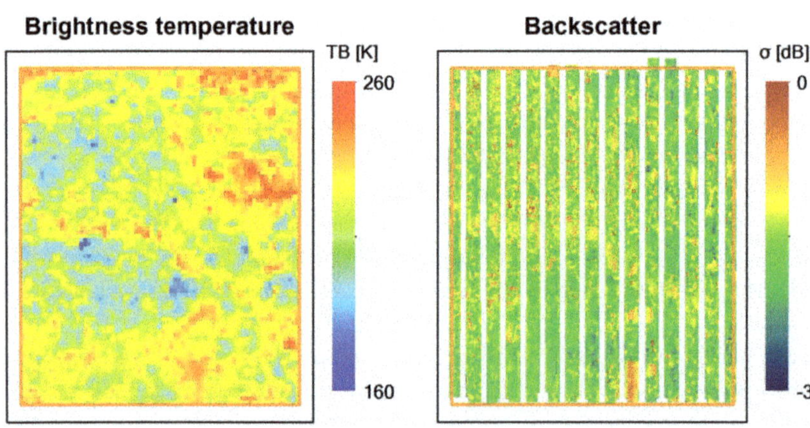

Figure 5 Maps of airborne brightness temperature normalized to 38.5° (left) and backscatter observations normalized to 40° (right) over the SMAP validation flight area on 11 May 2015. Source: Adapted from Ye et al. (2021).

5.3 Passive microwave remote sensing of soil moisture

Airborne soil moisture was retrieved from the dual-polarized PLMR brightness temperature observations using the L-MEB model (Wigneron et al., 2007). The VWC was estimated from the MODIS 250 m daily reflectance products (MOD09GQ) using the Gao et al. (2015) method. The required roughness and vegetation parameters in the L-MEB were adapted from previous studies in the Yanco area (Panciera et al., 2008, 2009) and literature (Grant et al., 2008; Wigneron et al., 2007). The surface and deep soil temperatures were obtained from monitoring stations and used to estimate effective temperature in Eqn. (14). The retrieved soil moisture data were validated using the ground 250 m spacing soil moisture measurements over all six focus areas. According to the comparison results shown in Fig. 6, an overall RMSE of 0.08 m^3/m^3 was achieved for SMAPEx-4/-5 soil moisture data at 1 km scale, while soil moisture was overestimated in SMAPEx-5 in some instances due to the presence of standing water.

Figure 7 shows the temporal variation of top 5 cm soil moisture retrieved from the 1 km PLMR brightness temperature observations across the SMAPEx-4 and -5, respectively. As the result of two rainfall events, 9 May and 18 May 2015, soil moisture over the flight area varied from a dry homogeneous condition to a heterogeneous spatially distributed pattern during the SMAPEx-4. The SMAPEx-5 started from a very wet and partly flooded condition on 8 September 2015, and experienced a drying out period to the end of the experiment on 26 September 2015. The 1 km PLMR-derived soil moisture data were averaged

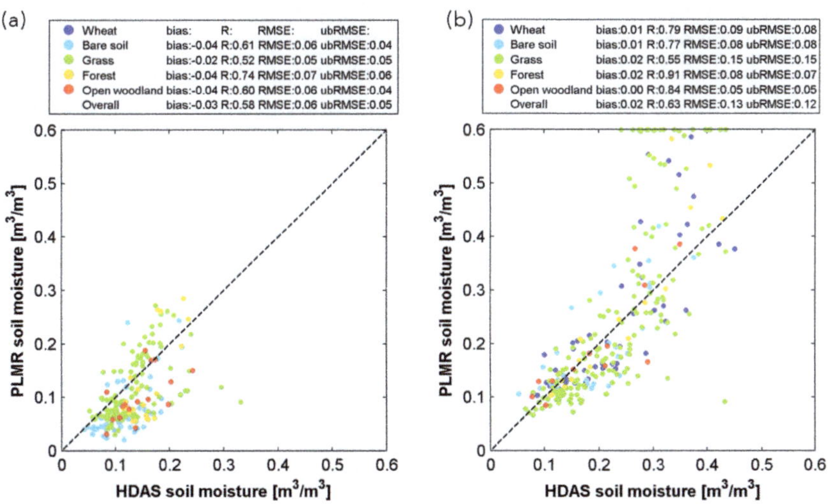

Figure 6 Comparison between top 5 cm soil moisture measurements and soil moisture retrieved from the 1 km PLMR brightness temperature observations during SMAPEx-4 (a) and -5 (b) using published parameters. Source: Adapted from Ye et al. (2021).

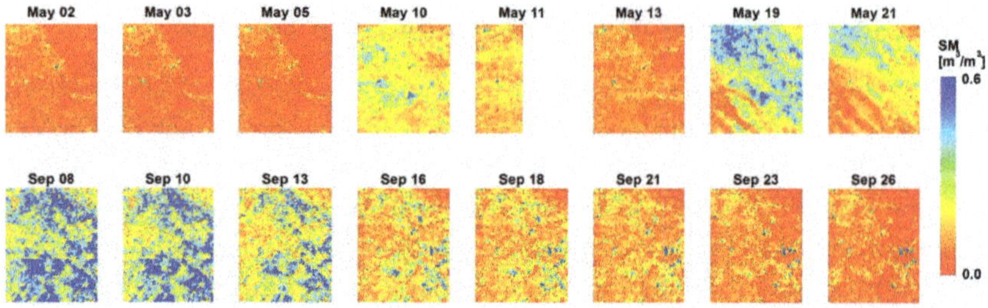

Figure 7 Maps of 1-kmPLMR-derived soil moisture over the SMAP validation flight area during the SMAPEx-4 (top) and -5 (bottom).

to the SMAP EASE grids at 3 km, 9 km, and 36 km, and then compared with SMAP soil moisture products at pixel levels as shown in Fig. 8. Taking the PLMR soil moisture as the reference, the SMAPL2 Passive, Active-Passive downscaled, and Active soil moisture products had high correlations with RMSEs of ≤0.04 m^3/m^3, 0.07 m^3/m^3, and 0.11 m^3/m^3 in both SMAPEx-4 and -5. Consequently, the SMAPEx-4 and -5 data sets demonstrate the capability of monitoring temporal and spatial variation of near-surface soil moisture using microwave remote sensing technique, and provided airborne brightness temperature and backscatter observations for the validation of SMAP.

5.4 Radar remote sensing of soil moisture

Radar data collected across the SMAPEx-5 campaign, covering three frequency bands (aircraft and satellite), L-, C-, and X-band with varying incidence angle and polarizations over time, were used for soil moisture retrieval here. Radar data were available for 15 days of the 3-week campaign, being a simulation of the dense data expected from combining recent and projected radar satellite missions. The available observations varied from day to day, with 5 and 10 days

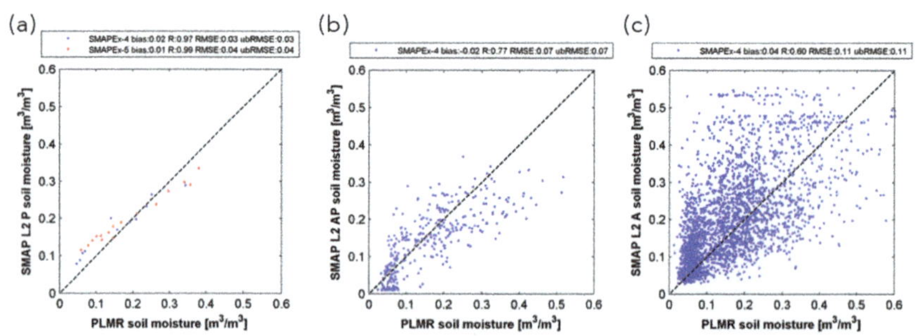

Figure 8 Comparison of SMAPEx-4 and -5 airborne PLMR soil moisture versus SMAPL2 Passive/radiometer (a), Active-Passive downscaled (b), and Active/radar (c) products. Source: Referred from Ye et al. (2021).

Table 2 The main components of the stochastic ensemble framework

Component	Description	Reference
Scattering models	Three frequency-specific look up tables (LUTs) built by NMM3D and distort Born approximation, representing the relationships between soil moisture, roughness, and radar observations.	Zhu et al. (2019b)
Change detection component	It implements an unsupervised change detection of the areas with roughness and/or vegetation changes, being a preprocessing procedure.	Zhu et al. (2019c)
Channel selection	It estimates the contribution of vegetation volume scattering and removes the channels dominated by volume scattering.	Zhu et al. (2019a)
Ensemble retrieval	It implements multiple soil moisture retrievals using random subsets of input radar data and get their ensemble average as the output.	Zhu et al. (2020)

having dual-frequency (L + C or C + X) and single-frequency data, respectively. A stochastic ensemble framework was proposed for soil moisture retrieval, with the main components summarized in Table 2.

Soil moisture retrieval was made at the 25 m pixel and paddock scales, respectively, using all available radar data. The 25-m pixel soil moisture maps for the YA4 area are depicted in Fig. 9 as an example, while the paddock scale

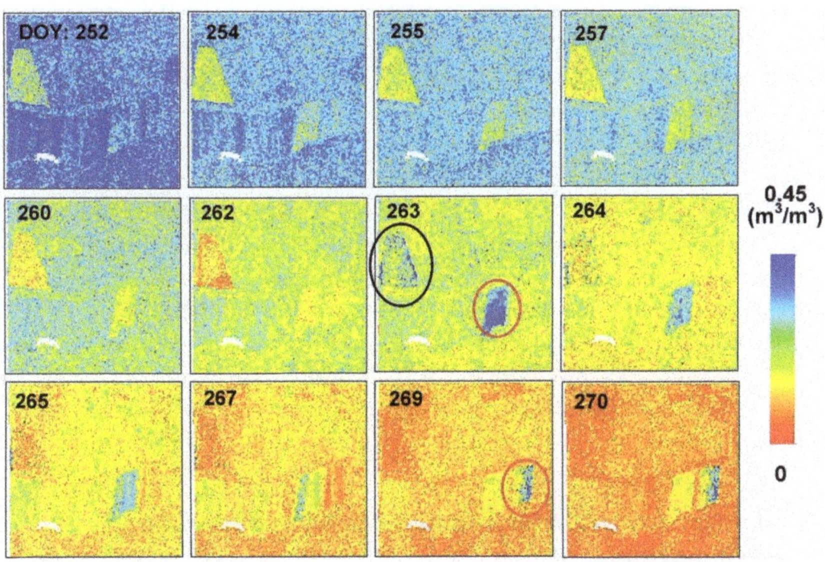

Figure 9 Retrieved soil moisture maps in YA4 with the day of year listed in top left. The paddocks in black circles are these with soil plowing, while the red circles are those with irrigation.

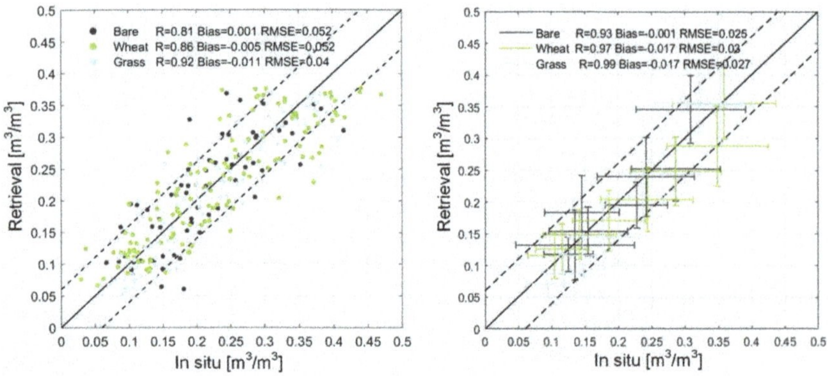

Figure 10 *In situ* versus retrieved soil moisture at the paddock scale (left panel) and daily average (right panel) using time series L-, C-, and X-band data. The dashed lines denote the ± 0.06 cm^3/cm^3.

and daily average comparison against corresponding ground measurements are shown in Fig. 10. The retrieved time series soil moisture maps agreed well with the dry down process observed during SMAPEx-5, with a faster dry down observed over bare soil paddocks. The observed cultivation activities in the circled paddocks were detected, with the sudden soil moisture increases being successfully recorded.

Moderate accuracy was achieved at the paddock scale, showing an RMSE of 0.04–0.06 m^3/m^3 and a correlation coefficient (R) of 0.8–0.9, reaching the 0.06 m^3/m^3 accuracy target of SMAP radar products. The retrieved soil moisture for wet conditions (larger than 0.4 m^3/m^3) was slightly underestimated. Such underestimations were also observed and ascribed to the decreased sensitivity of σ^0 in moist areas in other studies (Bai et al., 2016; Wang et al., 2011). Another interpretation could be the relatively low upper bound of soil moisture in the LUTs (~0.43 m^3/m^3) compared to the ground measurements during the first days of SMAPEx-5. As expected, the results were greatly improved for daily average values. The RMSE decreased to 0.025 m^3/m^3, 0.03 m^3/m^3, and 0.027 m^3/m^3 in bare, wheat, and grass, respectively. A good correlation (R: 0.93–0.99) was observed with negligible bias (<0.02 m^3/m^3). The retrieved accuracy was much better than the requirement of 0.05 m^3/m^3 suggested by Walker and Houser (2004) and World Meteorological Organization (available at http://www.wmosat.info/oscar/requirements).

5.5 Irrigator/tower/unmanned aerial vehicle-based remote sensing of soil moisture

Soil moisture is a temporal and spatially varying variable that plays an important role in the agriculture industry. Although the current level of irrigator technology supports the precise control of the volume of irrigation water, it is difficult to accurately estimate spatial irrigation water demand, due to the high spatial variability of soil moisture. Therefore, the efficiency of irrigation water use can be improved by introducing real-time soil moisture mapping to determine the spatial distribution of required irrigation water volume, and an irrigation-based soil moisture remote sensing experiment has been conducted to demonstrate this potential. The irrigator-based passive microwave soil moisture experiment was conducted over a 200 m by 600 m farm near Cora Lynn, in southeast of Australia (Fig. 11). The ELBARA III (the Eidgenössische Technische Hochschule [ETH] L-band radiometer for soil moisture research; Jonard et al., 2011; Schwank et al., 2010) was mounted on an irrigation boom with a 40° incidence angle and a 5 m footprint. Dual-polarized brightness temperature observations were collected over a 5 m by 600 m transect when the irrigator moved across the farm.

To validate soil moisture retrieval accuracy, point-based soil moisture measurements were collected at the same time using the Hydraprobe Data Acquisition System (HDAS) (Merlin et al., 2007; Panciera et al., 2006). Figure 12 shows the spatial distribution of soil moisture data measured using the HDAS and retrieved from the ELBARA III brightness temperature observations, respectively. Soil moisture was found to have a considerable variation of

Figure 11 Overview of the irrigator-based soil moisture remote sensing experiment (top) and picture of the irrigator with the ELBARA III mounted on the boom (bottom).

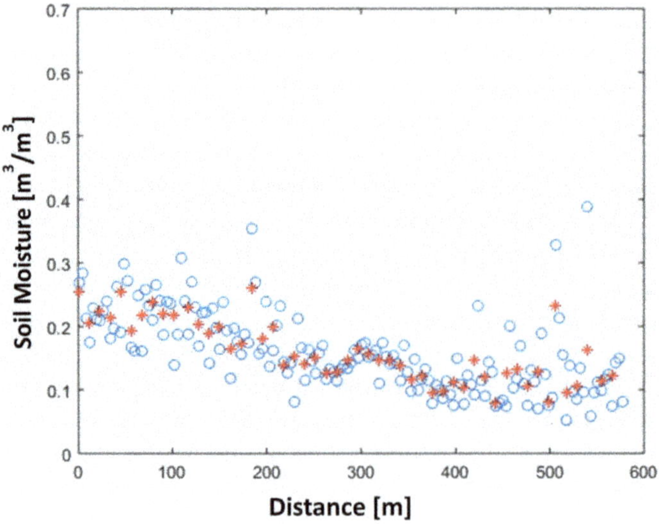

Figure 12 Soil moisture measurements along the ELBARA III transect. The red dots and blue circles indicate soil moisture data measured using the HDAS and retrieved from the ELBARA III brightness temperature observations respectively.

more than 0.2 m^2/m^2 within the same farm, although normally assumed as a homogeneous condition. This confirms the difficulty to accurately measure soil moisture at farm scales using conventional methods, and thus the potential of saving irrigation water by accounting for initial soil moisture using remote sensing techniques. The ELBARA III-derived soil moisture had a similar spatial pattern and a high correlation to the HDAS soil moisture measurements, demonstrating the applicability of boom-based passive microwave soil moisture remote sensing application in the agriculture industry.

6 Future trends in research

With the recent development of sensor technology, the sizes and costs of sensors have been reduced significantly, such that a series of portable sensors have been employed on different platforms including unmanned aerial vehicle (UAV). Compared with traditional aircraft (Bai and Purcell, 2018; Bellvert et al., 2015; Gerhards et al., 2018; Walker et al., 2012; Zhang et al., 2017) and satellite (Calera et al., 2017; Du et al., 2013; Helman et al., 2018) platforms, UAVs are small, inexpensive, easy, and autonomous, providing a new remote sensing platform with an improved spatial resolution and sampling flexibility in precision agriculture application (Vergara-Díaz et al., 2016; Zarco-Tejada et al., 2009, 2013). The optical remote sensing techniques has been tested for UAV-based surface soil moisture mapping (e.g. Hassan-Esfahani et al., 2014; Hoffmann et al., 2016; Lu et al., 2020; Turner et al., 2011). To overcome the limitation of optical techniques on soil moisture remote sensing, only a few passive microwave sensors have been developed for UAV-based soil moisture mapping (Acevo-Herrera et al., 2010; Houtz et al., 2020). However, it can be expected in the future that L-band passive microwave remote sensing with a high sensitivity to soil moisture will be used together with the optical technique on UAVs to map high-resolution soil moisture at farm scale.

The state-of-art L-band passive microwave soil moisture remote sensing technique is limited to measure water content in the top ~5 cm of the soil. According to theoretical and experimental studies (Laymon et al., 2001; Njoku and Entekhabi, 1996; Shi et al., 2005), the sensing depth of passive microwave sensors is a function of the wavelength. Thus a deeper layer of soil can be achieved by using radiometers at lower frequencies (Njoku et al., 2003; Njoku and Kong, 1977; Paloscia et al., 1993). In addition, the impacts of surface roughness and overlaying vegetation on soil moisture retrieval accuracy are expected to be reduced. Consequently, a radiometer with a doubled wavelength (40 cm) operating at P-band (750 MHz) has been developed to demonstrate the capability of measuring a deeper layer (~10 cm) soil moisture over more densely vegetated surfaces, and with overall improved accuracy. Currently, a series of tower-based and airborne field experiments are being conducted to explore the potential of P-band passive microwave soil moisture remote sensing. The initial results show a higher sensing depth, less sensitivity on vegetation and roughness, and higher accuracy for P-band than L-band soil moisture remote sensing (Shen et al., 2020; Ye et al., 2020).

With the increasing availability of 'Earth big data', the use of deep learning in soil characteristics mapping has attracted broad attention in recent years (Yuan et al., 2020). Deep learning models can accurately approximate the complicated nonlinear relationship between soil characteristics and various remote sensing observations, without requiring cumbersome physical modeling and the ill-posed inversion process (Ali et al., 2015; Notarnicola et al., 2008). A neural network has been successfully applied to global soil moisture mapping from passive microwave data (Rodriguez-Fernandez et al., 2015) and some regional-scale applications (Rodríguez-Fernández et al., 2017; Santi et al., 2016). However, the ground truth for training is either from climate models (e.g. the European Centre for Medium-Range Weather Forecasts model) or in situ measurements. Currently, the main limitation of deep learning is that the limited in situ measurements (Dorigo et al., 2021) cannot reach the requirement of training a deep and wide neural network, with the trained neural network being too simple to have satisfactory accuracy and generalization capability. Theories and/or techniques (e.g. transfer learning) that can be used to provide more samples and/or reduce the number of required ground samples are thus expected to be the key to further use of deep learning in soil mapping.

7 Where to look for further information

The following articles provide a good overview of the subject:

- Ulaby, F. T., Long, D. G., Blackwell, W. J., Elachi, C., Fung, A. K., Ruf, C., Sarabandi, K., Zebker, H. A. and Van Zyl, J. (2014). *Microwave Radar and Radiometric Remote Sensing*. Ann Arbor: The University of Michigan Press.

Key research in this area can be found at the following organizations:

- Department of Civil Engineering, Monash University.
- https://www.monash.edu/engineering/jeffwalker.
- https://www.researchgate.net/profile/Jeffrey-Walker-9.

8 References

Acevo-Herrera, R., Aguasca, A., Bosch-Lluis, X., Camps, A., Martínez-Fernández, J., Sánchez-Martín, N. and Pérez-Gutiérrez, C. (2010). Design and first results of an UAV-borne L-band radiometer for multiple monitoring purposes. *Remote Sensing* 2(7), 1662–1679.

Ali, I., Greifeneder, F., Stamenkovic, J., Neumann, M. and Notarnicola, C. (2015). Review of machine learning approaches for biomass and soil moisture retrievals from remote sensing data. *Remote Sensing* 7(12), 16398–16421.

Attema, E. P. W. and Ulaby, F. T. (1978). Vegetation modeled as a water cloud. *Radio Science* 13(2), 357–364.

Australian Bureau of Rural Science. (2001). *Mean Monthly and Mean Annual Rainfall and Temperature Data*. Melbourne, Australia: Bureau of Meteorology.

Baghdadi, N., Holah, N. and Zribi, M. (2006). Calibration of the integral equation model for SAR data in C-band and HH and VV polarizations. *International Journal of Remote Sensing* 27(4), 805–816.

Baghdadi, N., King, C., Chanzy, A. and Wigneron, J. P. (2002). An empirical calibration of the integral equation model based on SAR data, soil moisture and surface roughness measurement over bare soils. *International Journal of Remote Sensing* 23(20), 4325–4340.

Bai, X., He, B. and Li, X. (2016). Optimum surface roughness to parameterize advanced integral equation model for soil moisture retrieval in prairie area using Radarsat-2 data. IEEE Transactions on Geoscience and Remote Sensing 54, 2437–2449.

Bai, H. and Purcell, L. C. (2018). Aerial canopy temperature differences between fast-and slow-wilting soya bean genotypes. *Journal of Agronomy and Crop Science* 204(3), 243–251.

Beckmann, P. and Spizzichino, A. (1963). *The Scattering of Electromagnetic Waves From Rough Surfaces*. Norwood, MA: Artech House, Inc.

Bell, D., Menges, C., Ahmad, W. and Van Zyl, J. J. (2001). The application of dielectric retrieval algorithms for mapping soil salinity in a tropical coastal environment using airborne polarimetric SAR. *Remote Sensing of Environment* 75(3), 375–384.

Bellvert, J., Marsal, J., Girona, J. and Zarco-Tejada, P. J. (2015). Seasonal evolution of crop water stress index in grapevine varieties determined with high-resolution remote sensing thermal imagery. *Irrigation Science* 33(2), 81–93.

Burke, E. J., Shuttleworth, W. J. and French, A. N. (2001). Using vegetation indices for soil-moisture retrievals from passive microwave radiometry. *Hydrology and Earth System Sciences* 5(4), 671–678.

Calera, A., Campos, I., Osann, A., D'Urso, G. and Menenti, M. (2017). Remote sensing for crop water management: From ET modelling to services for the end users. *Sensors* 17(5), 1104.

Chen, K. S., Wu, Tzong-Dar, Tsang, Leung, Li, Qin, Shi, Jiancheng and Fung, A. K. (2003). Emission of rough surfaces calculated by the integral equation method with comparison to three-dimensional moment method simulations. *IEEE Transactions on Geoscience and Remote Sensing* 41(1), 90–101.

Chen, K. S., Yen, S. K. and Huang, W. P. (1995). A simple model for retrieving bare soil moisture from radar-scattering coefficients. *Remote Sensing of Environment* 54(2), 121–126.

Choudhury, B. J., Schmugge, T. J., Chang, A. and Newton, R. W. (1979). Effect of surface roughness on the microwave emission from soils. *Journal of Geophysical Research* 84(C9), 5699–5706.

Choudhury, B. J., Schmugge, T. J. and Mo, T. (1982). A parameterization of effective soil temperature for microwave emission. *Journal of Geophysical Research* 87(C2), 1301–1304.

D'urso, G. and Minacapilli, M. (2006). A semi-empirical approach for surface soil water content estimation from radar data without a-priori information on surface roughness. *Journal of Hydrology* 321(1–4), 297–310.

De Rosnay, P., Wigneron, J.-P., Holmes, T. and Calvet, J.-C. (2006). Parameterizations of the effective temperature for L-band radiometry. Inter-comparison and long term validation with SMOSREX field experiment. In: Mätzler, C., Rosenkranz, P. W., Battaglia, A. and Wigneron, J.-P. (Eds.), *Thermal Microwave Radiation - Applications for Remote Sensing Electromagnetic Waves Series*. London, UK: IEEE, pp. 312–324.

Dobson, M. C., Ulaby, F. T., Hallikainen, M. T. and El-Rayes, M. A. (1985). Microwave dielectric behavior of wet soil-Part II: Dielectric mixing models. *IEEE Transactions on Geoscience and Remote Sensing* GE-23(1), 35–46.

Dorigo, W., Himmelbauer, I., Aberer, D., Schremmer, L., Petrakovic, I., Zappa, L., Preimesberger, W., Xaver, A., Annor, F., Ardö, J., Baldocchi, D., Bitelli, M., Blöschl, G., Bogena, H., Brocca, L., Calvet, J., Camarero, J. J., Capello, G., Choi, M., Cosh, M. C., van de Giesen, N., Hajdu, I., Ikonen, J., Jensen, K. H., Kanniah, K. D., de Kat, I., Kirchengast, G., Kumar Rai, P., Kyrouac, J., Larson, K., Liu, S., Loew, A., Moghaddam, M., Martínez Fernández, J., Mattar Bader, C., Morbidelli, R., Musial, J. P., Osenga, E., Palecki, M. A., Pellarin, T., Petropoulos, G. P., Pfeil, I., Powers, J., Robock, A., Rüdiger, C., Rummel, U., Strobel, M., Su, Z., Sullivan, R., Tagesson, T., Varlagin, A., Vreugdenhil, M., Walker, J., Wen, J., Wenger, F., Wigneron, J. P., Woods, M., Yang, K., Zeng, Y., Zhang, X., Zreda, M., Dietrich, S., Gruber, A., van Oevelen, P., Wagner, W., Scipal, K., Drusch, M. and Sabia, R. (2021). The international soil moisture network: Serving earth system science for over a decade. *Hydrology and Earth System Sciences* 25(11), 5749–5804.

Du, L., Tian, Q., Yu, T., Meng, Q., Jancso, T., Udvardy, P. and Huang, Y. (2013). A comprehensive drought monitoring method integrating MODIS and TRMM data. *International Journal of Applied Earth Observation and Geoinformation* 23, 245–253.

Dubois, P. C. and Van Zyl, J. (1994). *An Empirical Soil Moisture Estimation Algorithm Using Imaging Radar*. New York: IEEE, vol. 1573, pp. 1573–1575.

Dubois, P. C., Van Zyl, J. and Engman, T. (1995). Measuring soil moisture with imaging radars. *IEEE Transactions on Geoscience and Remote Sensing* 33(4), 915–926.

Engman, E. T. (1990). Progress in microwave remote sensing of soil moisture. *Canadian Journal of Remote Sensing* 16(3), 6–14.

Engman, E. T. and Chauhan, N. (1995). Status of microwave soil moisture measurements with remote sensing. *Remote Sensing of Environment* 51(1), 189–198.

Entekhabi, D., Njoku, E. G., O'Neill, P. E., Kellogg, K. H., Crow, W. T., Edelstein, W. N., Entin, J. K., Goodman, S. D., Jackson, T. J., Johnson, J., Kimball, J., Piepmeier, J. R., Koster, R. D., Martin, N., McDonald, K. C., Moghaddam, M., Moran, S., Reichle, R., Shi, J. C., Spencer, M. W., Thurman, S. W., Tsang, L. and Van Zyl, J. (2010). The soil moisture active passive (SMAP) mission. *Proceedings of the IEEE* 98(5), 704–716.

Escorihuela, M. J., Kerr, Y. H., de Rosnay, P., Wigneron, J.-P., Calvet, J.-C. and Lemaître, F. (2007). A simple model of the bare soil microwave emission at L-band. *IEEE Transactions on Geoscience and Remote Sensing* 45(7), 1978–1987.

Fung, A. K. (1994a). *Microwave Scattering and Emission Models and Their Applications*. Norwood, MA: Artech House.

Fung, A. K. (1994b). *Microwave Scattering, Emission Model and Its Application*. Norwood, MA: Artech House.

Fung, A. K., Li, Z. and Chen, K. S. (1992). Backscattering from a randomly rough dielectric surface. *IEEE Transactions on Geoscience and Remote Sensing* 30(2), 356–369.

Gao, Y., Walker, J. P., Allahmoradi, M., Monerris, A., Ryu, D. and Jackson, T. J. (2015). Optical sensing of vegetation water content: A synthesis study. *IEEE Journal of Selected Topics in Applied Earth Observations and Remote Sensing* 8(4), 1456–1464.

Gao, Z., Xu, X., Wang, J., Yang, H., Huang, W. and Feng, H. (2013). A method of estimating soil moisture based on the linear decomposition of mixture pixels. *Mathematical and Computer Modelling* 58(3–4), 606–613.

Gerhards, M., Schlerf, M., Rascher, U., Udelhoven, T., Juszczak, R., Alberti, G., Miglietta, F. and Inoue, Y. (2018). Analysis of airborne optical and thermal imagery for detection of water stress symptoms. *Remote Sensing* 10(7), 1139.

Grant, J. P., Saleh-Contell, K., Wigneron, J.-P., Guglielmetti, M., Kerr, Y. H., Schwank, M., Skou, N. and Van de Griend, A. A. (2008). Calibration of the L-MEB model over a coniferous and a deciduous forest. *IEEE Transactions on Geoscience and Remote Sensing* 46(3), 808–818.

Hajnsek, I., Jagdhuber, T., Schön, H. and Papathanassiou, K. P. (2009). Potential of estimating soil moisture under vegetation cover by means of PolSAR. *IEEE Transactions on Geoscience and Remote Sensing* 47(2), 442–454.

Hallikainen, M. T., Ulaby, F. T., Dobson, M. C., El-Rayes, M. A. and Wu, L. K. (1985). Microwave dielectric behavior of wet soil-part 1: Empirical models and experimental observations. *IEEE Transactions on Geoscience and Remote Sensing* GE-23(1), 25–34.

Hassan-Esfahani, L., Torres-Rua, A., Ticlavilca, A. M., Jensen, A. and McKee, M. (2014). Topsoil moisture estimation for precision agriculture using unmmaned aerial vehicle multispectral imagery. In: *IEEE International Geoscience and Remote Sensing Symposium (IGARSS)*. New York: IEEE Publications, vol. 2014, pp. 3263–3266.

Helman, D., Bahat, I., Netzer, Y., Ben-Gal, A., Alchanatis, V., Peeters, A. and Cohen, Y. (2018). Using time series of high-resolution planet satellite images to monitor grapevine stem water potential in commercial vineyards. *Remote Sensing* 10(10), 1615.

Hoffmann, H., Jensen, R., Thomsen, A., Nieto, H., Rasmussen, J. and Friborg, T. (2016). Crop water stress maps for an entire growing season from visible and thermal UAV imagery. *Biogeosciences* 13(24), 6545–6563.

Hornbuckle, B. K. and England, A. W. (2005). Diurnal variation of vertical temperature gradients within a field of maize: Implications for satellite microwave radiometry. *IEEE Geoscience and Remote Sensing Letters* 2(1), 74–77.

Hornbuckle, B. K., England, A. W., De Roo, R. D., Fischman, M. A. and Boprie, D. L. (2003). Vegetation canopy anisotropy at 1.4 GHz. *IEEE Transactions on Geoscience and Remote Sensing* 41(10), 2211–2223.

Houtz, D., Naderpour, R. and Schwank, M. (2020). Portable L-band radiometer (PoLRa): Design and characterization. *Remote Sensing* 12(17), 2780.

Jackson, T. J., Hsu, A. Y., Van de Griend, A. and Eagleman, J. R. (2004). Skylab L-band microwave radiometer observations of soil moisture revisited. *International Journal of Remote Sensing* 25(13), 2585–2606.

Jackson, T. J. and Le Vine, D. E. (1996). Mapping surface soil moisture using an aircraft-based passive microwave instrument: Algorithm and example. *Journal of Hydrology* 184(1–2), 85–99.

Jackson, T. J. and O'Neill, P. E. (1987). Salinity effects on the microwave emission of soils. *IEEE Transactions on Geoscience and Remote Sensing* GE-25(2), 214–220.

Jackson, T. J. and Schmugge, T. J. (1989). Passive microwave remote sensing system for soil moisture: Some supporting research. *IEEE Transactions on Geoscience and Remote Sensing* 27(2), 225–235.

Jackson, T. J. and Schmugge, T. J. (1991). Vegetation effects on the microwave emission of soils. *Remote Sensing of Environment* 36(3), 203–212.

Jackson, T. J., Schmugge, T. J. and Wang, J. R. (1982). Passive microwave sensing of soil moisture under vegetation canopies. *Water Resources Research* 18(4), 1137–1142.

Jagdhuber, T., Hajnsek, I., Bronstert, A. and Papathanassiou, K. P. (2013). Soil moisture estimation under low vegetation cover using a multi-angular polarimetric decomposition. *IEEE Transactions on Geoscience and Remote Sensing* 51(4), 2201–2215.

Jonard, F., Weihermuller, L., Jadoon, K. Z., Schwank, M., Vereecken, H. and Lambot, S. (2011). Mapping field-scale soil moisture with L-band radiometer and ground-penetrating radar over bare soil. *IEEE Transactions on Geoscience and Remote Sensing* 49(8), 2863–2875.

Kankaku, Y., Sagisaka, M. and Suzuki, S. (2014). PALSAR-2 launch and early orbit status. In: *IEEE International Geoscience and Remote Sensing Symposium (IGARSS)*. New York: IEEE Publications, vol. 2014, pp. 3410–3412.

Kerr, Y. H., Waldteufel, P., Wigneron, J.-P., Delwart, S., Cabot, F., Boutin, J., Escorihuela, M. J., Font, J., Reul, N., Gruhier, C., Juglea, S. E., Drinkwater, M. R., Hahne, A., Martín-Neira, M. and Mecklenburg, S. (2010). The SMOS mission: New tool for monitoring key elements of the global water cycle. *Proceedings of the IEEE* 98(5), 666–687.

Khan, N. M., Rastoskuev, V. V., Sato, Y. and Shiozawa, S. (2005). Assessment of hydrosaline land degradation by using a simple approach of remote sensing indicators. *Agricultural Water Management* 77(1–3), 96–109.

Kim, S.-B., Tsang, L., Johnson, J. T., Huang, S., Van Zyl, J. J. and Njoku, E. G. (2012). Soil moisture retrieval using time-series radar observations over bare surfaces. *IEEE Transactions on Geoscience and Remote Sensing* 50(5), 1853–1863.

Kirdiashev, K. P., Chukhlantsev, A. A. and Shutko, A. M. (1979). Microwave radiation of the earth's surface in the presence of vegetation cover. *Radiotekhnika i Elektronika* 24, 256–264.

Kite, G. W. and Pietroniro, A. (1996). Remote sensing applications in hydrological modelling. *Hydrological Sciences Journal* 41(4), 563–591.

Lasne, Y., Paillou, P., Freeman, A., Farr, T., McDonald, K. C., Ruffie, G., Malezieux, J.-M., Chapman, B. and Demontoux, F. (2008). Effect of salinity on the dielectric properties of geological materials: Implication for soil moisture detection by means of radar remote sensing. *IEEE Transactions on Geoscience and Remote Sensing* 46(6), 1674–1688.

Laymon, C. A., Crosson, W. L., Jackson, T. J., Manu, A. and Tsegaye, T. D. (2001). Ground-based passive microwave remote sensing observations of soil moisture at S-band and L-band with insight into measurement accuracy. *IEEE Transactions on Geoscience and Remote Sensing* 39(9), 1844–1858.

Le Toan, T., Quegan, S., Davidson, M. W. J., Balzter, H., Paillou, P., Papathanassiou, K., Plummer, S., Rocca, F., Saatchi, S., Shugart, H. and Ulander, L. (2011). The BIOMASS mission: Mapping global forest biomass to better understand the terrestrial carbon cycle. *Remote Sensing of Environment* 115(11), 2850–2860.

Lievens, H., Verhoest, N., De Keyser, E., Vernieuwe, H., Matgen, P., Álvarez-Mozos, J. and De Baets, B. (2011). Effective roughness modelling as a tool for soil moisture retrieval from C-and L-band SAR. *Hydrology and Earth System Sciences Discussions* 7, 4995–5031.

Liu, H.-J., Zhang, Y.-Z., Zhang, X.-L., Zhang, B., Song, K.-S., Wang, Z.-M. and Tang, N. (2009). Quantitative analysis of moisture effect on black soil reflectance. *Pedosphere* 19(4), 532–540.

Liu, W., Baret, F., Gu, X., Tong, Q., Zheng, L. and Zhang, B. (2002). Relating soil surface moisture to reflectance. *Remote Sensing of Environment*, 81(2–3), 238–246.

Lu, F., Sun, Y. and Hou, F. (2020). Using UAV visible images to estimate the soil moisture of steppe. *Water* 12(9), 2334.

Martinez-Agirre, A., Álvarez-Mozos, J., Lievens, H. and Verhoest, N. E. C. (2017). Influence of surface roughness measurement scale on radar backscattering in different agricultural soils. *IEEE Transactions on Geoscience and Remote Sensing* 55(10), 5925–5936.

Martinez-Vazquez, A., Camps, A., Lopez-Sanchez, J. M., Vall-Llossera, M. and Monerris, A. (2009). Numerical simulation of the full-polarimetric emissivity of vines and comparison with experimental data. *Remote Sensing* 1(3), 300–317.

Mätzler, C., Rosenkranz, P. W., Battaglia, A. and Wigneron, J.-P. (2006). *Thermal Microwave Radiation: Applications for Remote Sensing Electromagnetic Waves Series*. London, UK: IET.

Meesters, A. G. C. A., De Jeu, R. A. M. and Owe, M. (2005). Analytical derivation of the vegetation optical depth from the microwave polarization difference index. *IEEE Geoscience and Remote Sensing Letters* 2(2), 121–123.

Merlin, O., Walker, J. P., Kalma, J. D., Kim, E. J., Hacker, J., Panciera, R., Young, R., Summerell, G., Hornbuckle, J., Hafeez, M. and Jackson, T. (2008). The NAFE'06 data set: Towards soil moisture retrieval at intermediate resolution. *Advances in Water Resources* 31(11), 1444–1455.

Merlin, O., Walker, J. P., Panciera, R., Young, R., Kalma, J. D. and Kim, E. J. (2007). Calibration of a soil moisture sensor in heterogeneous terrain with the national airborne field experiment (NAFE) data. In: Oxley, L. and Kulasiri, D. (Eds.), *MODSIM 2007 International Congress on Modelling and Simulation*. New Zealand: Modelling and Simulation Society of Australia, pp. 2604–2610.

Metternicht, G. and Zinck, J. A. (2008). *Spectral Behavior of Salt Types*. Boca Raton, FL: CRC Press.

Metternicht, G. I. and Zinck, J. A. (2003). Remote sensing of soil salinity: Potentials and constraints. *Remote Sensing of Environment* 85(1), 1–20.

Mironov, V. L., Dobson, M. C., Kaupp, V. H., Komarov, S. A. and Kleshchenko, V. N. (2004). Generalized refractive mixing dielectric model for moist soils. *IEEE Transactions on Geoscience and Remote Sensing* 42(4), 773–785.

Mo, T., Choudhury, B. J., Schmugge, T. J., Wang, J. R. and Jackson, T. J. (1982). A model for microwave emission from vegetation-covered fields. *Journal of Geophysical Research* 87(C13), 11229–11211,11237.

Mo, T. and Schmugge, T. J. (1987). A parameterization of the effect of surface roughness on microwave emission. *IEEE Transactions on Geoscience and Remote Sensing* GE-25(4), 481–486.

Monsivais-Huertero, A., Liu, P.-W. and Judge, J. (2018). Phenology-based backscattering model for corn at L-band. *IEEE Transactions on Geoscience and Remote Sensing* 56, 4989–5005.

Newton, R. and Rouse Jr., J. (1980). Microwave radiometer measurements of soil moisture content. *IEEE Transactions on Antennas and Propagation* 28(5), 680–686.

Njoku, E. G. and Entekhabi, D. (1996). Passive microwave remote sensing of soil moisture. *Journal of Hydrology* 184(1–2), 101–129.

Njoku, E. G., Jackson, T. J., Lakshmi, V., Chan, T. K. and Nghiem, S. V. (2003). Soil moisture retrieval from AMSR-E. *IEEE Transactions on Geoscience and Remote Sensing* 41(2), 215–229.

Njoku, E. G. and Kong, J. A. (1977). Theory for passive microwave remote sensing of near-surface soil moisture. *Journal of Geophysical Research* 82(20), 3108–3118.

Njoku, E. G., Wilson, W. J., Yueh, S. H., Dinardo, S. J., Li, F. K., Jackson, T. J., Lakshmi, V. and Bolten, J. (2002). Observations of soil moisture using a passive and active low-frequency microwave airborne sensor during SGP99. *IEEE Transactions on Geoscience and Remote Sensing* 40(12), 2659–2673.

Notarnicola, C., Angiulli, M. and Posa, F. (2008). Soil moisture retrieval from remotely sensed data: Neural network approach versus Bayesian method. *IEEE Transactions on Geoscience and Remote Sensing* 46(2), 547–557.

Oh, Y., Sarabandi, K. and Ulaby, F. T. (1992). An empirical model and an inversion technique for radar scattering from bare soil surfaces. *IEEE Transactions on Geoscience and Remote Sensing* 30(2), 370–381.

Oh, Y., Sarabandi, K. and Ulaby, F. T. (1994). *An Inversion Algorithm for Retrieving Soil Moisture and Surface Roughness From Polarimetric Radar Observation*. New York: IEEE Publications, pp. 1582–1584).

Owe, M., de Jeu, R. and Walker, J. P. (2001). A methodology for surface soil moisture and vegetation optical depth retrieval using the microwave polarization difference index. *IEEE Transactions on Geoscience and Remote Sensing* 39(8), 1643–1654.

Paloscia, S., Pampaloni, P., Chiarantini, L., Coppo, P., Gagliani, S. and Luzi, G. (1993). Multifrequency passive microwave remote sensing of soil moisture and roughness. *International Journal of Remote Sensing* 14(3), 467–483.

Panciera, R., Merlin, O., Young, R. and Walker, J. (2006). *The Hydraprobe Data Acquisition System (HDAS): User Guide*. Report. University of Melbourne.

Panciera, R., Walker, J. P., Jackson, T. J., Gray, D. A., Tanase, M. A., Ryu, D., Monerris, A., Yardley, H., Rudiger, C., Wu, X., Gao, Y. and Hacker, J. M. (2014). The soil moisture active passive experiments (SMAPEx): Towards soil moisture retrieval from the SMAP mission. *IEEE Transactions on Geoscience and Remote Sensing* 52(1), 490–507.

Panciera, R., Walker, J. P., Kalma, J. D., Kim, E. J., Hacker, J. M., Merlin, O., Berger, M. and Skou, N. (2008). The NAFE'05/CoSMOS data set: Toward SMOS soil moisture retrieval, downscaling, and assimilation. *IEEE Transactions on Geoscience and Remote Sensing* 46(3), 736–745.

Panciera, R., Walker, J. P., Kalma, J. D., Kim, E. J., Saleh, K. and Wigneron, J.-P. (2009). Evaluation of the SMOS L-MEB passive microwave soil moisture retrieval algorithm. *Remote Sensing of Environment* 113(2), 435–444.

Peischl, S., Walker, J. P., Rüdiger, C., Ye, N., Kerr, Y. H., Kim, E., Bandara, R. and Allahmoradi, M. (2012). The AACES field experiments: SMOS calibration and validation across the Murrumbidgee River catchment. *Hydrology and Earth System Sciences* 9, 2763–2795.

Periasamy, S. and Ravi, K. P. (2020). A novel approach to quantify soil salinity by simulating the dielectric loss of SAR in three-dimensional density space. *Remote Sensing of Environment* 251, 112059.

Petropoulos, G. P., Ireland, G. and Barrett, B. (2015). Surface soil moisture retrievals from remote sensing: Current status, products & future trends. *Physics and Chemistry of the Earth, Parts A/B/C* 83–84, 36–56.

Rice, S. O. (1951). Reflection of electromagnetic waves from slightly rough surfaces. *Communications on Pure and Applied Mathematics* 4(2–3), 351–378.

Rodriguez-Fernandez, N. J., Aires, F., Richaume, P., Kerr, Y. H., Prigent, C., Kolassa, J., Cabot, F., Jimenez, C., Mahmoodi, A. and Drusch, M. (2015). Soil moisture retrieval using neural networks: Application to SMOS. *IEEE Transactions on Geoscience and Remote Sensing* 53(11), 5991–6007.

Rodríguez-Fernández, N. J., de Souza, V., Kerr, Y. H., Richaume, P. and Al Bitar, A. (2017). Soil moisture retrieval using SMOS brightness temperatures and a neural network trained on in situ measurements. In: Bitar, A. (Ed.), *IEEE International Geoscience and Remote Sensing Symposium (IGARSS)*. New York: IEEE Publications, vol. 2017, pp. 1574–1577.

Rosen, P., Hensley, S., Shaffer, S., Edelstein, W., Kim, Y., Kumar, R., Misra, T., Bhan, R. and Sagi, R. (2017). The NASA-ISRO SAR (NISAR) mission dual-band radar instrument preliminary design. In: *IEEE International Geoscience and Remote Sensing Symposium (IGARSS)*. New York: IEEE Publications, vol. 2017, pp. 3832–3835.

Santi, E., Paloscia, S., Pettinato, S. and Fontanelli, G. (2016). Application of artificial neural networks for the soil moisture retrieval from active and passive microwave spaceborne sensors. *International Journal of Applied Earth Observation and Geoinformation* 48, 61–73.

Schmugge, T., O'Neill, P. E. and Wang, J. R. (1986). Passive microwave soil moisture research. *IEEE Transactions on Geoscience and Remote Sensing* GE-24(1), 12–22.

Schwank, M. and Mätzler, C. (2006). Air-to-soil transition model. In: Mätzler, C., Rosenkranz, P. W., Battaglia, A. and Wigneron, J.-P. (Eds.), *Thermal Microwave Radiation-Applications for Remote Sensing Electromagnetic Waves Series*. London, UK: IEEE, pp. 287–301.

Schwank, M., Wiesmann, A., Werner, C., Mätzler, C., Weber, D., Murk, A., Völksch, I. and Wegmüller, U. (2010). ELBARA II, an L-band radiometer system for soil moisture research. *Sensors* 10(1), 584–612.

Shen, X., Walker, J. P., Ye, N., Wu, X., Boopathi, N., Yeo, I.-Y., Zhang, L. and Zhu, L. (2020). Soil moisture retrieval depth of P-and L-band radiometry: Predictions and observations. *IEEE Transactions on Geoscience and Remote Sensing* 59, 6814–6822.

Shi, J., Chen, K. S., Li, Q., Jackson, T. J., O'Neill, P. E. and Tsang, L. (2002). A parameterized surface reflectivity model and estimation of bare-surface soil moisture with L-band radiometer. *IEEE Transactions on Geoscience and Remote Sensing* 40(12), 2674–2686.

Shi, J., Jiang, L., Zhang, L., Chen, K., Wigneron, J.-P. and Chanzy, A. (2005). A parameterized multifrequency-polarization surface emission model. *IEEE Transactions on Geoscience and Remote Sensing* 43(12), 2831–2841.

Shi, J., Wang, J., Hsu, A., O'Neili, P. and Engman, E. T. (1995). Estimation of soil moisture and surface roughness parameters using L-band SAR measurements. *Proceeding of IEEE Transaction on Geosicence and Remote Sensing* 501, 507–509.

Shoshany, M., Svoray, T., Curran, P. J., Foody, G. M. and Perevolotsky, A. (2000). The relationship between ERS-2 SAR backscatter and soil moisture: Generalization from a humid to semi-arid transect. *International Journal of Remote Sensing* 21(11), 2337–2343.

Shutko, A. M. and Reutov, E. M. (1982). Mixture formulas applied in estimation of dielectric and radiative characteristics of soils and grounds at microwave frequencies. *IEEE Transactions on Geoscience and Remote Sensing* 1, 29–32.

Sidike, A., Zhao, S. and Wen, Y. (2014). Estimating soil salinity in Pingluo County of China using QuickBird data and soil reflectance spectra. *International Journal of Applied Earth Observation and Geoinformation* 26, 156–175.

Singh, R. and Srivastav, S. (1990). Mapping of waterlogged and salt-affected soils using microwave radiometers. *Remote Sensing* 11, 1879–1887.

Smith, A. B., Walker, J. P., Western, A. W., Young, R. I., Ellett, K. M., Pipunic, R. C., Grayson, R. B., Siriwardena, L., Chiew, F. H. S. and Richter, H. (2012). The Murrumbidgee soil moisture monitoring network data set. *Water Resources Research* 48(7), 1-6. doi:10.1029/2012WR011976.

Taghadosi, M. M., Hasanlou, M. and Eftekhari, K. (2019). Soil salinity mapping using dual-polarized SAR Sentinel-1 imagery. *International Journal of Remote Sensing* 40(1), 237–252.

Torres, R., Snoeij, P., Geudtner, D., Bibby, D., Davidson, M., Attema, E., Potin, P., Rommen, B., Floury, N., Brown, M., Traver, I. N., Deghaye, P., Duesmann, B., Rosich, B., Miranda, N., Bruno, C., L'Abbate, M., Croci, R., Pietropaolo, A., Huchler, M. and Rostan, F. (2012). GMES Sentinel-1 mission. *Remote Sensing of Environment* 120, 9–24.

Turner, D., Lucieer, A. and Watson, C. (2011). Development of an unmanned aerial vehicle (UAV) for hyper resolution vineyard mapping based on visible, multispectral, and thermal imagery. In: *Proceedings of the 34th International Symposium on Remote Sensing of Environment*, p. 4.

Ulaby, F. T., Dubois, P. C. and Van Zyl, J. (1996). Radar mapping of surface soil moisture. *Journal of Hydrology* 184(1-2), 57-84.

Ulaby, F. T., Long, D. G., Blackwell, W. J., Elachi, C., Fung, A. K., Ruf, C., Sarabandi, K., Zebker, H. A. and Van Zyl, J. (2014). *Microwave Radar and Radiometric Remote Sensing*. Ann Arbor: The University of Michigan Press.

Ulaby, F. T., Moore, R. K. and Fung, A. K. (1981). *Microwave Remote Sensing: Active and Passive. Volume I-Microwave Remote Sensing Fundamentals and Radiometry*. Boston, MA: Artech House.

Ulaby, F. T., Moore, R. K. and Fung, A. K. (1982). *Microwave Remote Sensing: Active and Passive. Volume II-Radar Remote Sensing and Surface Scattering and Emission Theory*. Boston, MA: Artech House.

Ulaby, F. T., Moore, R. K. and Fung, A. K. (1986). *Microwave Remote Sensing: Active and Passive. Volume III-Volume Scattering and Emission Theory, Advanced Systems and Applications*. Boston, MA: Artech House.

Ulaby, F. T. and Wilson, E. A. (1985). Microwave attenuation properties of vegetation canopies. *IEEE Transactions on Geoscience and Remote Sensing* GE-23(5), 746-753.

Van de Griend, A. A. and Wigneron, J.-P. (2004). The b-factor as a function of frequency and canopy type at H-polarization. *IEEE Transactions on Geoscience and Remote Sensing* 42(4), 786-794.

Van Zyl, J. J., Chapman, B. D., Dubois, P. and Shi, J. (1993). The effect of topography on SAR calibration. *IEEE Transactions on Geoscience and Remote Sensing* 31(5), 1036-1043.

Vergara-Díaz, O., Zaman-Allah, M. A., Masuka, B., Hornero, A., Zarco-Tejada, P., Prasanna, B. M., Cairns, J. E. and Araus, J. L. (2016). A novel remote sensing approach for prediction of maize yield under different conditions of nitrogen fertilization. *Frontiers in Plant Science* 7, 666.

Wagner, W., Lemoine, G., Borgeaud, M. and Rott, H. (1999a). A study of vegetation cover effects on ERS scatterometer data. *IEEE Transactions on Geoscience and Remote Sensing* 37(2), 938-948.

Wagner, W., Lemoine, G. and Rott, H. (1999b). A method for estimating soil moisture from ERS scatterometer and soil data. *Remote Sensing of Environment* 70(2), 191-207.

Walker, J. P., Dumedah, G., Monerris, A., Gao, Y., Rüdiger, C., Wu, X., Panciera, R., Merlin, O., Pipunic, R., Ryu, D. and Zreda, M. (2012). High resolution soil moisture mapping. In: *5th Global Workshop on Digital Soil Mapping*, Sydeny, Australia.

Walker, J. P. and Houser, P. R. (2004). Requirements of a global near-surface soil moisture satellite mission: accuracy, repeat time, and spatial resolution. *Advances in Water Resources* 27, 785-801.

Walker, J. P., Houser, P. R. and Willgoose, G. R. (2004). Active microwave remote sensing for soil moisture measurement: A field evaluation using ERS-2. *Hydrological Processes* 18(11), 1975-1997.

Wang, J., Ding, J., Yu, D., Ma, X., Zhang, Z., Ge, X., Teng, D., Li, X., Liang, J., Lizaga, I., Chen, X., Yuan, L. and Guo, Y. (2019). Capability of Sentinel-2 MSI data for monitoring and mapping of soil salinity in dry and wet seasons in the Ebinur Lake region, Xinjiang, China. *Geoderma* 353, 172-187.

Wang, J. R. (1983). Passive microwave sensing of soil moisture content: The effects of soil bulk density and surface roughness. *Remote Sensing of Environment* 13(4), 329-344.

Wang, J. R. and Choudhury, B. J. (1981). Remote sensing of soil moisture content, over bare field at 1.4 GHz frequency. *Journal of Geophysical Research* 86(C6), 5277-5282.

Wang, J. R., O'Neill, P. E., Jackson, T. J. and Engman, E. T. (1983). Multifrequency measurements of the effects of soil moisture, soil texture, and surface roughness. *IEEE Transactions on Geoscience and Remote Sensing* GE-21(1), 44–51.

Wang, J. R. and Schmugge, T. J. (1980). An empirical model for the complex dielectric permittivity of soils as a function of water content. *IEEE Transactions on Geoscience and Remote Sensing* GE-18(4), 288–295.

Wang, L. and Qu, J. J. (2009). Satellite remote sensing applications for surface soil moisture monitoring: A review. *Frontiers of Earth Science in China* 3(2), 237–247.

Wang, S., Li, X., Han, X. and Jin, R. (2011). Estimation of surface soil moisture and roughness from multi-angular ASAR imagery in the Watershed Allied Telemetry Experimental Research (WATER). *Hydrology and Earth System Sciences* 15, 1415–1426.

Wegmuller, U. and Mätzler, C. (1999). Rough bare soil reflectivity model. *IEEE Transactions on Geoscience and Remote Sensing* 37(3), 1391–1395.

Weidong, L., Baret, F., Xingfa, G., Qingxi, T., Lanfen, Z. and Bing, Z. (2002). Relating soil surface moisture to reflectance. *Remote Sensing of Environment* 81(2–3), 238–246.

Wigneron, J.-P., Chanzy, A., Calvet, J. C. and Bruguier, N. (1995). A simple algorithm to retrieve soil moisture and vegetation biomass using passive microwave measurements over crop fields. *Remote Sensing of Environment* 51(3), 331–341.

Wigneron, J.-P., Kerr, Y. H., Waldteufel, P., Saleh, K., Escorihuela, M.-J., Richaume, P., Ferrazzoli, P., De Rosnay, P., Gurney, R., Calvet, J.-C., Grant, J. P., Guglielmetti, M., Hornbuckle, B., Mätzler, C., Pellarin, T. and Schwank, M. (2007). L-band microwave emission of the biosphere (L-MEB) model: Description and calibration against experimental data sets over crop fields. *Remote Sensing of Environment* 107(4), 639–655.

Wigneron, J.-P., Laguerre, L. and Kerr, Y. H. (2001). A simple parameterization of the L-band microwave emission from rough agricultural soils. *IEEE Transactions on Geoscience and Remote Sensing* 39(8), 1697–1707.

Wigneron, J.-P., Pardé, M., Waldteufel, P., Chanzy, A., Kerr, Y., Schmidl, S. and Skou, N. (2004). Characterizing the dependence of vegetation model parameters on crop structure, incidence angle, and polarization at L-band. *IEEE Transactions on Geoscience and Remote Sensing* 42(2), 416–425.

Ye, N., Walker, J. P. and Rüdiger, C. (2015). A cumulative distribution function method for normalizing variable-angle microwave observations. *IEEE Transactions on Geoscience and Remote Sensing* 53(7), 3906–3916.

Ye, N., Walker, J. P., Wu, X., de Jeu, R., Gao, Y., Jackson, T. J., Jonard, F., Kim, E., Merlin, O., Pauwels, V. R. N., Renzullo, L. J., Rudiger, C., Sabaghy, S., von Hebel, C., Yueh, S. H. and Zhu, L. (2021). The soil moisture active passive experiments: Validation of the SMAP products in Australia. *IEEE Transactions on Geoscience and Remote Sensing* 59(4), 2922–2939.

Ye, N., Walker, J. P., Yeo, I.-Y., Jackson, T. J., Kerr, Y., Kim, E., McGrath, A., PopStefanija, I., Goodberlet, M. and Hills, J. (2020). Toward P-band passive microwave sensing of soil moisture. *IEEE Geoscience and Remote Sensing Letters* 18, 504–508.

Yuan, Q., Shen, H., Li, T., Li, Z., Li, S., Jiang, Y., Xu, H., Tan, W., Yang, Q., Wang, J., Gao, J. and Zhang, L. (2020). Deep learning in environmental remote sensing: Achievements and challenges. *Remote Sensing of Environment* 241, 111716.

Zarco-Tejada, P. J., Berni, J. A. J., Suárez, L., Sepulcre-Cantó, G., Morales, F. and Miller, J. R. (2009). Imaging chlorophyll fluorescence with an airborne narrow-band multispectral camera for vegetation stress detection. *Remote Sensing of Environment* 113(6), 1262–1275.

Zarco-Tejada, P. J., Guillén-Climent, M. L., Hernández-Clemente, R., Catalina, A., González, M. R. and Martín, P. (2013). Estimating leaf carotenoid content in vineyards using high resolution hyperspectral imagery acquired from an unmanned aerial vehicle (UAV). *Agricultural and Forest Meteorology* 171–172, 281–294.

Zhang, H., Han, M., Chávez, J. L. and Lan, Y. (2017). Improvement in estimation of soil water deficit by integrating airborne imagery data into a soil water balance model. *International Journal of Agricultural and Biological Engineering* 10, 37–46.

Zhao, W. and Li, Z.-L. (2013). Sensitivity study of soil moisture on the temporal evolution of surface temperature over bare surfaces. *International Journal of Remote Sensing* 34(9–10), 3314–3331.

Zhu, L., Walker, J. P. and Shen, X. (2020). Stochastic ensemble methods for multi-SAR-mission soil moisture retrieval. *Remote Sensing of Environment* 251, 112099.

Zhu, L., Walker, J. P., Tsang, L., Huang, H., Ye, N. and Rüdiger, C. (2019a). A multi-frequency framework for soil moisture retrieval from time series radar data. *Remote Sensing of Environment* 235, 111433.

Zhu, L., Walker, J. P., Tsang, L., Huang, H., Ye, N. and Rüdiger, C. (2019b). Soil moisture retrieval from time series multi-angular radar data using a dry down constraint. *Remote Sensing of Environment* 231, 111237.

Zhu, L., Walker, J. P., Ye, N. and Rüdiger, C. (2019c). Roughness and vegetation change detection: A pre-processing for soil moisture retrieval from multi-temporal SAR imagery. *Remote Sensing of Environment* 225, 93–106.

Zhu, L., Walker, J. P., Ye, N. and Rudiger, C. (2016). The effect of radar configuration on effective correlation length. In: *2016 International Conference on Electromagnetics in Advanced Applications (ICEAA)*. New York: IEEE Publications, pp. 820–823.

Part 2

Advances in proximal sensing technologies

Chapter 4

Advances in using proximal spectroscopic sensors to assess soil health

Kenneth A. Sudduth and Kristen S. Veum, USDA-ARS, USA

1 Introduction

Due to the sensitivity of the soil ecosystem and its role in global carbon cycling, multiple national and international initiatives are focused on soil carbon sequestration, soil carbon markets, and monitoring of soil health across soils and climates. Assessment of soil health involves determining how well a soil is performing its biological, chemical, and physical functions by measuring soil properties referred to as soil health indicators. Applied soil health research is focused on the needs of 'on-farm' soil health assessment and interpretation and requires tools and strategies that are affordable and accessible. The high cost and labor requirements of most currently available soil health assessments render them impractical for large-scale efforts. Ideally, an on-farm soil health assessment would be easy to measure, applicable to field conditions, affordable, and capture a wide range of soil functions that vary spatially and temporally (Doran and Parkin, 1996; Andrews et al., 2004; Karlen et al., 2017). Thus, development of rapid, low-cost methods for in-field assessment of soil health is a priority. Soil spectroscopy techniques offer high-density spatial and temporal soil

http://dx.doi.org/10.19103/AS.2022.0107.10
Published by Burleigh Dodds Science Publishing Limited, 2023.

information to help achieve agronomic and sustainability goals related to soil health. This chapter highlights the successful application of soil spectroscopy to soil health assessment, identifies novel tools and strategies that are advancing the science of soil health assessment through soil spectroscopy, and outlines areas for future research.

2 Soil spectroscopy methods

Diffuse reflectance spectroscopy (DRS) is a method that has been successfully used to estimate soil organic carbon (SOC), soil organic matter (SOM), and numerous other properties related to soil health (see reviews by Malley et al., 2004; Viscarra Rossel et al., 2006; Stenberg et al., 2010). Spectral analysis of soils is generally carried out in the visible (Vis; 400–700 nm), near-infrared [NIR; 750–2500 nm (13500–4000 cm^{-1})], or mid-infrared [MIR; 2500–23 500 nm (4000–450 cm^{-1})] range of the electromagnetic spectrum (Soriano Disla et al., 2014; Nocita et al., 2015; Johnson et al., 2019). The DRS method is based on the diffusion of incoming radiation in soil prior to being reflected and detected. This interaction of light with soil constituents produces spectral features that can subsequently be used to develop chemometric models for estimation of soil properties (Malley et al., 2004).

Spectra are generally quantified in per cent reflectance or decimal reflectance (R) as a function of wavelength. Often, reflectance will be transformed to apparent absorbance: $A = \log_{10}(1/R)$ as that quantity is more linearly related to the concentration of an absorbing compound. The fundamental vibrations of chemical bonds excited by radiation occur in the MIR region. However, instrumentation to quantify MIR DRS is more expensive, less robust to field deployment, and generally requires additional sample pretreatment beyond that required for Vis-NIR DRS. Therefore, most soil spectral analysis occurs in the NIR or Vis-NIR spectral range where overtones of the fundamental vibrations exist but become progressively weaker at shorter wavelengths. These overtones are broad and overlap with one another, such that few well-defined spectral features are found in the Vis-NIR region. This renders the spectra difficult to interpret directly because they are not amenable to assignment or quantification as individual peaks. However, Vis-NIR spectra still contain useful information on organic and inorganic soil constituents. Absorptions in the Vis region are often associated with iron-containing minerals (Sherman and Waite, 1985), and due to the inherent 'darkness' of SOM, it can also broadly affect Vis absorbance. In the NIR region, absorptions result from a variety of bonds containing carbon, nitrogen, hydrogen, and oxygen atoms, making quantification of organic compounds, clay minerals, and water possible (Fig. 1).

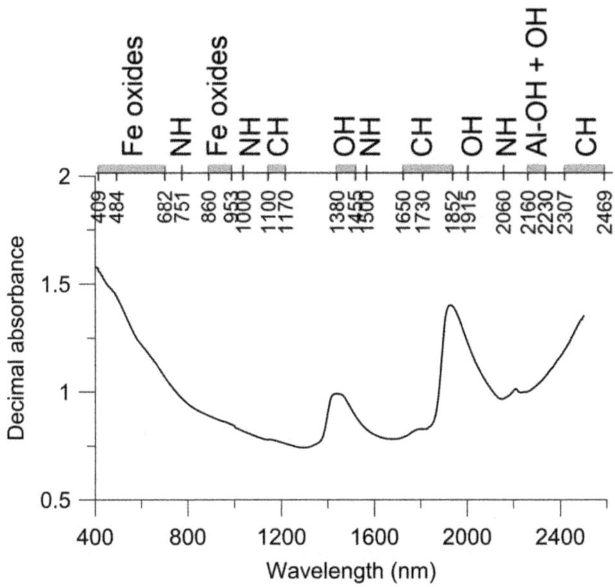

Figure 1 Typical absorbance spectrum of moist soil, including specific absorption bands for soil chemical bonds (after Katuwal et al., 2018).

2.1 Calibration approaches

Early approaches to statistical analysis of spectral datasets often involved a multiple, stepwise regression approach to select informative wavebands from the spectrum (Hruschka, 1987). This generally followed one or more spectral pretreatments to remove interference and/or linearize the relationship between spectral data and the response variable. First and second derivatives of the spectra and ratios of reflectance or absorbance at selected bands were common. Presently, the wavelength selection approach has generally been replaced by the full-spectrum approaches described later. An exception is where wavelengths are being selected with the intention of developing a discrete-waveband sensor, which could be expected to be less complex and more rugged than a full-spectrum instrument (Lee et al, 2009; Zhou et al., 2022).

The full-spectrum calibration uses multivariate techniques such as partial least squares regression (PLSR) or principal components regression (PCR; Viscarra Rossel and Behrens, 2010) that can handle highly correlated, high-dimensional data where the number of predictor variables is greater than the number of observations (Haaland and Thomas, 1988). Also, as signals in the Vis-NIR region are weaker than vibrations in the MIR region, multivariate techniques such as PLSR can overcome these challenges (Viscarra Rossel and Behrens, 2010). More recently, Veum et al. (2018) applied Bayesian covariate-assisted techniques to estimation of SOC, total nitrogen, and soil texture (i.e. clay, silt, and sand content). In addition, a range of mathematical pretreatments or spectral transformations may be required, with no consensus on best practices in the literature (Stenberg et al., 2010). Some have combined PCR or PLSR with feature selection methods to retain only the most informative spectral variables for calibration. For example, Zhou et al. (2022) used backward interval PLSR (Zou et al., 2007) to select the most informative spectral regions for estimating multiple soil properties. They reported improved accuracy compared to standard PLSR analysis for some, but not all, variables.

Various machine learning approaches have also been used to develop prediction models, allowing better representation of nonlinearities than PCR and PLSR. Some of the machine learning techniques include neural networks (Zhou et al., 2022), random forests (Viscarra Rossel and Behrens, 2010), and boosted regression trees (Brown et al., 2006). Pei et al. (2019) compared PLSR, neural networks, and random forests for estimating 11 soil physical and chemical properties. They reported that best results were obtained by PLSR in six cases, by neural networks in three cases, and by random forests in two cases. When considering multiple studies comparing calibration methods, no single algorithm, including PLSR, neural networks, or other machine learning algorithms, has consistently provided best results (Igne et al., 2010; Pei et al., 2019; Viscarra Rossel and Behrens, 2010).

Another approach to spectral estimation of soil properties relies on soil spectral libraries. These libraries have been created at a variety of scales, ranging from national (e.g. Wijewardane et al., 2016b) to global (Viscarra Rossel et al, 2016). The libraries contain soil property data, spectra, and associated metadata, which can be used with various modeling methods to estimate soil properties for spectra that are not in the library. However, these models often perform poorly at a local scale (i.e. field or farm), even when the libraries are inclusive of sites from the locality or have been augmented or 'spiked' with local samples (Wetterlind and Stenberg, 2010). Methods to improve local representations using soil spectral libraries are a current research topic. For example, Shen et al. (2022) used a process called 'deep transfer learning' to use a global soil spectral library (Viscarra Rossel et al., 2016) for estimating field-scale variations in soil carbon, finding that the approach provided improved results compared to locally calibrated models.

2.2 Lab-based soil spectroscopy

Early soil spectroscopy investigations were laboratory-based, due to the complexity and cost of spectrophotometers. Generally, soils were obtained in the field, air- or oven-dried, crushed to pass a sieve (usually 2 mm), and then scanned. A variety of complex, expensive spectrophotometers were used, but over time the portable Vis-NIR (350–2500 nm) ASD FieldSpec spectrometer (Analytical Spectral Devices, now a part of Malvern Panalytical, Malvern, UK) became almost a de-facto standard in soil research. A standard protocol for lab-based spectroscopy was proposed by Viscarra Rossel et al. (2016).

2.3 Field systems

In contrast to more traditional laboratory approaches, on-the-go sensor technology has the potential to provide high-resolution, in situ data quickly at low cost (Hummel et al., 1996). In situ sensing can also provide measurements at a high spatial or temporal density at relatively low cost (Adamchuk et al., 2004), leading to greater overall accuracy in mapping soil variability. Lab-based analyses have two sources of error: (1) analysis error due to sub-sampling and analytical determination and (2) sampling error due to point-to-point variation in soils. With traditional soil testing, analysis error is expected to be relatively low, but sampling error can be substantial due to limited sampling density across space and time. Sensors can provide a spatiotemporal sampling intensity several orders of magnitude greater than traditional methods, thus reducing sampling error and potentially reducing overall error even if analysis errors are higher.

The process of collecting soil samples in the field and sample preparation for lab-based DRS requires soil disturbance and considerable labor and processing time. These issues are especially important when dense spatial sampling is desired, for example, to estimate soil health across fields and landscapes. Thus, in situ DRS instruments have been developed to estimate soil properties in real time after the required calibrations are established. Both continuous spectrum and discrete spectral band instruments have been the subject of research and development. In the continuous spectrum approach, Sudduth and Hummel (1993) designed a mobile NIR spectrophotometer (Fig. 2) and tested it in the laboratory and field. Christy (2008) developed a mobile spectrometer-based system with a range of 920–1718 nm and used it to map spatial differences in SOM content. Mouazen et al. (2007, 2014) developed a soil sensing device with a measuring range of 306–1711 nm and reported results for estimation of total carbon, soil moisture content, pH, and phosphorus.

In-field soil sensors based on discrete spectral bands have typically used light-emitting diodes, laser diodes, or tungsten halogen lamps as light sources (An et al., 2014; Zhou et al., 2019). Compared with a spectrometer that acquires data for hundreds to thousands of spectral bands, soil sensors based on discrete spectral bands are relatively inexpensive. Shonk et al. (1991) developed a prototype real-time SOM sensor that used red light-emitting diodes as the light source. An et al. (2014) developed a total nitrogen sensor based on six discrete

Figure 2 Mobile near-infrared (1700–2420 nm) spectrophotometer developed in the late 1980s for estimating in-field variations in soil organic matter.

NIR wavebands, and Zhou et al. (2019) developed an in situ total nitrogen and soil moisture sensor with eight discrete NIR bands. Kweon and Maxton (2013) reported on a commercialized two-band (660 nm and 940 nm) sensor designed to estimate SOM, showing good agreement with laboratory-measured SOM across six test fields. Although these discrete-band sensors could be effective in estimating a single soil property, the critical sensing wavelengths for multiple soil properties are generally not coincident (Lee et al., 2009; Zhou et al., 2022). Therefore, the continuous-spectrum approach is likely more appropriate when the goal is to estimate multiple soil properties for a soil health assessment.

2.4 Moisture and other interfering factors

Prediction of multiple soil properties using DRS has been successful on air-dried surface samples (Sudduth and Hummel, 1993; Chang et al., 2001; Veum et al., 2015b), air-dried whole-profile soil samples (Lee et al., 2009), and moist samples (Sudduth and Hummel, 1993; Morgan et al., 2009; Veum et al., 2015b) scanned in the laboratory. Some studies have compared predictions obtained at different soil moisture levels, with some reporting better results with dry soil (Sudduth and Hummel, 1993; Stevens et al., 2006) and others reporting better results with moist soil (Fystro, 2002; Nocita et al., 2013). It is known that Vis-NIR spectra are particularly sensitive to soil moisture, decreasing the estimation accuracy of Vis-NIR collected in the field. In particular, hydroxyl bands from soil moisture mask some spectral features produced by SOC (Mouazen et al., 2007). It has been suggested

Published by Burleigh Dodds Science Publishing Limited, 2023.

that including a wide range of soil moisture contents in calibration models may mitigate problems associated with soil moisture variation (Sudduth and Hummel, 1993). Moreover, in a field setting, spectra are sensitive to other environmental conditions (e.g. temperature and soil structure) further decreasing prediction accuracy and the ultimate utility of spectra collected in the field (Sudduth and Hummel, 1993; Mouazen et al., 2007; Morgan et al., 2009; Reeves, 2010; Minasny et al., 2011; Ji et al., 2015).

Various techniques have been applied to account for moisture and other environmental factors to improve model performance, including external parameter orthogonalization (EPO), direct standardization, and global moisture modeling (Wijewardane et al., 2016a). The EPO algorithm removes variation due to external factors by projecting the soil spectra orthogonal to the space of unwanted variation (Roger et al., 2003). Studies have successfully applied EPO for the estimation of soil properties, including soil carbon (Minasny et al., 2011; Ge et al., 2014; Veum et al., 2018) and clay content (Ge et al., 2014; Ackerson et al., 2017; Veum et al., 2018). Alternatively, the direct standardization approach derives a transfer matrix to characterize differences between corresponding field and laboratory spectra and has successfully been used to predict SOM using a portable spectrometer (Ji et al., 2015). With the global moisture modeling technique, a secondary variable with a relationship to the primary variable is intentionally manipulated, resulting in a more robust calibration model (Kawano et al., 1995). This approach, akin to spiking, has been applied to datasets that span large geographical regions or use combined spectral libraries to estimate soil carbon and clay content (Brown, 2007; Wetterlind and Stenberg, 2010). The ultimate goal is to leverage existing spectral libraries collected on dry, laboratory-processed soils for prediction of soil properties from spectra collected in situ under variable environmental conditions.

2.5 Soil profile information

Soil health research has focused on surface soils (~0–15 cm) that are critical in plant growth and development and represent the zone of primary agricultural management. Surface soils are the interface with the atmosphere, and alteration of the upper soil horizons through management can impact many biological, physical, and chemical soil properties and processes. In addition, there are practical limitations in soil sampling equipment and budget constraints that result in an emphasis on surface soil horizons in most soil health assessments. However, vertical stratification and anisotropy in the soil profile affect infiltration, storage, and movement of water (Bouma et al., 1977), nutrient availability (Jobbágy and Jackson, 2001), carbon storage (Harrison et al., 2011), and microbial activity (Fontaine et al., 2007). For example, the subsurface argillic horizon characteristic of claypan soils impacts water partitioning, runoff, lateral movement of water through the soil, and microbial community structure and function (Hsiao et al., 2018).

Vertical stratification of soil health indicators is frequently observed under no-till systems. Adoption of no-till is one of the most important strategies for Conservation Agriculture, and it can improve soil health (Nunes et al., 2018; Veum et al., 2014). However, long-term no-till systems managed without other conservation practices (e.g. diversified cropping systems and permanent soil cover by mulching and/or cover crops) can also result in strong stratification of chemical, physical, and biological soil health indicators, with a high concentration of nutrients and SOC within the uppermost topsoil layer (~0–8 cm) and low nutrient availability and high soil compaction below that depth (Deubel et al., 2011; Houx et al., 2011; Nunes et al., 2019a; Powlson and Jenkinson, 1981). These characteristics may restrict plant root growth, decrease the absorption and translocation of water and nutrients to plants, and ultimately limit crop yield. Furthermore, soil compaction also reduces infiltration capacity, leading to increased runoff, soil erosion, and transport of pesticides and nutrients. Therefore, in-field, profile estimation of soil properties could provide vertical spatial data to inform soil health status and site-specific constraints (e.g. subsurface soil compaction) that are essential for informed management and enhanced crop production.

Previous laboratory studies have demonstrated successful estimation of profile soil properties with DRS. For example, Dalal and Henry (1986) estimated profile SOC, total nitrogen, and soil moisture with $R^2 > 0.84$. Others used NIR DRS to estimate profile soil moisture ($R^2 = 0.87$) and SOC ($R^2 > 0.77$) in the laboratory (Hummel et al., 2001), and to estimate several profile soil properties, including clay, calcium, cation exchange capacity, and SOC with $R^2 \geq 0.80$ (Lee et al., 2009). Another approach has been in-field spectrometry on intact, extracted soil cores (Kusumo, et al., 2008; Viscarra Rossel, et al., 2017). With this approach, SOC was successfully predicted at depth resolutions as small as 1 cm (Hedley et al., 2015; Roudier et al., 2015).

Acquisition of in situ profile soil information has been investigated by several groups, as it offers many potential benefits over laboratory analysis of profile samples in terms of efficiency, timeliness, and expense. A penetrometer foreoptic developed for a prototype NIR spectrophotometer estimated soil moisture with $R^2 = 0.90$ (Hummel et al., 2004). In field studies, prototype penetrometer foreoptics coupled to commercial spectrometers estimated clay content with 20–25% greater error compared to laboratory spectra on dried and sieved samples (Poggio et al., 2017; Ackerson et al., 2017). A commercial penetrometer instrument, the Veris P4000 (Veris Technologies, Salina, Kansas, USA) can be deployed in the field for profile VNIR DRS data collection (Fig. 3). Data from this instrument have been used to estimate multiple soil properties. In one study, the cross-validation R^2 of bulk density, SOC, moisture, clay, silt,

Figure 3 Veris P4000 instrument capable of sensing soil spectra (343–2202 nm), soil apparent electrical conductivity, and penetration resistance to a depth of 1 m.

and sand content were found to be 0.32, 0.67, 0.40, 0.65, 0.61, and 0.38, respectively (Cho et al., 2017b). In another study, soil texture was estimated with errors of ~6% for clay and silt, 10–11% for sand, and SOM with errors of 0.3–0.5% (Wetterlind et al., 2015). The SOC estimates from in situ Veris P4000 data were less accurate (R^2 from 0.78 to 0.90) than estimates from P4000 data obtained in the laboratory on dry soil samples (R^2 from 0.93 to 0.96; Hodge and Sudduth, 2012). Veum et al. (2018) reported similar results, with the root mean square error of prediction smaller (0.19%) for SOC estimates based on laboratory DRS than for estimates based on in situ DRS (0.26%).

2.6 Auxiliary data and sensor data fusion

Soil health reflects an integration of physical, chemical, and biological functions, and it is not currently feasible to provide a complete soil health assessment with a single soil sensing approach. In particular, models for important chemical and physical aspects of soil health have been less successful than the DRS models for biological indicators. One technique to improve estimation of inconsistent target soil health indicators is to incorporate auxiliary variables into the estimation models that may be directly or indirectly related to the target soil property or may be less expensive to measure. This can be accomplished by combining simple laboratory measurements with reflectance spectra (Brown et al., 2006; Kinoshita et al., 2012; Morgan et al., 2009; Veum et al., 2015b, 2018). Although this may overcome some of the limitations of using reflectance spectra alone, this approach still requires collection and analysis of soil in the laboratory.

Another approach combines or 'fuses' data from complementary sensors and avoids laboratory soil analysis altogether (Adamchuk et al., 2011; Cho et al., 2017b; Kusumo et al., 2008; Roudier et al., 2015; Veum et al., 2017). In one example, an in-field core-scanning system that included gamma-ray attenuation and digital imaging along with NIR spectroscopy was used to estimate multiple soil profile properties (Viscarra Rossel et al., 2017). In addition, the combination of reflectance spectra with soil apparent electrical conductivity (EC_a) and penetration resistance (or soil strength) sensor data, as implemented in the Veris P4000 instrument, has demonstrated improved estimates of multiple soil properties (Pei et al., 2019; Veum et al., 2017). These studies illustrate some potential for rapid quantification of soil health by fusing auxiliary laboratory data or data obtained from complementary sensors. However, critical gaps remain in the development and application of sensors for nutrient availability, soil structure, and advanced soil biological indicators.

3 Estimation of soil health indicators and indices

3.1 Soil biological properties

Within the developing science of soil health, soil biology is playing a leading role in advancing our understanding of ecosystem function as well as agricultural productivity and sustainability (Lehman et al., 2015). Therefore, it is critical to gain a better understanding of soil biological function across space and time. A wide range of soil biological properties usually quantified by traditional laboratory methods are potential candidates for spectroscopic estimation. These range from standard SOM and SOC measurements to carbon and nitrogen mineralization assays (e.g. Doran, 1987; Franzluebbers et al., 1996), soil enzyme assays (e.g. Dick, 1994; Eivazi and Tabatabai, 1988), a rapid carbon oxidation test (Culman et al., 2012; Weil et al., 2003), a rapid protein extraction (Hurisso et al., 2018), and analyses of the microbial community via genomics (Manter et al., 2017) or high-throughput lipid analysis (Buyer and Sasser, 2012).

DRS has been successfully used to estimate SOC or SOM for some time in the laboratory as an alternative to wet chemistry methods (see reviews by Stenberg et al., 2010; Viscarra Rossel et al., 2006). In addition to SOC and SOM, spectra in this range have been used to successfully estimate several other biological soil health indicators across a range of soils and ecosystems, including total nitrogen, β-glucosidase activity, oxidizable carbon, microbial biomass-carbon, microbial lipids, and soil respiration (Cécillon et al., 2009; Chang et al., 2001; Chaudhary et al., 2012; Kinoshita et al., 2012; Palmborg and Nordgren, 1993, 1996; Pietikäinen and Fritze, 1995; Sudduth and Hummel, 1993; Veum et al., 2014, 2015b; Zornoza et al., 2008).

Published by Burleigh Dodds Science Publishing Limited, 2023.

Despite the relative success of these studies, estimation of these soil properties may be only achieved by proxy through correlation with the spectral signature of SOC or SOM. For example, good estimates may be the result of changes in surrogate (i.e. proxy or highly correlated) soil properties instead of directly detecting changes in the property of interest when using NIR spectroscopy methods (Reeves, 2010; Vågen et al., 2006). Organic carbon bonds and inorganic mineral bonds are the primary sources of spectral features (see Cécillon et al., 2009; Viscarra Rossel et al., 2006), and when a model relies on an indirect, proxy measurement, performance is dependent upon the relationship between the target soil property and the proxy measurement.

3.2 Soil physical properties

Physical soil properties, such as bulk density, aggregate stability, and water-filled pore space, reflect multiple soil functions such as infiltration capacity, resistance to erosion, and microbial habitat (Angers, 1992; Franzluebbers, 1999; Logsdon and Karlen, 2004). In-field estimation of these important physical soil health indicators using reflectance spectra has been less consistent than for soil biological properties, although some studies have demonstrated success with soil texture and aggregate stability (Bogrekci and Lee, 2005; Chang et al., 2001; Vågen et al., 2006) and bulk density (Cho et al., 2017a).

In some cases, spectral data have been supplemented with data from other sensor variables, including soil strength (Hemmat and Adamchuk, 2008) and EC_a (Corwin and Lesch, 2005). Soil strength is highly correlated with other dynamic soil physical properties, such as bulk density, macroporosity, aggregate stability, moisture, and degree of compaction, as well as inherent soil properties such as texture (Chung et al., 2006; Nunes et al., 2019a), and root growth (Nunes et al., 2019b). While vertically operating cone penetrometers are the standard device for measuring soil strength, horizontal, on-the-go sensors have also been developed to improve data collection efficiency (Chung et al., 2006; Hemmat and Adamchuk, 2008). Soil apparent electrical conductivity is affected by numerous attributes, including soil texture, mineralogy, cation exchange capacity, and moisture (Doolittle and Brevik, 2014; McNeill, 1992; Sudduth et al., 2013). In non-saline soils, texture variations generally have the largest effect, so EC_a has been used to map soil texture differences (Anderson-Cook et al., 2002; Kitchen et al., 1996). However, the lack of a unique relationship between EC_a values and specific soil properties, mainly due to the multiple properties that can affect EC_a in any given situation, has hindered its use for quantitative assessment of soil properties in a practical setting. While good results are often obtained in research, it is difficult to elucidate protocols that will transfer those results to practice.

3.3 Soil chemical properties

Chemical soil health indicators primarily reflect nutrient availability for crop growth and are typically represented by laboratory measurements of soil pH, electrical conductivity (paste), extractable phosphorus, and extractable potassium (Smith and Doran, 1996; Staggenborg et al., 2007; Wienhold et al., 2009). Precision, in-field soil health assessments could provide high-resolution information for improved fertility management decisions (e.g. right time, right place, right rate) and would enhance the sustainability of production systems by reducing environmental impacts and improving profitability. However, sensing of available phosphorus and potassium using reflectance spectra has been met with limited success. A few researchers (e.g. Bogrekci and Lee, 2005; Daniel et al., 2003) reported consistently good ($R^2 > 0.8$) results for both phosphorus and potassium, but other studies (e.g. Ge et al., 2007; Lee et al., 2009; Viscarra Rossel et al., 2006) reported consistently poor ($R^2 < 0.5$) results. This lack of consistency has been attributed to the fact that spectral estimation of phosphorus and potassium relies on the covariation of nutrient concentrations with other, optically active soil constituents (Stenberg et al., 2010). For a more accurate estimation of soil nutrients, La et al. (2016) supplemented spectral data with laboratory data from electrochemical sensors, improving phosphorus and potassium estimates to $R^2 \geq 0.95$. Additional information on the electrochemical sensing of nutrients can be found in reviews by Kim et al. (2009) and Sinfield et al. (2010).

3.4 Soil health indices

The ability to estimate multiple soil health indicators and provide a soil health score to land managers without expensive soil sampling and analysis would support multiple national and international sustainability initiatives related to carbon sequestration, carbon markets, and soil health. To this end, many studies have related Vis-NIR reflectance to individual soil health indicators as described earlier or suites of soil health measurements. In addition, Vis-NIR has been applied to the direct estimation of soil health indices. The general aim of a soil health index is to translate measured values of soil health indicators into a more user-friendly, interpretable classification or score, typically related to an outcome such as erosion, crop productivity, or water quality. A wide range of soil health scoring frameworks are available depending on the application and the outcome of interest, and Vis-NIR has been used to estimate soil health indices in ecosystems across the globe. Vis-NIR spectra have been applied by Vågen et al. (2006) to estimate a 3-category soil fertility index based on 10 soil health indicators in Madagascar, by Cohen et al. (2006) to classify 3 tiers of soil health based on 17 indicators in the USA, and by Kinoshita et al. (2012) to estimate a 3-category soil quality index for soils in Western Kenya.

Few studies, however, have evaluated the simultaneous estimation of biological, physical, and chemical soil health indicators with the goal of estimating a comprehensive soil health index (as opposed to a categorical soil health index). Veum et al. (2015a) estimated soil health scores for soils from the Midwestern USA based on the Soil Management Assessment Framework (SMAF). The SMAF index scores measured values using a set of non-linear curves with thresholds for 'more is better', 'less is better', or mid-point optima (Andrews et al., 2004). Veum et al. (2015b) found that their Vis-NIR models successfully estimated the biological components of the SMAF score but performed poorly for the chemical and physical components of the SMAF scores. Due to the non-linear nature of the SMAF scoring curves, the distribution of the scores likely impacted model performance, particularly for the chemical and nutrient scores. In particular, the mid-point optima scoring curves for soil pH and phosphorus performed poorly in the models. Overall, the nonlinearity of the SMAF scoring algorithms presented a challenge for the PLSR approach. Thus, non-linear techniques may provide more robust estimation models. However, a single soil health dataset is unlikely to exhibit ideal distributional characteristics across all soil health indicators simultaneously, which may reduce the ability to estimate a comprehensive soil health index using Vis-NIR spectra alone.

Adding auxiliary variables to Vis-NIR spectra may improve model performance, particularly for estimation of a soil health index where multiple soil health indicators are represented. Veum et al. (2015b) added selected auxiliary variables, alone and in combination, to improve estimation of the SMAF soil health score. Bulk density, water-filled pore space, pH, aggregate stability, and/or phosphorus measurements all resulted in improved estimation of the overall SMAF score, suggesting that Vis-NIR sensors in conjunction with other sensors or simple field and/or laboratory analyses may improve index estimates. Selecting auxiliary variables that reflect aspects of soil health that are not represented well by Vis-NIR, such as physical, chemical, or nutrient indicators, may enhance estimation. As noted earlier, no single sensor has demonstrated the ability to estimate a comprehensive soil health index. This challenge represents an ideal opportunity for the application of sensor fusion technology. For example, Veum et al. (2017) noted that the fusion of EC_a and cone index (penetrometer) data with Vis-NIR improved model estimates of the physical SMAF category and subsequently the overall SMAF soil health score, while chemical and fertility-related soil health indicators remained poorly estimated.

Overall, current research supports using sensor fusion to improve estimation of soil health indicators and SMAF scores. Modern agriculture is challenged in attaining a balance of agronomic productivity with long-term sustainability, and sensor fusion technology along with novel data analysis techniques has the potential to provide high-resolution soil health assessment results for more informed management decisions. Current research points to the need for improved sensors for chemical and physical soil health indicators and better techniques to handle the challenges of varying environmental conditions for an effective in-field approach to soil health assessment.

4 Case study: combining spectra and auxiliary sensor data for improved soil health estimation

A series of related studies illustrate the potential and limitations of soil spectroscopy for estimating variables important in quantifying soil health. These studies focused on the Central Claypan Region in northeastern Missouri, USA. The soils found at these sites, characterized as claypan soils, are primarily of the Mexico-Putnam association (fine, montmorillonitic, mesic Udollic Ochraqualfs). These soils were formed in moderately fine textured loess over a fine textured pedisediment. Surface textures range from a silt loam to a silty clay loam. The subsoil claypan horizon(s) are silty clay loam, silty clay, or clay and may commonly contain as much as 50–60% montmorillonitic clay. This abrupt increase in clay content over a few centimeters is the distinguishing factor of claypan soils, and the strong stratification between topsoil and high-clay subsoil contributes to poor soil water holding capacity, excessive surface runoff, and in many years, insufficient soil water for optimum crop growth.

4.1 Laboratory Vis-NIR estimation of soil organic carbon

The first in this sequence of studies (Chaudhary et al., 2012) had the goal of evaluating laboratory-based Vis-NIR DRS as a tool for discriminating differences in SOC among systems in a long-term cropping systems experiment. Surface soil samples were collected from three different grain cropping systems and three perennial grass systems in the experiment. To provide an independent calibration dataset, other samples were collected from fields under various management systems located within a few kilometers of the experimental site. Spectral data were collected in the laboratory with an ASD FieldSpec Pro FR spectrometer between 400 nm and 2500 nm on both field-moist and oven-dry soil samples. Laboratory SOC data were provided by dry combustion analysis. Models estimating SOC were established with PLSR on the calibration dataset, and those models were then applied to the plot data with good results ($R^2 \geq 0.81$; Fig. 4). Spectral analysis of oven-dry samples provided better results than with field-moist samples, with both able to detect SOC differences among cropping systems. However, neither was as successful in detecting differences as dry combustion lab data.

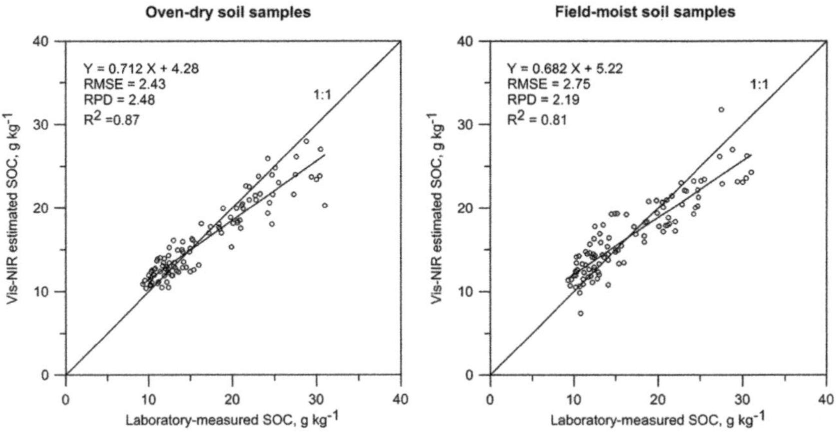

Figure 4 Visible and near-infrared (Vis-NIR) spectroscopy-estimated vs. laboratory-measured values of soil organic carbon (SOC) for oven-dry and field-moist plot soil samples (after Chaudhary et al., 2012).

4.2 Laboratory Vis-NIR estimation of soil health indicators and scoring functions

This study (Veum et al., 2015b) employed the same experimental sites and soil samples as Chaudhary et al. (2012). The goal was to evaluate laboratory Vis-NIR spectroscopy as a tool to estimate soil health indicator variables and soil health scores. The indicator variables and scoring categories were taken from those used in the SMAF (Andrews et al., 2004). This soil health index combines values of multiple indicator variables into calculated scores for biological, physical, and chemical categories, as well as an overall SMAF score. Biological variables were SOC, β-glucosidase, microbial biomass-carbon, and mineralizable nitrogen. Physical variables were bulk density, water-filled pore space, and water-stable aggregates. The chemical category included extractable phosphorus and potassium, pH, and electrical conductivity (paste).

Laboratory Vis-NIR spectroscopy was effective in estimating biological indicators but not those in the physical or chemical categories (Fig. 5). When estimating scoring functions, similar results were found, with the biological SMAF score well estimated ($R^2 = 0.76$) but not the scores for the other categories ($R^2 \leq 0.27$). Although the overall SMAF score was reasonably well-estimated ($R^2 = 0.69$), we expected this could be improved with a better estimation of physical and chemical indicators through the incorporation of data from auxiliary sensors.

4.3 Improving Vis-NIR soil health estimates with auxiliary data

To improve DRS performance for SMAF soil health scores, Veum et al. (2017) combined laboratory Vis-NIR spectral data with auxiliary data collected in the

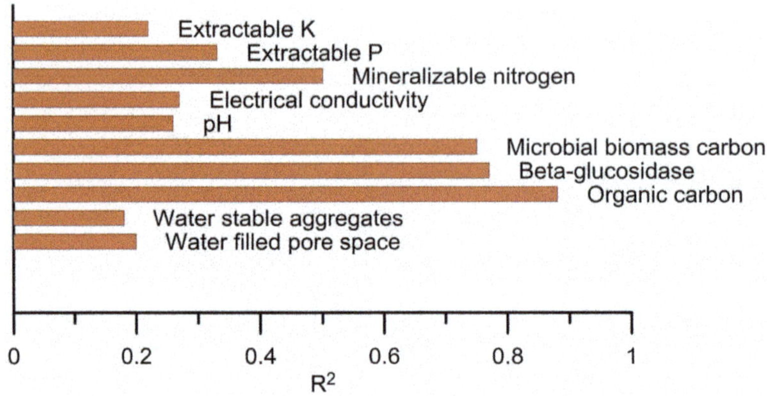

Figure 5 Accuracy of laboratory visible and near-infrared spectroscopy applied to oven-dry soil samples for estimating selected soil health indicators.

field. Soil samples were collected at 2 depths (0–5 cm and 5–15 cm) at 108 locations across a 10-ha research site including different cropping systems and landscape positions. Samples were oven-dried, and spectra (400–2500 nm) were obtained in the laboratory with an ASD FieldSpec Pro FR spectrometer. Near each sample collection point, a Veris Profiler 3000 instrument (Veris Technologies, Salina, KS, USA) was used to obtain in situ EC_a and penetration resistance data, which were then averaged over the soil sample depth increments. Models were created by PLSR with spectra alone and with spectra, EC_a, and penetration resistance data.

Results showed that adding auxiliary variables improved model estimates for all SMAF category scores and the overall SMAF score (Fig. 6). Substantial improvement was seen in the physical score, which then translated to improvement in the overall score. Improvements to biological and chemical scores were minor. While the biological SMAF score was well estimated with spectra alone (Fig. 6), the chemical score was not. Including other auxiliary data, such as that from electrochemical sensors (La et al., 2016) could be one approach to improve models for chemical scores.

4.4 Profile in situ soil sensing using Vis-NIR spectra and auxiliary data

Studies described by Cho et al. (2017b) and Pei et al. (2019) compared soil property estimates by Vis-NIR spectra alone to those from datasets that also included EC_a and penetration resistance (Fig. 7). Data were collected with a Veris P4000 instrument over multiple field sites to a depth of approximately 1 m. Spectra and auxiliary variables were averaged over depth increments corresponding with soil horizon delineations and calibration soil samples were

Published by Burleigh Dodds Science Publishing Limited, 2023.

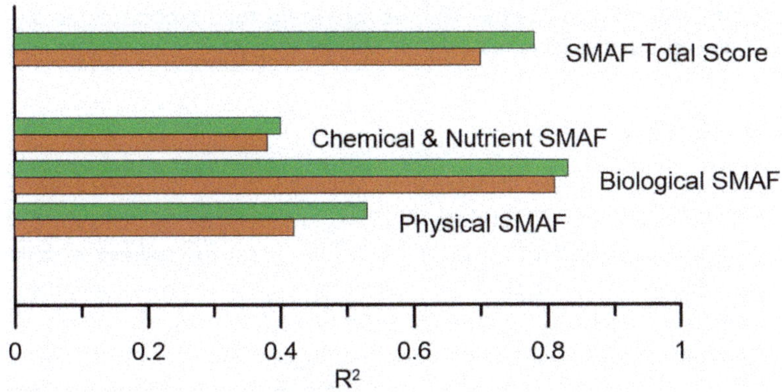

Figure 6 Improvement in Soil Management Assessment Framework (SMAF) total and categorical chemical, biological, and physical scores by including auxiliary variables of soil apparent electrical conductivity (EC_a) and penetration resistance along with Vis-NIR spectra (green bars), as compared to spectra alone (brown bars).

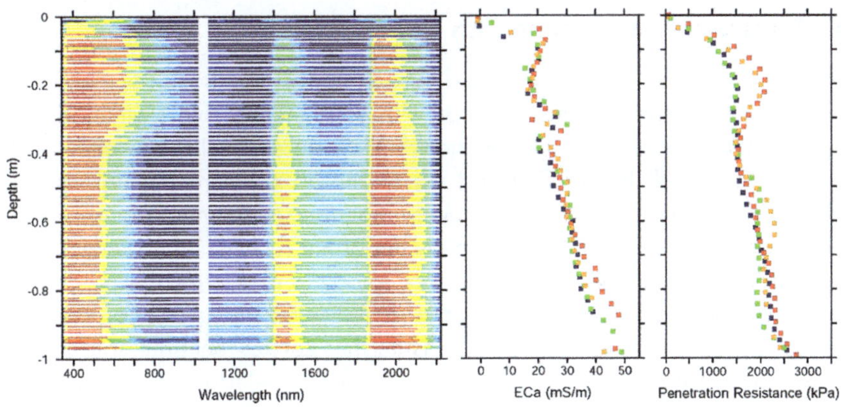

Figure 7 Soil spectra, apparent electrical conductivity (EC_a), and penetration resistance data obtained by four vertical probings to 1 m at a claypan field sample site. Spectral data are shown in absorbance units, with highest absorbance in red and lowest in blue.

obtained over those same horizons and analyzed in the laboratory. In both studies, only slight improvement was seen when adding the auxiliary variables. Pei et al. (2019) reported that root mean square error decreased by more than 5% for only 1 of the 11 soil properties evaluated. It should be noted that only three soil properties - SOC, pH, and extractable potassium - were common between these studies and the SMAF-oriented evaluations described earlier. This is due in part to the concentration of SOM and associated soil health indicators in the near-surface horizons of most soils. However, profile knowledge of keystone soil health indicators, such as SOC, and inherent soil properties, such as soil texture, can contribute to better soil management recommendations and soil health assessment interpretation.

5 Conclusion

Soil spectroscopy has the potential to provide rapid, cost-effective, high-resolution data to support critical initiatives related to soil health. Current research results support the use of Vis-NIR sensors for in-field assessment and quantification of biological soil health indicators, even under conditions of variable soil moisture content. Important soil physical or chemical soil health indicators may need to be supplied by auxiliary data collected using complementary sensors, field test kits, or simple laboratory measurements. Overall, soil spectroscopy using sensor-based technology in the field has the potential to reliably assess soil health for improved sustainability, profitability, and environmental protection.

6 Future trends in research

Achieving the goal of rapid, in-field assessment of soil health will allow for the creation of comprehensive maps with soil function layers gleaned from in-field, sensor estimates of soil health indicators. Ideally, the soil health indicator data will be transformed via on-the-go software into interpretive maps for producers along with recommendations for management. Despite the successes in this area to date, many important research and development needs remain, including new sensor technology to measure soil strength, soil nutrients, and biological characteristics directly and accurately. For example, the development of novel in-field sensors that collect real-time soil process data (e.g. soil enzyme activity) will expand our knowledge of soil variability to improve crop production and better protect soil and water resources. Integration of advanced sensor platforms and data analysis techniques will almost certainly be required. Sensor optimization to improve data quality and reliability is needed, and methods to address a range of environmental conditions will be necessary to move additional sensor techniques from the laboratory to the field.

Following data collection, improved soil health interpretation frameworks are also necessary to provide scientifically sound guidance to landowners and producers. Increased accessibility to soil health information, interpretation, and management recommendations via user-friendly software applications will allow landowners to make science-based decisions. As interpretation tools are expanded at the national and global scale, data-driven soil health management information will help optimize management decisions to provide economic and environmental benefits.

7 Where to look for further information

Two books that provide thorough coverage of spectroscopy uses in agriculture are Williams and Norris (1987) and Roberts et al. (2004). Both cover spectroscopy basics, analysis methods, and applications. A chapter in the second book (Malley et al., 2004) provides coverage of soil applications. Good reviews specific to soil applications include the articles by Viscarra Rossel et al. (2006), Stenberg et al. (2010), and Nocita et al. (2015).

Current advances in soil spectroscopy are often presented at the International, European, and Asian Conferences on Precision Agriculture, Information about and links to proceedings papers from these three biennial conferences can be found at the website of the International Society of Precision Agriculture (www.ispag.org). International journals that publish soil spectroscopy research include *Biosystems Engineering, Computers and Electronics in Agriculture, European Journal of Soil Science, Geoderma, Journal of the ASABE* (previously *Transactions of the ASABE*), *Precision Agriculture,* and *Soil Science Society of America Journal.*

For details on laboratory methods for soil health analysis, a recent two-volume set published by the Soil Science Society of America serves as excellent references: Approaches to Soil Health Analysis, Volume 1 (Karlen et al., 2021a) and Laboratory Methods for Soil Health Analysis, Volume 2 (Karlen et al., 2021b).

8 References

Ackerson, J. P., Morgan, C. L. S. and Ge, Y. (2017). Penetrometer-mounted VisNIR spectroscopy: Application of EPO-PLS to in situ VisNIR spectra, *Geoderma* 286, 131–138.

Adamchuk, V. I., Hummel, J. W., Morgan, M. T. and Upadhyaya, S. K. (2004). On-the-go soil sensors for precision agriculture, *Comput. Electron. Agric.* 44(1), 71–91.

Adamchuk, V. I., Viscarra Rossel, R. A., Sudduth, K. A. and Schulze Lammers, P. (2011). Sensor fusion for precision agriculture. In: Thomas, C. (Ed.), *Sensor Fusion - Foundation and Applications*. Tech, Rijeka, Croatia, pp. 27–40.

An, X., Li, M., Zheng, L., Liu, Y. and Sun, H. (2014). A portable soil nitrogen detector based on NIRS, *Precis. Agric.* 15(1), 3–16.

Anderson-Cook, C. M., Alley, M. M., Roygard, J. K. F., Khosla, R., Noble, R. B. and Doolittle, J. A. (2002). Differentiating soil types using electromagnetic conductivity and crop yield maps, *Soil Sci. Soc. Am. J.* 66(5), 1562–1570.

Andrews, S. S., Karlen, D. L. and Cambardella, C. A. (2004). The soil management assessment framework: A quantitative soil quality evaluation method, *Soil Sci. Soc. Am. J.* 68(6), 1945–1962.

Angers, D. A. (1992). Changes in soil aggregation and organic carbon under corn and alfalfa, *Soil Sci. Soc. Am. J.* 56(4), 1244–1249.

Bogrekci, I. and Lee, W. S. (2005). Improving phosphorus sensing by eliminating soil particle size effect in spectral measurement, *Trans. ASAE* 48, 1971–1978.

Bouma, J., Jongerius, A., Boersma, O., Jager, A. and Schoonderbeek, D. (1977). The function of different types of macropores during saturated flow through four swelling soil horizons, *Soil Sci. Soc. Am. J.* 41(5), 945–950.

Brown, D. J. (2007). Using a global VNIR soil-spectral library for local soil characterization and landscape modeling in a 2nd-order Uganda watershed, *Geoderma* 140(4), 444–453.

Brown, D. J., Shepherd, K. D., Walsh, M. G., Mays, M. D. and Reinsch, T. G. (2006). Global soil characterization with VNIR diffuse reflectance spectroscopy, *Geoderma* 132(3-4), 273–290.

Buyer, J. S. and Sasser, M. (2012). High throughput phospholipid fatty acid analysis of soils, *Appl. Soil Ecol.* 61, 127–130.

Cécillon, L., Barthès, B. G., Gomez, C., Ertlen, D., Genot, V., Hedde, M., Stevens, A. and Brun, J. J. (2009). Assessment and monitoring of soil quality using near-infrared reflectance spectroscopy (NIRS), *Eur. J. Soil Sci.* 60(5), 770–784.

Chang, C. W., Laird, D. A., Mausbach, M. J. and Hurburgh, C. R. (2001). Near-infrared reflectance spectroscopy-principal components regression analysis of soil properties, *Soil Sci. Soc. Am. J.* 65(2), 480–490.

Chaudhary, V. P., Sudduth, K. A., Kitchen, N. R. and Kremer, R. J. (2012). Reflectance spectroscopy detects management and landscape differences in soil carbon and nitrogen, *Soil Sci. Soc. Am. J.* 76(2), 597–606.

Cho, Y., Sheridan, A. H., Sudduth, K. A. and Veum, K. S. (2017a). Comparison of field and laboratory VNIR spectroscopy for profile soil property estimation, *Trans. ASABE* 60(5), 1503–1510.

Cho, Y., Sudduth, K. A. and Drummond, S. T. (2017b). Profile soil property estimation using a VIS-NIR-EC-force probe, *Trans. ASABE* 60(3), 683–692.

Christy, C. D. (2008). Real-time measurement of soil attributes using on-the-go near infrared reflectance spectroscopy, *Comput. Electron. Agric.* 61(1), 10–19.

Chung, S. O., Sudduth, K. A. and Hummel, J. W. (2006). Design and validation of an on-the-go soil strength profile sensor, *Trans. ASABE* 49(1), 5–14.

Cohen, M. J., Dabral, S., Graham, W. D., Prenger, J. P. and Debusk, W. F. (2006). Evaluating ecological condition using soil biogeochemical parameters and near infrared reflectance spectra, *Environ. Monit. Assess.* 116(1-3), 427–457.

Corwin, D. L. and Lesch, S. M. (2005). Apparent soil electrical conductivity measurements in agriculture, *Comput. Electron. Agric.* 46(1-3), 11–43.

Culman, S. W., Snapp, S. S., Freeman, M. A., Schipanski, M. E., Beniston, J., Lal, R., Drinkwater, L. E., Franzluebbers, A. J., Glover, J. D., Grandy, A. S., Lee, J., Six, J., Maul, J. E., Mirksy, S. B., Spargo, J. T. and Wander, M. M. (2012). Permanganate oxidizable carbon reflects a processed soil fraction that is sensitive to management, *Soil Sci. Soc. Am. J.* 76(2), 494–504.

Dalal, R. C. and Henry, R. J. (1986). Simultaneous determination of moisture, organic carbon, and total nitrogen by near-infrared reflectance spectrophotometry, *Soil Sci. Soc. Am. J.* 50(1), 120–123.

Daniel, K. W., Tripathi, N. K. and Honda, K. (2003). Artificial neural network analysis of laboratory and in situ spectra for the estimation of macronutrients in soils of Lop Buri (Thailand), *Soil Res.* 41(1), 47–59.

Deubel, A., Hofmann, B. and Orzessek, D. (2011). Long-term effects of tillage on stratification and plant availability of phosphate and potassium in a loess chernozem, *Soil Till. Res.* 117, 85–92.

Dick, R. P. (1994). Soil enzyme activity as an indicator of soil quality. In: Doran, J. W., Coleman, D. C., Bezdicek, D. F. and Stewart, B. A. (Eds). *Defining Soil Quality for a Sustainable Environment*, Soil Science Society of America and American Society of Agronomy, Madison, WI, pp. 107–124.

Doolittle, J. A. and Brevik, E. C. (2014). The use of electromagnetic induction techniques in soils studies, *Geoderma* 223–225, 33–45.

Doran, J. W. (1987). Microbial biomass and mineralizable nitrogen distributions in no-tillage and plowed soils, *Biol. Fertil. Soils* 5(1), 68–75.

Doran, J. W. and Parkin, T. B. (1996). Quantitative indicators of soil quality: A minimum data set. In: Doran, J. W. and Jones, A. J. (Eds.), *Methods for Assessing Soil Quality*. Soil Science Society of America, Madison, WI, pp. 25–37.

Eivazi, F. and Tabatabai, M. A. (1988). Glucosidases and galactosidases in soils, *Soil Biol. Biochem.* 20(5), 601–606.

Fontaine, S., Barot, S., Barré, P., Bdioui, N., Mary, B. and Rumpel, C. (2007). Stability of organic carbon in deep soil layers controlled by fresh carbon supply, *Nature* 450(7167), 277–280.

Franzluebbers, A. J. (1999). Microbial activity in response to water-filled pore space of variably eroded southern Piedmont soils, *Appl. Soil Ecol.* 11(1), 91–101.

Franzluebbers, A. J., Haney, R. L., Hons, F. M. and Zuberer, D. A. (1996). Determination of microbial biomass and nitrogen mineralization following rewetting of dried soil, *Soil Sci. Soc. Am. J.* 60(4), 1133–1139.

Fystro, G. (2002). The prediction of C and N content and their potential mineralization in heterogeneous soil samples using VIS-NIR spectroscopy and comparative methods, *Plant Soil* 246(2), 139–149.

Ge, Y., Morgan, C. L. S. and Ackerson, J. P. (2014). VisNIR spectra of dried ground soils predict properties of soils scanned moist and intact, *Geoderma* 221–222, 61–69.

Ge, Y., Thomasson, J. A., Morgan, C. L. and Searcy, S. W. (2007). VNIR diffuse reflectance spectroscopy for agricultural soil property determination based on regression-kriging, *Trans. ASABE* 50(3), 1081–1092.

Haaland, D. M. and Thomas, E. V. (1988). Partial least-squares methods for spectral analyses. 1. Relation of other quantitative calibration methods and the extraction of qualitative information, *Anal. Chem.* 60(11), 1193–1202.

Harrison, R., Footen, P. and Strahm, B. (2011). Deep soil horizons: Contribution and importance to soil carbon pools and in assessing whole-ecosystem response to management and global change, *For. Sci.* 57, 67–76.

Hedley, C., Roudier, P. and Maddi, L. (2015). VNIR soil spectroscopy for field analysis, *Commun. Soil Sci. Plant Anal.* 46(Suppl. 1), 104–121.

Hemmat, A. and Adamchuk, V. I. (2008). Sensor systems for measuring soil compaction: Review and analysis, *Comput. Electron. Agric.* 63(2), 89–103.

Hodge, A. M. and Sudduth, K. A. (2012). Comparison of two spectrometers for profile soil carbon sensing, *American Society of Agricultural and Biological Engineers Annual International Meeting*, ASABE, St. Joseph, MI, Paper No. 121338240.

Houx, J. H., Wiebold, W. J. and Fritschi, F. B. (2011). Long-term tillage and crop rotation determines the mineral nutrient distributions of some elements in a Vertic Epiaqualf, *Soil Till. Res.* 112(1), 27–35.

Hruschka, W. R. (1987). Data analysis: Wavelength selection methods. In: Williams, P. and Norris, K. (Eds.), *Near-Infrared Technology in the Agricultural and Food Industries*, American Association of Cereal Chemists, St. Paul, MN, pp. 35–56.

Hsiao, C.-J., Sassenrath, G. F., Zeglin, L. H., Hettiarachchi, G. M. and Rice, C. W. (2018). Vertical changes of soil microbial properties in claypan soils, *Soil Biol. Biochem.* 121, 154–164.

Hummel, J. W., Ahmad, I. S., Newman, S. C., Sudduth, K. A. and Drummond, S. T. (2004). Simultaneous soil moisture and cone index measurement, *Trans. ASAE* 47, 607–618.

Hummel, J. W., Gaultney, L. D. and Sudduth, K. A. (1996). Soil property sensing for site-specific crop management, *Comput. Electron. Agric.* 14(2–3), 121–136.

Hummel, J. W., Sudduth, K. A. and Hollinger, S. E. (2001). Soil moisture and organic matter prediction of surface and subsurface soils using an NIR soil sensor, *Comput. Electron. Agric.* 32(2), 149–165.

Hurisso, T. T., Moebius-Clune, D. J., Culman, S. W., Moebius-Clune, B. N., Thies, J. E. and van Es., H. M. (2018). Soil protein as a rapid soil health indicator of potentially available organic nitrogen, *Agric. Environ. Lett.* 3, 180006.

Igne, B., Reeves, J. B., III, McCarty, G., Hively, W. D., Lund, E. and Hurburgh, C. R., Jr. (2010). Evaluation of spectral pretreatments, partial least squares, least squares support vector machines and locally weighted regression for quantitative spectroscopic analysis of soils, *J. Near Infrared Spectrosc.* 18(3), 167–176.

Ji, W., Viscarra Rossel, R. A. and Shi, Z. (2015). Accounting for the effects of water and the environment on proximally sensed vis–NIR soil spectra and their calibrations, *Eur. J. Soil Sci.* 66(3), 555–565.

Jobbágy, E. G. and Jackson, R. B. (2001). The distribution of soil nutrients with depth: Global patterns and the imprint of plants, *Biogeochemistry* 53(1), 51–77.

Johnson, J. M., Vandamme, E., Senthilkumar, K., Sila, A., Shepherd, K. D. and Saito, K. (2019). Near-infrared, mid-infrared or combined diffuse reflectance spectroscopy for assessing soil fertility in rice fields in sub-Saharan Africa, *Geoderma* 354, 113840.

Karlen, D. L., Goeser, N. J., Veum, K. S. and Yost, M. A. (2017). On-farm soil health evaluations: Challenges and opportunities, *J. Soil Water Conserv.* 72(2), 26A–31A.

Karlen, D. L., Stott, D. E. and Mikha, M. M. (Eds.). (2021a). *Approaches to Soil Health Analysis* (Vol. 1), Wiley, Hoboken, NJ.

Karlen, D. L., Stott, D. E. and Mikha, M. M. (Eds.). (2021b). *Laboratory Methods for Soil Health Analysis* (Vol. 2), Wiley, Hoboken, NJ.

Katuwal, S., Knadel, M., Moldrup, P., Norgaard, T., Greve, M. H. and deJonge, L. W. (2018). Visible-near-infrared spectroscopy can predict mass transport of dissolved chemicals through intact soil, *Sci. Rep.* 8(1), 11188.

Kawano, S., Abe, H. and Iwamoto, M. (1995). Development of a calibration equation with temperature compensation for determining the Brix value in intact peaches, *J. Near Infrared Spectrosc.* 3(4), 211–218.

Kim, H. J., Sudduth, K. A. and Hummel, J. W. (2009). Soil macronutrient sensing for precision agriculture, *J. Environ. Monit.* 11(10), 1810–1824.

Kinoshita, R., Moebius-Clune, B. N., van Es, H. M., Hively, W. D. and Bilgilis, A. V. (2012). Strategies for soil quality assessment using visible and near-infrared reflectance spectroscopy in a Western Kenya chronosequence, *Soil Sci. Soc. Am. J.* 76(5), 1776–1788.

Kitchen, N. R., Sudduth, K. A. and Drummond, S. T. (1996). Mapping of sand deposition from 1993 midwest floods with electromagnetic induction measurements, *J. Soil Water Conserv.* 51, 336–340.

Kusumo, B. H., Hedley, C. B., Hedley, M. J., Hueni, A., Tuohy, M. P. and Arnold, G. C. (2008). The use of diffuse reflectance spectroscopy for in situ carbon and nitrogen analysis of pastoral soils, *Soil Res.* 46(7), 623–635.

Kweon, G. and Maxton, C. (2013). Soil organic matter sensing with an on-the-go optical sensor, *Biosys. Eng.* 115, 66–81.

La, W. J., Sudduth, K. A., Kim, H. J. and Chung, S. O. (2016). Fusion of spectral and electrochemical sensor data for estimating soil macronutrients, *Trans. ASABE* 59(4), 787–794.

Lee, K. S., Lee, D. H., Sudduth, K. A., Chung, S. O., Kitchen, N. R. and Drummond, S. T. (2009). Wavelength identification and diffuse reflectance estimation for surface and profile soil properties, *Trans. ASAE* 52, 683–695.

Lehman, R., Cambardella, C., Stott, D., Acosta-Martinez, V., Manter, D., Buyer, J., Maul, J., Smith, J., Collins, H., Halvorson, J., Kremer, R., Lundgren, J., Ducey, T., Jin, V. and Karlen, D. (2015). Understanding and enhancing soil biological health: The solution for reversing soil degradation, *Sustainability* 7(1), 988–1027.

Logsdon, S. D. and Karlen, D. L. (2004). Bulk density as a soil quality indicator during conversion to no-tillage, *Soil Till. Res.* 78(2), 143–149.

Malley, D. F., Martin, P. D. and Ben-Dor, E. (2004). Application in analysis of soils. In: Roberts, C. A., Workman, J., Jr. and Reeves, J. B., III (Eds). Near-Infrared Spectroscopy in Agriculture, *American Society of Agronomy*. Crop Science Society of America and Soil Science Society of America, Madison, WI, pp. 729–784.

Manter, D. K., Delgado, J. A., Blackburn, H. D., Harmel, D., Pérez de León, A. A. and Honeycutt, C. W. (2017). Opinion: Why we need a national living soil repository, *Proc. Natl. Acad. Sci. U. S. A.* 114(52), 13587–13590.

McNeill, J. D. (1992). Rapid, accurate mapping of soil salinity by electromagnetic ground conductivity meters. In: Topp, G. C., Reynolds, W. D. and Green, R. E. (Eds.), *Advances in Measurement of Soil Physical Properties: Bringing Theory into Practice*, Soil Science Society of America, Madison, WI, pp. 209–229.

Minasny, B., McBratney, A. B., Bellon-Maurel, V., Roger, J. M., Gobrecht, A., Ferrand, L. and Joalland, S. (2011). Removing the effect of soil moisture from NIR diffuse reflectance spectra for the prediction of soil organic carbon, *Geoderma* 167–168, 118–124.

Morgan, C. L. S., Waiser, T. H., Brown, D. J. and Hallmark, C. T. (2009). Simulated in situ characterization of soil organic and inorganic carbon with visible near-infrared diffuse reflectance spectroscopy, *Geoderma* 151(3–4), 249–256.

Mouazen, A. M., Alhwaimel, S. A., Kuang, B. and Waine, T. (2014). Multiple on-line soil sensors and data fusion approach for delineation of water holding capacity zones for site specific irrigation, *Soil Till. Res.* 143, 95–105.

Mouazen, A. M., Maleki, M. R., De Baerdemaeker, J. and Ramon, H. (2007). On-line measurement of some selected soil properties using a VIS–NIR sensor, *Soil Till. Res.* 93(1), 13–27.

Nocita, M., Stevens, A., Noon, C. and van Wesemael, B. (2013). Prediction of soil organic carbon for different levels of soil moisture using Vis-NIR spectroscopy, *Geoderma* 199, 37–42.

Nocita, M., Stevens, A., van Wesemael, B., Aitkenhead, M., Bachmann, M., Barthès, B., Ben Dor, E., Brown, D. J., Clairotte, M., Csorba, A., Dardenne, P., Demattê, J. A. M., Genot, V., Guerrero, C., Knadel, M., Montanarella, L., Noon, C., Ramirez-Lopez, L., Robertson, J., Sakai, H., Soriano-Disla, J. M., Shepherd, K. D., Stenberg, B., Towett, E. K., Vargas, R. and Wetterlind, J. (2015). Soil spectroscopy: An alternative to wet chemistry for soil monitoring. In: Sparks, D. L. (Ed.), *Advances in Agronomy* (Vol. 132), Academic Press, Burlington, MA, pp. 139–159.

Nunes, M. R., Karlen, D. L., Denardin, J. E. and Cambardella, C. A. (2019a). Corn root and soil health indicator response to no-till production practices, *Agric. Ecosyst. Environ.* 285.

Nunes, M. R., Pauletto, E. A., Denardin, J. E., Suzuki, L. E. A. S. and van Es, H. M. (2019b). Dynamic changes in compressive properties and crop response after chisel tillage in a highly weathered soil, *Soil Till. Res.* 186, 183–190.

Nunes, M. R., van Es, H. M., Schindelbeck, R., James Ristow, A. J. and Ryan, M. (2018). No-till and cropping system diversification improve soil health and crop yield, *Geoderma* 328, 30–43.

Palmborg, C. and Nordgren, A. (1993). Modelling microbial activity and biomass in forest soil with substrate quality measured using near infrared reflectance spectroscopy, *Soil Biol. Biochem.* 25(12), 1713–1718.

Palmborg, C. and Nordgren, A. (1996). Partitioning the variation of microbial measurements in forest soils into heavy metal and substrate quality dependent parts by use of near infrared spectroscopy and multivariate statistics, *Soil Biol. Biochem.* 28(6), 711–720.

Pei, X., Sudduth, K. A., Veum, K. S. and Li, M. (2019). Improving in-situ estimation of soil profile properties using a multi-sensor probe, *Sensors (Basel)* 19(5), 1011.

Pietikäinen, J. and Fritze, H. (1995). Clear-cutting and prescribed burning in coniferous forest: Comparison of effects on soil fungal and total microbial biomass, respiration activity and nitrification, *Soil Biol. Biochem.* 27(1), 101–109.

Poggio, M., Brown, D. J. and Bricklemyer, R. S. (2017). Comparison of Vis-NIR on in situ, intact core, and dried, sieved soil to estimate clay content at field to regional scales, *Eur. J. Soil Sci.* 68(4), 434–448.

Powlson, D. S. and Jenkinson, D. S. (1981). A comparison of the organic matter, biomass, adenosine triphosphate and mineralizable nitrogen contents of ploughed and direct-drilled soils, *J. Agric. Sci.* 97(3), 713-721.

Reeves, J. B. (2010). Near- versus mid-infrared diffuse reflectance spectroscopy for soil analysis emphasizing carbon and laboratory versus on-site analysis: Where are we and what needs to be done?, *Geoderma* 158(1–2), 3–14.

Roberts, C. A., Workman, J., Jr. and Reeves, J. B., III. (2004). *Near-Infrared Spectroscopy In Agriculture*, American Society of Agronomy, Crop Science Society of America, and Soil Science Society of America, Madison, WI.

Roger, J. M., Chauchard, F. and Bellon Maurel, V. (2003). EPO-PLS external parameter orthogonalisation of PLS application to temperature-independent measurement of sugar content of intact fruits, *Chemom. Intellig Lab. Syst.* 66(2), 191–204.

Roudier, P., Hedley, C. B. and Ross, C. W. (2015). Prediction of volumetric soil organic carbon from field-moist intact soil cores, *Eur. J. Soil Sci.* 66(4), 651–660.

Shen, Z., Ramirez-Lopez, L., Behrens, T., Cui, L., Zhang, M., Walden, L., Wetterlind, J., Shi, Z., Sudduth, K. A., Baumann, P., Song, Y., Catambay, K. and Viscarra Rossel, R. A. (2022). Deep transfer learning of global spectra for local soil carbon monitoring, *ISPRS J. Photogramm.* 188, 190–200.

Sherman, D. M. and Waite, T. D. (1985). Electronic spectra of Fe^{3+} oxides and oxide hydroxides in the near IR to near UV, *Am. Mineralogist* 70, 1262–1269.

Shonk, J. L., Gaultney, L. D., Schulze, D. G. and Van, G. E. (1991). Spectroscopic sensing of soil organic matter content, *Trans. ASAE* 34, 1978–1984.

Sinfield, J. V., Fagerman, D. and Colic, O. (2010). Evaluation of sensing technologies for on-the-go detection of macronutrients in cultivated soils, *Comput. Electron. Agric.* 70(1), 1-18.

Smith, J. L. and Doran, J. W. (1996). Measurement and use of pH and electrical conductivity for soil quality analysis. In: Doran, J. W. and Jones, A. J. (Eds.), *Methods for Assessing Soil Quality*, Soil Science Society of America, Madison, WI, pp. 169-185.

Soriano Disla, J. M., Janik, L. J., Viscarra Rossel, R. A., Macdonald, L. M. and McLaughlin, M. J. (2014). The performance of visible, near-, and mid-infrared reflectance spectroscopy for prediction of soil physical, chemical, and biological properties, *Appl. Spectrosc. Rev.* 49(2), 139-186.

Staggenborg, S. A., Carignano, M. and Haag, L. (2007). Predicting soil pH and buffer pH in situ with a real-time sensor, *Agron. J.* 99(3), 854-861.

Stenberg, B., Viscarra Rossel, R. A., Mouazen, A. M. and Wetterlind, J. (2010). Visible and near infrared spectroscopy in soil science. In: Sparks, D. L. (Ed.), *Advances in Agronomy* (Vol. 107), Academic Press, Burlington, MA, pp. 163-215.

Stevens, A., van Wesemael, B., Vandenschrick, G., Touré, S. and Tychon, B. (2006). Detection of carbon stock change in agricultural soils using spectroscopic techniques, *Soil Sci. Soc. Am. J.* 70(3), 844-850.

Sudduth, K. A. and Hummel, J. W. (1993). Soil organic matter, CEC, and moisture sensing with a prototype NIR spectrometer, *Trans. ASAE* 36, 1571-1582.

Sudduth, K. A., Myers, D. B., Kitchen, N. R. and Drummond, S. T. (2013). Modeling soil electrical conductivity-depth relationships with data from proximal and penetrating ECa sensors, *Geoderma* 199, 12-21.

Vågen, T.-G., Shepherd, K. D. and Walsh, M. G. (2006). Sensing landscape level change in soil fertility following deforestation and conversion in the highlands of Madagascar using Vis-NIR spectroscopy, *Geoderma* 133(3-4), 281-294.

Veum, K. S., Goyne, K. W., Kremer, R. J., Miles, R. J. and Sudduth, K. A. (2014). Biological indicators of soil quality and soil organic matter characteristics in an agricultural management continuum, *Biogeochemistry* 117(1), 81-99.

Veum, K. S., Kremer, R. J., Sudduth, K. A., Kitchen, N. R., Lerch, R. N., Baffaut, C., Stott, D. E., Karlen, D. L. and Sadler, E. J. (2015a). Conservation effects on soil quality indicators in the Missouri Salt River Basin, *J. Soil Water Conserv.* 70(4), 232-246.

Veum, K. S., Parker, P. A., Sudduth, K. A. and H Holan, S. (2018). Predicting profile soil properties with reflectance spectra via Bayesian covariate assisted external parameter orthogonalization, *Sensors (Basel)* 18(11), 3869.

Veum, K. S., Sudduth, K. A., Kremer, R. J. and Kitchen, N. R. (2015b). Estimating a soil quality index with VNIR reflectance spectroscopy, *Soil Sci. Soc. Am. J.* 79(2), 637-649.

Veum, K. S., Sudduth, K. A., Kremer, R. J. and Kitchen, N. R. (2017). Sensor data fusion for soil health assessment, *Geoderma* 305, 53-61.

Viscarra Rossel, R. A., Behrens, T., Ben-Dor, E., Brown, D. J., Demattê, J. A. M., Shepherd, K. D., Shi, Z., Stenberg, B., Stevens, A., Adamchuk, V., Aïchi, H., Barthès, B. G., Bartholomeus, H. M., Bayer, A. D., Bernoux, M., Böttcher, K., Brodský, L., Du, C. W., Chappell, A., Fouad, Y., Genot, V., Gomez, C., Grunwald, S., Gubler, A., Guerrero, C., Hedley, C. B., Knadel, M., Morrás, H. J. M., Nocita, M., Ramirez-Lopez, L., Roudier, P., Campos, E. M. R., Sanborn, P., Sellitto, V. M., Sudduth, K. A., Rawlins, B. G., Walter, C., Winowiecki, L. A., Hong, S. Y. and Ji, W. (2016). A global spectral library to characterize the world's soil, *Earth Sci. Rev.* 155, 198-230.

Viscarra Rossel, R. A., Lobsey, C. R., Sharman, C., Flick, P. and McLachlan, G. (2017). Novel proximal sensing for monitoring soil organic C stocks and condition, *Environ. Sci. Technol.* 51(10), 5630–5641.

Viscarra Rossel, R. A., Walvoort, D. J. J., McBratney, A. B., Janik, L. J. and Skjemstad, J. O. (2006). Visible, near infrared, mid infrared or combined diffuse reflectance spectroscopy for simultaneous assessment of various soil properties, *Geoderma* 131(1–2), 59–75.

Viscarra Rossel, R. A. and Behrens, T. (2010). Using data mining to model and interpret soil diffuse reflectance spectra, *Geoderma* 158(1–2), 46–54.

Weil, R. R., Islam, K. R., Stine, M. A., Gruver, J. B. and Samson-Liebig, S. E. (2003). Estimating active carbon for soil quality assessment: A simplified method for laboratory and field use, *Am. J. Alt. Agric.* 18, 3–17.

Wetterlind, J., Piikki, K., Stenberg, B. and Söderström, M. (2015). Exploring the predictability of soil texture and organic matter content with a commercial integrated soil profiling tool, *Eur. J. Soil Sci.* 66(4), 631–638.

Wetterlind, J. and Stenberg, B. (2010). Near-infrared spectroscopy for within-field soil characterization: Small local calibrations compared with national libraries spiked with local samples, *Eur. J. Soil Sci.* 61(6), 823–843.

Wienhold, B. J., Karlen, D. L., Andrews, S. S. and Stott, D. E. (2009). Protocol for indicator scoring in the soil management assessment framework (SMAF), *Renew. Agric. Food Syst.* 24(4), 260–266.

Wijewardane, N. K., Ge, Y. and Morgan, C. L. S. (2016a). Prediction of soil organic and inorganic carbon at different moisture contents with dry ground VNIR: A comparative study of different approaches, *Eur. J. Soil Sci.* 67(5), 605–615.

Wijewardane, N. K., Ge, Y., Wills, S. and Loecke, T. (2016b). Prediction of soil carbon in the conterminous United States: Visible and near infrared reflectance spectroscopy analysis of the rapid carbon assessment project, *Soil Sci. Soc. Am. J.* 80(4), 973–982.

Williams, P. and Norris, K. (1987). *Near-Infrared Technology in the Agricultural and Food Industries*, American Association of Cereal Chemists, St. Paul, MN.

Zhou, P., Sudduth, K. A., Veum, K. S. and Li, M. (2022). Extraction of reflectance spectra features for estimation of surface, subsurface, and profile soil properties, *Comput. Electron. Agric.* 196, 106845.

Zhou, P., Zhang, Y., Yang, W., Li, M., Liu, Z. and Liu, X. (2019). Development and performance test of an in-situ soil total nitrogen-soil moisture detector based on near-infrared spectroscopy, *Comput. Electron. Agric.* 160, 51–58.

Zornoza, R., Guerrero, C., Mataix-Solera, J., Scow, K. M., Arcenegui, V. and Mataix-Beneyto, J. (2008). Near infrared spectroscopy for determination of various physical, chemical and biochemical properties in Mediterranean soils, *Soil Biol. Biochem.* 40(7), 1923–1930.

Zou, X., Zhao, J. and Li, Y. (2007). Selection of the efficient wavelength regions in FT-NIR spectroscopy for determination of SSC of "Fuji" apple based on BiPLS and FiPLS models, *Vib. Spectrosc.* 44(2), 220–227.

Chapter 5

Advances in using proximal ground penetrating radar sensors to assess soil health

Katherine Grote, Missouri University of Science and Technology, USA

1 Introduction

Geophysical methods can provide important supplemental information to conventional soil characterization techniques. Most conventional techniques are point measurements and encompass a relatively small volume of soil. In fields with significant soil heterogeneity, conventional methods are often inadequate to characterize soil at the field scale (Bittelli, 2011; Vischel et al., 2008; Huisman et al., 2001). Additionally, these methods are often invasive or destructive (Castro et al., 2015), so it is difficult to determine whether the sensor installation has affected the soil behavior or to monitor temporal changes (Danjon and Reubens, 2008). Remote sensing methods provide continuous coverage over a field, but the resolution is typically too coarse and the sampling depth too shallow to be appropriate for many agricultural applications (Zhang et al., 2014a,b; Robinson et al., 2008). Also, remote sensing data may be limited by vegetation (Famiglietti et al., 2008). Geophysical techniques can provide information on soil properties

http://dx.doi.org/10.19103/AS.2022.0107.11

at a scale in between that of conventional measurements and remote sensing. Some geophysical techniques can quickly provide many measurements over a field, and the sampling depth of some techniques extends through the root zone depth or deeper. Geophysical techniques are also non-destructive and thus can be used to monitor changes in soil properties with time.

Ground penetrating radar (GPR) is a high-frequency electromagnetic geophysical technique that can be used to characterize soil properties, map soil structure, and provide information on other parameters of interest, such as drainage tile location or root density. GPR is a versatile geophysical technique, and data can be acquired with antennas in contact with the ground, held above the ground, or in boreholes. GPR data can also be acquired quickly; thousands of measurements can be collected across a field within a few hours.

2 Electromagnetic parameters and ground penetrating radar surveying and data processing

2.1 Electromagnetic parameters

GPR uses high-frequency electromagnetic energy, usually in the 10 MHz to 2 GHz range (Jol, 2009). The electromagnetic attribute most often measured with GPR is the electromagnetic velocity (v), which is influenced by the dielectric permittivity (ε), electrical conductivity (σ), magnetic permeability (μ), and frequency of the GPR antenna (ω). The dielectric permittivity is usually expressed as the relative dielectric permittivity, ε_r, which is the dielectric permittivity normalized by the permittivity of free space (ε_0). Similarly, the magnetic permeability is usually expressed as the relative magnetic permeability, μ_r, which is 1 for non-magnetic materials. The relationship between these parameters and the electromagnetic velocity is:

$$v = c\sqrt{\frac{2}{\mu_r \varepsilon_r}}\left[\sqrt{1+\left(\frac{\sigma}{\omega\varepsilon}\right)^2}+1\right]^{-0.5} \tag{1}$$

where c is the speed of an electromagnetic wave in a vacuum (Reynolds, 2011). For the frequency range used for GPR, electromagnetic energy is primarily capacitive. Within this frequency range, the frequency dependence of the velocity becomes negligible, and for less electrically conductive environments (electrical conductivity less than ~100 mS/m), σ can also be neglected (Reppert et al., 2000). If the material is non-magnetic, as is true for most soils, the electromagnetic velocity measured by GPR can be written as:

$$v = \frac{c}{\sqrt{\varepsilon_r}} = \frac{0.3\,\text{m/ns}}{\sqrt{\varepsilon_r}}. \tag{2}$$

As shown in Equation 2, for most geologic environments, the most important parameter for determining the electromagnetic velocity is the dielectric permittivity. The dielectric permittivity is a measure of how easily a material can become polarized in the presence of an applied electric field. Materials with high natural polarity, such as a liquid water molecule, can easily change position (become polarized) when an electric field is applied; these materials have high dielectric permittivity. Materials with less polarized molecules, such as most minerals, have lower permittivities. The relative dielectric permittivity of most minerals ranges from 3 to 8, while the permittivities of air and water are ~1 and ~80, respectively (Robinson, 2004). Since the permittivity of air is similar to that of minerals, and since the range of permittivity for minerals is small compared to the permittivity of water, the dielectric permittivity of most soils is primarily determined by how much water is in the soil pores. Thus, mineralogy is relatively insignificant, and most contrasts in dielectric permittivity in agricultural soils can be related to contrasts in soil water content. It should be noted, however, that contrasts in soil water content can be caused by changes in mineralogy, so these parameters are not necessarily independent.

While the electrical conductivity and magnetic permeability are not evaluated as frequently as the dielectric permittivity in GPR applications, they influence the attenuation of the GPR energy. Therefore, they are important in evaluating the likely success of GPR surveys. Electrical conductivity is the ease with which an electrical current can flow through a material; when the soil has a very high electrical conductivity, more of the electromagnetic energy becomes conductive, as opposed to capacitive, which increases attenuation. Similarly, magnetic permeability is high in materials that can easily become magnetized, and electromagnetic energy is also attenuated in these soils (Cassidy, 2009). Most soils have low concentrations of magnetic minerals, so electrical conductivity is usually a much greater concern for agricultural applications. Electrical conductivity is usually high in soils with a large clay fraction and in saline soils, so GPR penetration depth is often limited in these soils (Doolittle and Butnor, 2009). The mineralogy of the clay and the types and concentrations of salts in saline soils also affect the attenuation of the GPR signal; Doolittle et al. (2007) have created a map of soils in the United States that are considered suitable for GPR surveying. While this map is useful in determining the applicability of sites for standard engineering or archeological GPR applications, it should be noted that the shallow depth of many agricultural applications (<0.5 m in some cases) allows GPR to be successfully applied even at sites generally deemed unsuitable for other applications that require a greater penetration depth.

2.2 Ground penetrating radar data acquisition methods

GPR instruments consist of a transmitting antenna, which emits electromagnetic energy, and a receiving antenna. The receiving antenna records a time series of voltages (the signal amplitude), which can be interpreted to derive information about the surface. Data can be acquired with the antennas on the ground, in the air, or lowered down a borehole. For agricultural applications, data are most frequently collected with the antennas on the ground, so this chapter will provide more background on this data acquisition method than the other methods.

For all types of GPR data acquisition, the choice of antenna frequency is important. Lower frequency antennas have less attenuation and can penetrate deeper into the subsurface. However, lower frequency antennas also have larger wavelengths, which reduces the resolution. Lower resolution results in less precise determination of the location of targets and an increase in the minimum size of targets that can be detected. Both the survey objectives and the subsurface conditions must be considered when selecting an appropriate GPR frequency.

2.2.1 Ground-coupled ground penetrating radar

GPR transmitters emit electromagnetic energy outward in a spherical wavefront. The spherical wavefront can be represented as ray paths, and for ground-coupled transmitters, there are four main paths that return electromagnetic energy to the receiving antenna (Fig. 1). The first path is energy that travels directly through the air from the transmitting to the receiving antenna, which is referred to as the airwave. The electromagnetic velocity of air is very high, and

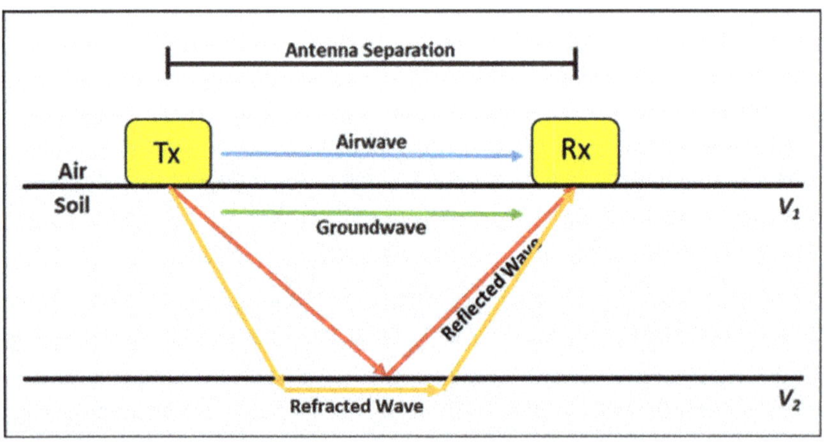

Figure 1 GPR energy can be represented as ray paths. The airwave (blue) and groundwave (green) travel directly between the transmitting antenna (Tx) and receiving antenna (Rx), while the reflected wave (red) and refracted wave (orange) are generated when the energy encounters a subsurface interface.

the travel path is short, so the airwave is the first energy detected by the receiver. The next energy detected is the groundwave, which travels directly between the transmitting and receiving antennas in the shallow subsurface. Groundwaves are evanescent, so this energy can only be measured when the antennas are in contact with the ground. The third travel path is the reflected wave, which occurs when subsurface materials have contrasting electromagnetic velocities. Reflections are the energy most often used in conventional GPR applications. The fourth travel path is the refracted wave, which can return to the surface when the electromagnetic velocity of the lower layer is greater than that of the overlying layer. While refractions can be observed in some GPR datasets, they are usually less useful than the other types of energy and are usually not analyzed.

Each GPR measurement creates a time series of amplitudes called a trace (Fig. 2). Amplitudes can be positive or negative, and each GPR 'event' (airwave, groundwave, reflection, or refraction) is composed of a wavelet with both positive and negative amplitudes. Ground-coupled GPR data are usually displayed in a radargram, in which sequential traces are plotted in a series; the vertical axis of a radargram is time (or depth, if the velocity is known), while the horizontal axis is the distance between traces (Fig. 3). There are two main methods for acquiring ground-coupled GPR data. The first method is

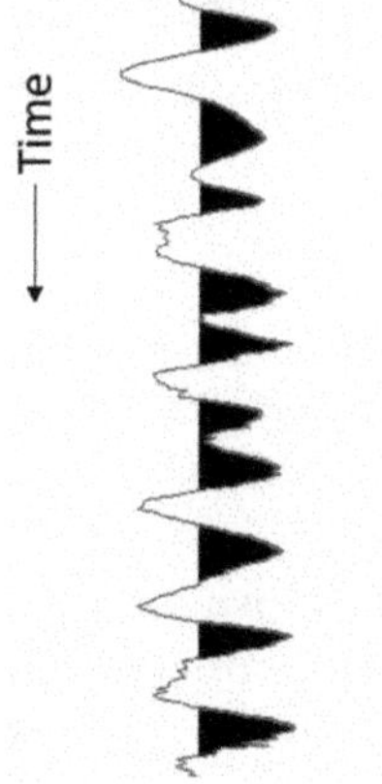

Figure 2 Each GPR trace is a time series of amplitudes, where amplitudes can be either positive or negative.

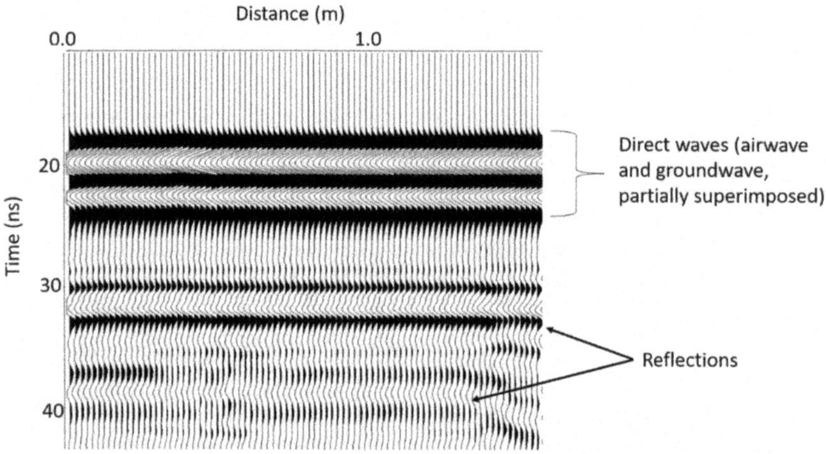

Figure 3 A radargram is a series of traces, where the amplitude is displayed as the variation from a vertical line or with different colors.

common-offset profiling (aka fixed-offset profiling), where the transmitting and receiving antennas are held a fixed distance apart and pulled along a traverse. This is the most common data acquisition method for covering large areas and is the simplest to process and interpret data. When reflections occur from units with different electromagnetic properties, common-offset data can show a pseudosection of the subsurface interfaces between the units (Fig. 4). If the electromagnetic velocity of the soil is known or can be estimated, the depth to the interface can be estimated using the arrival time of the reflection in the radargram.

The second data acquisition mode for ground-coupled data is variable-offset profiling. In this mode, the distance between the transmitting and receiving antennas is increased for each trace. One method of increasing the distance between antennas is to move both the transmitting and the receiving antenna by the same amount after each measurement, which is called a common-midpoint (CMP) survey (Fig. 5). Another method is to keep one antenna at a constant location and to move the other antenna progressively further away, which is a wide-angle reflection and refraction (WARR) survey (Fig. 6). The advantage of variable-offset surveys is that the electromagnetic velocity can be determined directly by analyzing the change in travel time of each event as a function of antenna separation distance. The velocities of the airwave and groundwave can be determined from a simple linear fit of the GPR event, while the velocity of reflections can be determined by fitting a hyperbola to the event. Most GPR software packages will calculate the electromagnetic velocity for different events directly from the radargram. For the airwave and groundwave, CMP and WARR surveys are equally valid, but for reflections, WARR surveys can

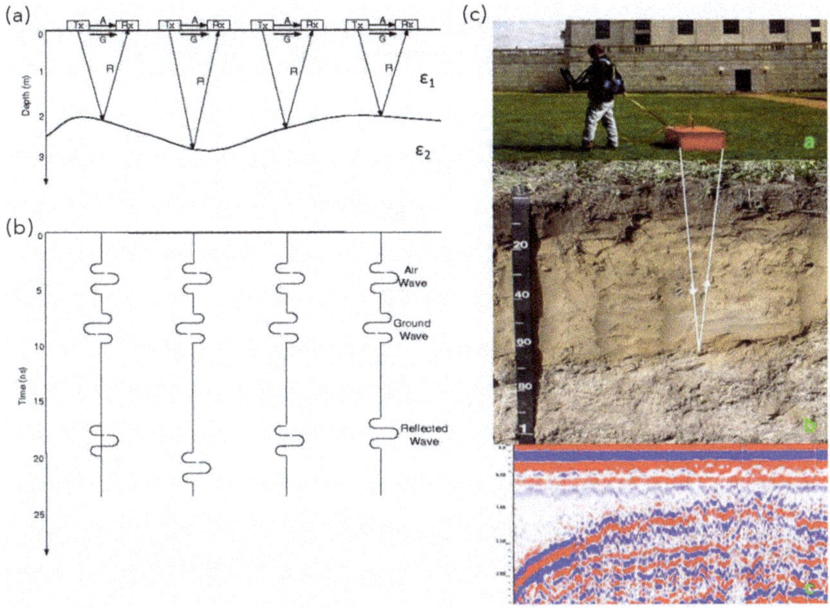

Figure 4 (a) Common-offset data are acquired when the transmitting and receiving antennas are kept at a constant separation distance and pulled along the traverse. (b) The radargram shows a pseudosection of the reflected interface. (c) Common-offset data acquired over soil horizons. Source: Image courtesy of James Doolittle (2021).

provide inaccurate velocity measurements if the reflective interface is dipping. Therefore, CMP surveys are preferred for most applications.

The design of GPR systems varies. The design of GPR systems varies. Some systems use a single antenna housing for both transmitting and receiving the GPR signal; these systems are referred to as monostatic, and it is not possible to perform variable-offset surveys with a single system. The design of GPR systems varies. Some systems use a single antenna housing for both transmitting and receiving the GPR signal; these systems are referred to as monostatic transmitting and receiving the signal, so variable-offset surveys are straightforward. Bi-static antennas are usually placed in a sled that keeps them in a set position for common-offset profiling.

2.2.2 Air-launched ground penetrating radar

GPR data can also be acquired if the antennas are held a short distance (~0.2–2 m) above the ground. Air-launched data can be acquired quickly, but usually have poor depth penetration, so only the shallowest soil can be analyzed. Most air-launched data are acquired using horn antennas, which are often larger than ground-coupled units, have a larger bandwidth, and are usually higher frequency than what is used for most ground-coupled soil studies. Air-launched

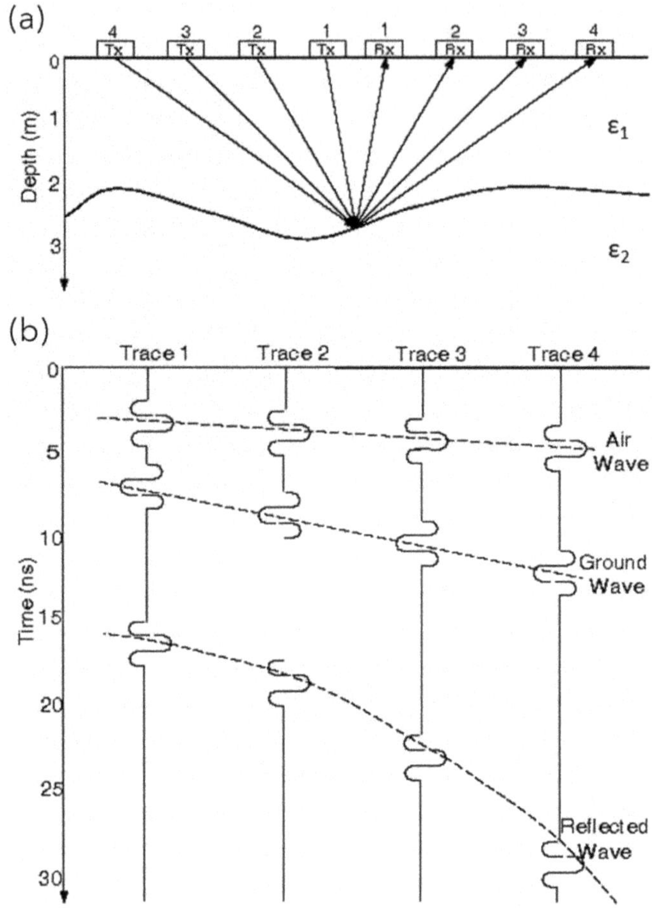

Figure 5 (a) Common-midpoint data are acquired when the transmitting and receiving antennas are both moved further apart for each measurement. The numbers above the antennas indicate the measurement number. (b) The electromagnetic velocity can be determined for each GPR event.

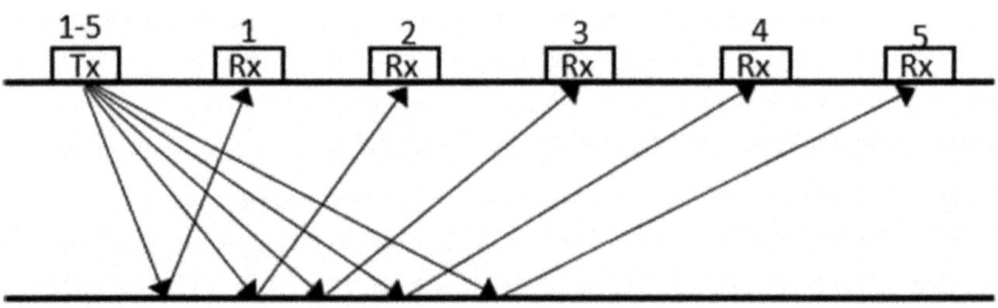

Figure 6 WARR data are acquired when one antenna is held in a fixed position and the other antenna is moved further away. For this example, the receiving antenna is moved for each measurement.

antennas are generally mounted on a vehicle for rapid data acquisition, and usually only common-offset data are acquired.

2.2.3 Borehole ground penetrating radar

GPR data can also be acquired in boreholes using antennas designed for this purpose. For conventional borehole surveying, at least two boreholes are needed, and the transmitting and receiving antennas are lowered into separate boreholes. The receiver measures the magnitude and arrival time of the energy that passes between the boreholes. Data can be acquired either in transillumination mode, by keeping the transmitting and receiving antennas at the same depth for each measurement (Fig. 7a), or in tomographic mode (Fig. 7b). In tomographic mode, one antenna is held at a fixed position while the other antenna is moved along the length of the borehole, then the fixed position is changed and the measurement is again repeated along the length of the borehole. The tomographic mode can provide the highest resolution data of any type of GPR surveying but is time consuming to collect.

A newer method of borehole surveying uses one borehole, and the transmitting antenna is kept on the ground surface. In this method, the transmitter is placed at different distances from the borehole while the receiver is lowered to different depths (Tronicke and Hamann, 2014; Strobach

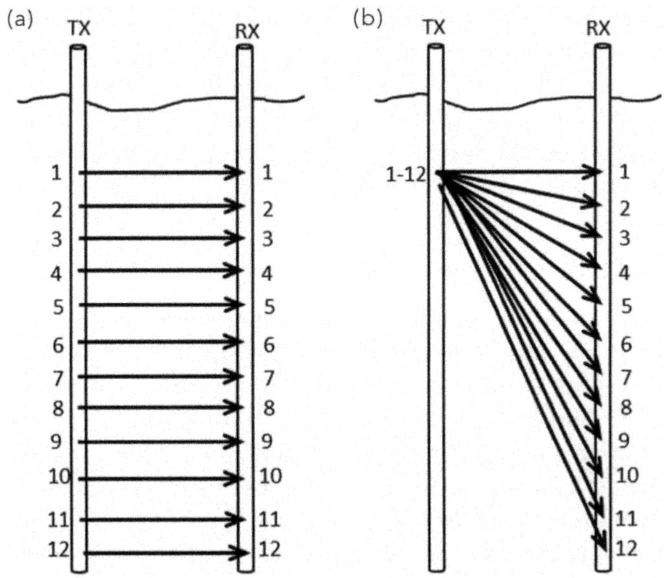

Figure 7 (a) Borehole data acquired in transillumination mode. (b) Data acquired in tomographic mode. Only the first fixed measurement for tomography is shown in this schematic.

et al., 2014). This method allows greater depth penetration than surface-based techniques and better characterization of shallow depths than conventional borehole measurements, although the transmitter and receiver locations must be chosen carefully to avoid interference from the direct wave.

Although borehole methods provide very detailed information about the subsurface, they miss two of the main advantages of GPR surveying, because they are intrusive and cannot be acquired quickly or over large areas. Therefore, borehole data are not commonly used for agricultural applications and will not be discussed further in this chapter.

2.3 Data processing and interpretation

GPR data processing and interpretation can range from very simple qualitative approaches to complex numerical modeling. Most GPR users employ a relatively simple processing procedure that involves removing noise outside of the desired frequency range through a bandpass or dewow filter and sometimes employing gain when needed to strengthen weak signals. This simple processing procedure is well suited to applications of GPR that primarily use the dielectric permittivity or electromagnetic velocity for data analysis or interpretation.

GPR data can be used to determine the structure or thickness of soil layers and/or to evaluate physical properties of the soil based upon the measured electromagnetic properties. The parameter most often analyzed to estimate soil properties is the dielectric permittivity or electromagnetic velocity. There are several methods used to estimate the dielectric permittivity; travel time methods and reflection amplitude analysis are the most commonly used methods, while early time analysis, full waveform inversion, and guided wave analysis have been used in some studies.

2.3.1 Travel time methods

The simplest method for determining the electromagnetic velocity using travel time methods is to use a variable-offset survey, since this requires no special processing or prior knowledge of the subsurface. Variable-offset surveys are also useful for interpreting common-offset data and are recommended for quality control/data verification for most large surveys. However, variable-offset surveys require two antennas and are time-consuming to perform and analyze, so other methods are often used.

A common method for estimating velocity, especially for engineering or archeological applications, is to analyze diffraction hyperbolas from isolated 'point' reflectors in the subsurface, such as pipes (Fig. 8). The slope of the legs of the hyperbola is controlled by the subsurface velocity, so the relationship between the arrival time of the hyperbola reflection and the antenna position can be used to estimate the velocity (Olhoeft,2000). Most GPR processing programs include this functionality, but the accuracy of this method is limited due to errors that are introduced if the GPR antenna passes over a buried object obliquely. Additionally, this method is spatially limited, as it provides the velocity only over the buried object and thus fails to capture the variations in velocity that typically occur in natural soils.

A third method for estimating velocity is to measure the time needed for a reflection to return to the surface for an object or layer at a known depth (Goodman et al., 2009; Lunt et al., 2005; Grote et al., 2002). Data processing for this method is very straightforward and requires only a few parameters, including the depth of the reflector (*d*), the separation distance of the antennas (*S*), and the travel time of the reflected wave (t_r). The velocity (*v*) can then be calculated as:

$$v = \frac{2\sqrt{d^2 + (0.5\,S)^2}}{t_r} \quad (3)$$

The travel time of the reflected wave can be calculated using the arrival time of the reflection and the time at which the energy leaves the transmitter (zero time). A simple method for choosing the zero time (t_z) is given by:

$$t_z = t_a - \frac{S}{c} \quad (4)$$

where t_a is the arrival time of the airwave, and the second term in this equation accounts for the time needed for the airwave to travel between the antennas.

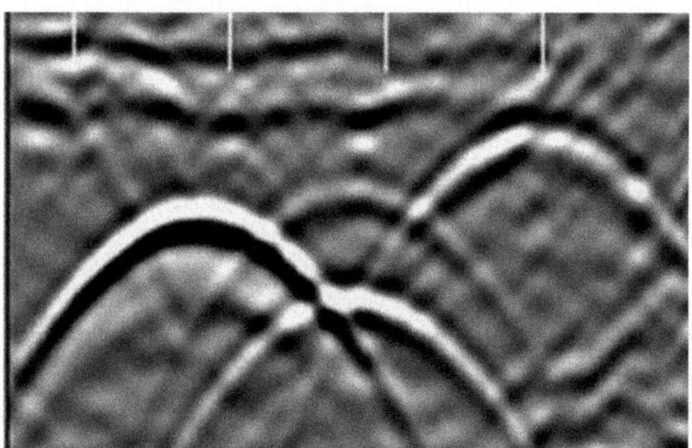

Figure 8 Point reflectors, such as pipes, create reflection hyperbolas that can be analyzed for velocity estimation. Source: Image courtesy of James Doolittle (2021).

The velocity estimated using reflectors at a known depth can be very accurate, but this method is also limited by logistical constraints. One problem arises if the user wishes to map the velocity over large areas, as it is rare to have a continuous reflector at a known depth over extended areas. Another problem is that even for small targets, the process of installing a target at a known depth or determining the depth of a natural horizon disturbs the soil, which can influence velocity measurements.

The last commonly used method for estimating the velocity based on travel time is to use the groundwave. Groundwaves can be analyzed using variable-offset data, as discussed previously, or common-offset data. Common-offset groundwave data can be used to estimate the velocity when the separation distance between the transmitting and receiving antennas is known and the arrival time of the airwave (t_a) and groundwave (t_g) is measured in the GPR data:

$$v = \frac{S}{t_{gw} - t_a + \frac{S}{c}}. \tag{5}$$

Advantages of common-offset groundwave data are that they occur without a subsurface interface, so this technique can be widely applied without prior knowledge of the subsurface, data can be collected very quickly, data processing is straightforward and can be done using most commercially available processing software, and different sampling depths are possible based upon the antenna frequency chosen. Studies that focused on the sampling depth of the groundwave show that higher frequencies have a shallower sampling depth than lower frequencies. Numerical modeling suggests that the sampling depth ranges from 0.13 m for 900 MHz antennas to 0.85 m for 100 MHz antennas (Galagedara et al., 2005b). Experimental studies performed in a laboratory show that the sampling depth ranges from 0.12 m for 1000 MHz antennas to 0.30 m for 250 MHz antennas (Grote et al., 2010).

Although groundwaves are a versatile method for estimating the electromagnetic velocity over depths of interest for many agricultural applications, there are limitations to this method. Careful survey planning is needed to acquire common-offset groundwave data, as the antenna separation distance must be chosen carefully. If the separation distance is too large, the groundwave may attenuate before it reaches the receiver, and thus cannot be detected. If the separation is too small, the airwave arrives only a short time before the groundwave. The airwave and groundwave wavelets may be superimposed, so the arrival time of the groundwave cannot be determined. Variable-offset surveys are needed to determine the optimal antenna separation prior to common-offset surveys. Another challenge occurs if there are reflections from shallow layers, as these early-arriving reflections may also superimpose with the groundwave. Finally, data quality may be poor in very dry soils, as the groundwave velocity is high under these conditions and it may be impossible to find an antenna separation distance that will avoid superposition with the airwave.

2.3.2 Reflection amplitude analysis

Another method for analyzing GPR data is to consider the amplitude of reflections. Quantitative amplitude analysis is generally performed for air-launched data and is usually done only for the reflection from the air-ground interface. This interface is analyzed because a large fraction of the energy returns to the receiver when a reflection occurs from the air-ground interface, so little energy remains to be transmitted into the subsurface. The simplest method for estimating the dielectric permittivity from the amplitude of the surface reflection is:

$$\varepsilon = \left(\frac{1 + \frac{A_s}{A_m}}{1 - \frac{A_s}{A_m}} \right)^2 \tag{6}$$

where A_s is the amplitude from the air-soil interface and A_m is the amplitude from a perfect reflector. In practice, a metal plate is used as the calibrating (perfect) reflector. The antennas should be held the same distance off the ground for the calibration measurement and subsequent measurements of the soil surface, and the metal plate should be larger than the footprint of the GPR energy.

Most air-launched GPR surveys are performed using high-frequency antennas (>900 MHz), primarily because the resolution is higher, but a side benefit is the size of the metal plate needed for calibration is smaller. Air-launched GPR surveys are used extensively in the transportation engineering industry but have been used much less frequently for agricultural applications. One of the main reasons air-launched methods are not as popular for agricultural applications is the shallow penetration depth; reflections from the ground surface penetrate less than 5 cm into the subsurface (Escorihuela et al., 2010; Serbin and Or, 2003), which is insufficient for many agricultural applications. Also, the accuracy of air-launched methods can be significantly decreased if the ground surface is rough/irregular (surface undulations greater than ~ 4 cm), if there are vertical changes in permittivity in the soil profile, or if the surface is covered by vegetation (Serbin and Or, 2005; Huisman et al., 2003). The main advantages of air-launched methods are the rapid data acquisition and, for the simple amplitude analysis used in Equation 6, ease of data processing.

2.3.3 Full waveform inversion

Full waveform inversion is an advanced processing technique based on numerical modeling of the GPR signal in the frequency domain. The main advantage of this technique is that it can provide an estimate of the electrical conductivity as well as the dielectric permittivity. Different approaches have been applied for full waveform inversion, including solving Maxwell's equations using the finite element or finite different methods (Durand and Slodic, 2011; Ernst et al., 2007) and assuming planar soil layers and using Green's function (Tran et al., 2015). Most studies that use full waveform inversion have been conducted with air-launched data (Tran et al., 2015; Ardekani, 2013; Weihermüller et al., 2007; Lambot et al., 2004), although one study also applied this technique to the ground-coupled data by analyzing the signal from groundwaves in WARR surveys (Busch et al., 2014).

Although full waveform inversion uses more of the GPR signal information than either travel time or amplitude analysis, and is therefore more robust, it currently has some significant limitations. Some of these limitations are related to the method of analysis. The technique is complex and requires considerable expertise, and there is currently no commercially available software to conduct full waveform inversion, so the technique cannot be applied by non-expert users. Another problem is the computational power needed to perform finite element or finite difference modeling (Tran et al., 2015), although this problem is usually overcome as computational power continues to increase. Other limitations are physical in nature. Most full waveform inversion is done using air-launched data, which have a very shallow sampling depth. This method also does not work well in soil with high electrical conductivity, so it may not work well in soils with high silt or clay content (Ardekani, 2013; Minet et al., 2012) or in soils with high water content (Mourmeaux et al., 2014) or high salinity. Soil roughness can also influence the results (Lambot et al., 2006). Thus, although this technique has potential for maximizing the information that can be extracted from the GPR signal, significant progress must be made before it can be adopted at large scales.

2.3.4 Early time analysis

As described previously, the airwave and groundwave are direct waves that arrive early in the radargram. Groundwaves can be analyzed using travel time methods for many soil moisture conditions, but groundwave analysis may not be possible if the soil is very dry and the airwave and groundwave cannot be sufficiently separated by increasing the antenna separation. Early time analysis focuses on the amplitude of the airwave and groundwave when the antenna offset is fairly small and the airwave and groundwave are superimposed.

Under these conditions, travel time techniques will not be successful, but the amplitude of the superimposed airwave and groundwave can be related to the dielectric permittivity and electrical conductivity of the soil (Pettinelli et al., 2014; Di Matteo et al., 2013; Pettinelli et al., 2007). Pettinelli et al. (2014) show similar levels of correlation between amplitude and dielectric permittivity and between amplitude and electrical conductivity, but a modeling study suggests that permittivity has a greater effect on the early time amplitude (Di Matteo et al., 2013). The portion of the signal that appears to be optimal for early time analysis is the first positive half cycle of the early time signal, as this early signal has less superposition with reflections from shallow objects (Di Matteo et al., 2013). Analyses of early time data use the maximum amplitude observed during this first half cycle for correlation with electromagnetic parameters. Using this maximum amplitude, the amplitude was observed to increase as permittivity decreased (Pettinelli et al., 2014; Ferrara et al., 2013).

Advantages of the early time analysis method are that it can be used in very dry soils or in very clayey soils where groundwave travel time methods may not be successful. Also, this method could be used in conjunction with reflection surveys, where the antenna separation may intentionally be kept small for optimal transmission of reflected energy. In this application, early time analysis may provide information about the shallow subsurface (as determined by groundwave penetration depth), while the reflection survey could provide information about deeper layers.

The biggest disadvantage of the early time analysis method is that it is a semi-empirical method, and amplitudes must be calibrated for different sites. This requires determination of the dielectric permittivity through some other means for use in calibration. Di Matteo et al. (2013) suggest that the amplitudes can be related to permittivity using either a linear relationship or a linear with inverse amplitude relationship, although they found a slightly better correlation with the linear relationship. Also, Ferrara et al. (2013) note that the sensitivity of the early time technique decreases for permittivity values greater than ~12, so this method may not work well for wetter soils.

2.3.5 Guided waves

Most GPR data processing techniques assume a homogeneous medium, but under certain conditions, the dielectric permittivity can have large contrasts within a layered structure. The most likely occurrence in an agricultural setting is when recent precipitation or irrigation has penetrated only a short distance into the soil, resulting in a thin very wet layer overlying a dry layer. In this circumstance, the GPR energy can become trapped in this thin layer; the multiple and superimposing reflections that occur in this layer are referred to as guided waves (Arcone et al., 2003). Traditional travel time and amplitude analysis methods are inappropriate for guided waves, but the techniques used for analyzing seismic Rayleigh or Love waves can be applied to variable-offset GPR data (van der Kruk et al., 2010).

The possibility of a thin layer of very different permittivity that acts as a GPR waveguide should be considered when planning GPR surveys. Although it is possible to analyze guided wave data, this technique is not easy to apply. A significant limitation is that the data processing is time-consuming compared to other GPR methods and cannot be performed using commercially available GPR software. Another limitation is that variable-offset surveys are needed to analyze guided waves. Finally, the conditions that generate guided waves may not be continuous over large areas. Therefore, it is important to recognize when guided waves might interfere with other types of data acquisition, such as groundwaves, but guided waves have not yet been used for large-scale agricultural applications.

3 Soil structure

One of the most useful agricultural applications of GPR is to map soil structure. A very common application is to map the depth of the soil-bedrock interface (Novakova et al., 2013; Sucre et al., 2011). Another common application is mapping the thickness of soil horizons (Zhang et al., 2014a,b; Novakova et al., 2013; Simeoni et al., 2009; van Dam et al., 2003; Freeland et al., 1998; Collins et al., 1990; Collins and Doolittle, 1987). Interfaces between soil layers or between soil and bedrock are generally mapped using GPR reflections, usually from ground-coupled GPR systems, as shown in Fig. 9. GPR is most successful in mapping soil horizons when they have differences in water content, which can be caused by variations in soil texture or organic matter in different soil layers (Zajícová and Chuman, 2019). Zhang et al. (2014a,b) noted that the reflections

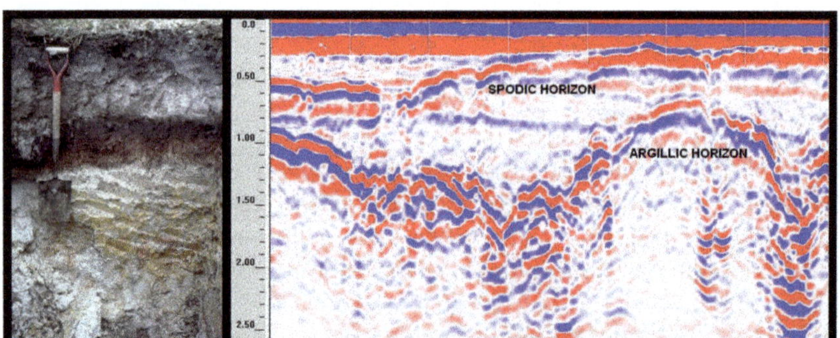

Figure 9 Ground-coupled, common-offset GPR reflection data that show interfaces between different soil horizons. Source: Image courtesy of James Doolittle (2021).

from interfaces between soil horizons were generally clearest under wet conditions, as the water-holding capacity of the horizons differed, and greater contrasts in water content (and thus dielectric permittivity) occurred under wet conditions. However, they also noted that preferential flow along root paths could obscure reflections under wet conditions, and the reflection from the soil-bedrock interface was usually clearer under drier conditions. The sensitivity of GPR to water content also makes it a good tool for mapping the water table, especially in coarser-grained soils without a well-developed capillary fringe, as shown in Fig. 10.

One type of soil mapping that has received special attention is mapping of the organic layer (Collins and Doolittle, 1987). Li et al (2015) and Winkelbauer et al. (2011) mapped the organic-mineral interface with fairly high accuracy (~4% and 15% error in the depth estimate for these studies, respectively). Mapping the organic layer can be difficult, as organic material tends to have high attenuation (Fig. 11), so the higher frequency GPR systems (which provide

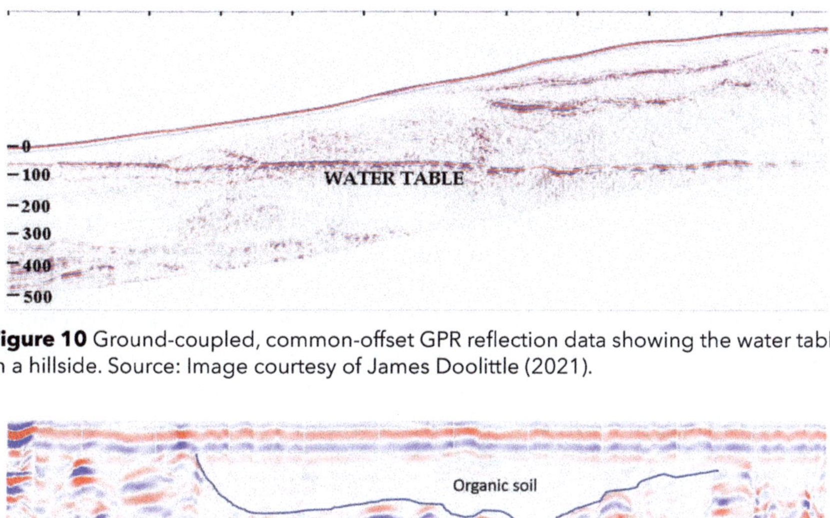

Figure 10 Ground-coupled, common-offset GPR reflection data showing the water table in a hillside. Source: Image courtesy of James Doolittle (2021).

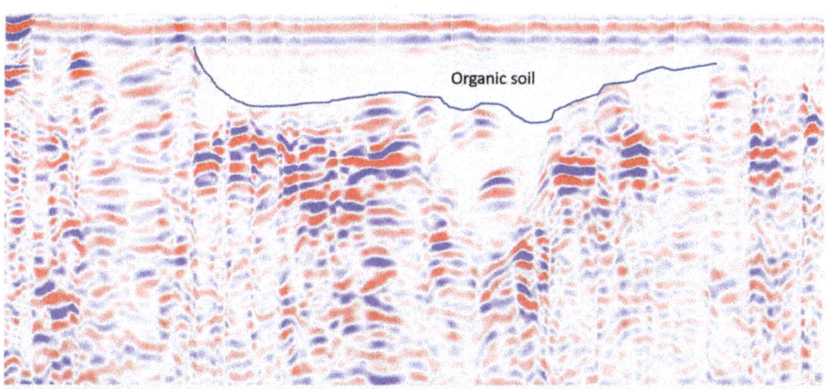

Figure 11 Organic soil tends to have high attenuation, so low signal amplitudes may be observed in these units.

higher accuracy depth measurements) may not achieve sufficient penetration depths in deeper organic soils. Additionally, researchers have obtained differing results for the dielectric permittivity of organic layers, with results ranging from ~1 to 60, depending on the soil moisture and the study considered (André et al., 2015; Winkelbauer et al., 2011). Winkelbauer et al. (2011) also determined that GPR could not detect differences between different organic horizons.

In addition to standard organic soil layer mapping, GPR has been used extensively to map the depth of peat layers (William and Comas, 2016; Plado et al., 2011; Kettridge et al., 2008). Peat layers can have very high water content, depending on the plant species and degree of decomposition, and have dielectric permittivities ranging from 50 to 70 (Zajícová and Chuman, 2019). Most mineral-based soils have much lower permittivity, even under conditions of full saturation, so the interface between peat and mineral soils is relatively easy to detect using GPR reflection techniques (Slater and Reeve, 2002). However, mapping the depth of peat poses some significant challenges, as peat is usually very attenuative of the GPR signal, so lower frequency antennas must be used to penetrate to the peat-mineral soil interface, and the lower frequency decreases the resolution of the depth estimate. Additionally, the dielectric permittivity of peat varies both as a function of the saturation of the peat and the state of peat decomposition (Plado et al., 2011), which makes accurate determination of the depth to the peat-mineral interface more difficult. Researchers have attempted to map different peat layers as a function of decomposition, but with only partial success, as these layers do not always generate reflections, and reflections are sometimes generated from interfaces not related to peat layers (Karuss and Berzins, 2015; Kettridge et al., 2012). Also, the wet conditions in many peat bogs can make GPR data acquisition challenging, so data collection is sometimes performed when the bog is frozen, but snow cover can complicate data interpretation (Plado et al., 2011). The challenges in acquiring accurate dielectric permittivity data and in interpreting GPR reflections make GPR a good tool for providing supplemental information to manual probing or other geophysical techniques, but GPR is generally not suited as a stand-alone tool for peat mapping (Zajícová and Chuman, 2019).

4 Soil water content

4.1 Petrophysical relationships

One of the applications for which GPR is best suited is the estimation of soil water content. Since the dielectric permittivity is so strongly influenced by the amount of pore water in soil, estimation of the volumetric water content (VWC) is relatively straightforward once the permittivity has been measured. To estimate permittivity from water content, a site-specific empirical relationship can be developed, or more commonly, a previously established petrophysical relationship can be used. The three types of petrophysical relationships most often used for VWC estimation are empirical relationships, volumetric mixing models, and effective medium approximations. The most commonly applied empirical relationship is a model developed by Topp et al. (1980):

$$VWC = -5.3\times10^{-2} + 2.92\times10^{-2}\varepsilon - 5.5\times10^{-4}\varepsilon^2 + 4.3\times10^{-6}\varepsilon^3 \quad (7)$$

which was developed using a range of agricultural soils. Other researchers (Jacobsen and Schjønning, 1993; Roth et al., 1992) have developed similar empirical relationships, but Topp's equation has been applied the most widely in GPR studies. Petrophysical relationships based upon volumetric mixing models use the proportion of different phases of the soil (usually air, water, and soil solids) and the dielectric permittivity of each phase. The most common mixing model is the complex refraction index model (CRIM), which is given by:

$$VWC = \frac{\sqrt{\varepsilon} - (1-n)\sqrt{\varepsilon_s} - n\sqrt{\varepsilon_a}}{\sqrt{\varepsilon_w} - \sqrt{\varepsilon_a}} \quad (8)$$

where n is the porosity and ε_a, ε_s, and ε_w are the dielectric permittivities of air, soil (mineral) solids, and water, respectively (Roth et al., 1990). Other researchers have suggested four-phase models to include the bound water phase (Dobson et al., 1985), but the three-phase CRIM model generally produces adequate results (Robinson et al., 1999). The last type of petrophysical model, effective medium approximations, takes into account both the proportions of air, water, and soil solids and the structure of the soil, such as grain shape, pore geometry, or pore water distribution (Steelman and Endres, 2011). Thus, effective medium approximations are similar to the volumetric mixing models in that they require knowledge of the porosity and the permittivity of soil solids, but they also require additional information about the soil structure.

Of the different types of petrophysical relationships, empirical relationships are used most frequently, as they are simple to apply and do not require knowledge of the soil properties beyond the dielectric permittivity. For greatest accuracy, empirical relationships should be applied to soils similar to the ones used to develop the relationship. Volumetric mixing models and effective medium approximations can represent different soil conditions more effectively than simple empirical relationships, but in practice, the parameters needed to apply these models are usually not known over large areas, especially for heterogeneous soils. Thus, the application of these models often requires simplifying assumptions, such as a uniform porosity over an entire field, that may reduce the accuracy of the model.

Table 1 Studies of soil moisture using ground penetrating radar

Data acquisition method	Data analyzed	Data processing method	Reference
Ground-coupled, variable offset	Reflection from a subsurface object or interface	Travel time	Greaves et al. (1996)
Ground-coupled, variable offset	Reflection from a subsurface object or interface	Travel time	van Overmeeren et al. (1997)
Ground-coupled, variable offset	Reflection from a subsurface object or interface	Travel time	Nakashima et al. (2001)
Ground-coupled, variable offset	Reflection from a subsurface object or interface	Travel time	Garambois et al. (2002)
Ground-coupled, common offset	Reflection from a subsurface object or interface	Travel time	Grote et al. (2002)
Ground-coupled, common offset	Reflection from a subsurface object or interface	Travel time	Stoffgregen et al. (2002)
Ground-coupled, common offset	Reflection from a subsurface object or interface	Travel time	Grote et al. (2005)
Ground-coupled, common offset	Reflection from a subsurface object or interface	Travel time	Lunt et al. (2005)
Ground-coupled, common offset	Reflection from a subsurface object or interface	Travel time	Wollschläger and Roth (2005)
Ground-coupled, variable offset	Reflection from a subsurface object or interface	Travel time	Turesson (2006)
Ground-coupled, variable offset	Reflection from a subsurface object or interface	Travel time	Bradford (2008)
Ground-coupled, common offset	Reflection from a subsurface object or interface	Travel time	Weiler et al. (1998)
Ground-coupled, variable offset	Reflection from a subsurface object or interface	Travel time	Pan et al. (2012)
Ground-coupled, variable offset	Groundwave	Travel time	Huisman et al. (2001)
Ground-coupled, variable offset	Groundwave	Travel time	Galagedera et al. (2003)
Ground-coupled, variable offset	Groundwave	Travel time	Grote et al. (2003)
Ground-coupled, variable offset	Groundwave	Travel time	Steelman and Endres (2010)
Ground-coupled, variable offset	Groundwave	Travel time	Lu et al. (2017)
Ground-coupled, common offset	Groundwave	Travel time	Grote et al. (2003)
Ground-coupled, common offset	Groundwave	Travel time	Galagedera et al. (2005a)
Ground-coupled, common offset	Groundwave	Travel time	Weihermüller et al. (2007)

Data acquisition method	Data analyzed	Data processing method	Reference
Ground-coupled, common offset	Groundwave	Travel time	Pallavi et al. (2011)
Ground-coupled, common offset	Groundwave	Travel time	Ardekani (2013)
Ground-coupled, common offset	Groundwave	Travel time	Lu et al. (2017)
Ground-coupled, common offset	Groundwave	Travel time	Kummode et al. (2020)
Ground-coupled, common offset	Surface reflection	Amplitude analysis	Ardekani (2013)
Air-launched	Surface reflection	Amplitude analysis	Redman et al. (2002)
Air-launched	Surface reflection	Full waveform inversion	Lambot et al. (2006, 2008)
Air-launched	Surface reflection	Full waveform inversion	Weihermüller et al. (2007)
Air-launched	Surface reflection	Full waveform inversion	Jadoon et al. (2010)
Air-launched	Surface reflection	Full waveform inversion	Minet et al. (2010, 2011, 2012)
Air-launched	Surface reflection	Full waveform inversion	Slob et al. (2010)
Air-launched	Surface reflection	Full waveform inversion	Jonard et al. (2011)
Air-launched	Surface reflection	Full waveform inversion	Ardekani (2013)
Ground-coupled, common offset	Airwave and groundwave	Early time amplitude analysis	Ferrara et al. (2013)
Ground-coupled, common offset	Airwave and groundwave	Early time amplitude analysis	Pettinelli et al. (2014)
Ground-coupled, common offset	Airwave and groundwave	Early time amplitude analysis	Algeo et al. (2016)
Ground-coupled, common offset	Airwave and groundwave	Early time amplitude analysis	Oimbe et al. (2018)

4.2 Water content estimation

Researchers have used all combinations of GPR data acquisition and data processing methods which were described in Section 2 to estimate the soil water content. Table 1 lists some of the studies that have been performed for different data acquisition and processing methods. As seen from Table 1, water content estimation using GPR has a very well-established history, and new studies continue to be produced. Review articles summarizing studies using GPR for water content estimation are given in Liu et al. (2019), Zajícová and Chuman (2019), Klotzsche et al. (2018), Paz et al. (2017), and Huisman et al. (2003).

5 Soil density/compaction

GPR has been used to estimate soil density in agricultural soils and pavement aggregates primarily by relating dielectric permittivity to soil density. Since the bulk dielectric permittivity of a soil is a function of the fractions and permittivities of the different components (typically soil solids, water, and air), and since the permittivities of air and soil are relatively similar, the impact of water content must be considered when estimating soil density based on permittivity. If the soil is quite dry (VWC less than 0.05%), increases in soil density result in a decrease in the volume of air, so the permittivity increases with increasing soil density (Benedetto and Tosti, 2013). However, if the soil is wetter, the bulk permittivity is more likely to be influenced by the presence of water in the pores, and an increase in density (reduction of pore space) causes a reduction in the volume of soil water, which results in a negative correlation between bulk permittivity and soil density (Olhoeft, 2000).

Few studies have been performed to quantitatively determine the density of agricultural soils with GPR. More work has been done in engineered soils underlying pavements, since aggregate density is an essential parameter for assessing pavement condition. One study (Plati and Loizos, 2013) used air-launched GPR to estimate the density of the upper asphalt-aggregate layer. This study used reflection amplitudes to estimate the dielectric permittivity and a mixing model (Al-Qadi et al., 2010) to relate bulk permittivity to the density of this layer, but they did not evaluate the accuracy of this technique using conventionally acquired density measurements.

Most GPR-based studies of compaction in agricultural soils have been semi-quantitative, in that they determine the location or depth of compaction, but do not determine the density of the soil. Jonard et al. (2013) acquired ground-coupled GPR data at a field where three tilling techniques were used in different plots. These researchers determined the depth to the bottom of the tilled layer, but did not estimate the soil density. Similarly, Muñiz et al. (2016) used GPR to map the depth of the compacted layer in an agricultural field. Akinsunmade et al. (2019) acquired GPR data before and after a tractor was driven over the survey area, and they observed that the amplitude of the GPR signal increased after the tractor had passed over the soil, suggesting that amplitudes could be used to indicate compaction. Freeland et al. (1998) also noticed changes in the GPR signal when data were collected over highly compacted areas.

Wang et al (2016) performed a study to quantitatively estimate soil density using a test pit that was 2.4 m × 1.1 m and 0.8 m deep. Different soils were placed in layers within the pit, and GPR surveys were acquired before and after infiltration into the pit. Soil properties were characterized after the infiltration using conventional methods. These researchers found that the relationship between soil density and dielectric permittivity was not significant if the soil water content was not independently known, but partial correlation analysis showed that permittivity was correlated to bulk density if the water content was considered in the analysis. The results for this study were limited, and the authors note that the data were inadequate for establishing a statistical relationship between compaction or soil density and GPR signals.

A large challenge in using GPR for measuring soil compaction is the interdependence of soil density and soil water content. Since GPR is sensitive to both, but especially to water content, studies that correlate GPR attributes to compaction (i.e. before and after tilling or compaction by farm equipment) but do not consider water content may be difficult to apply for large-scale soil characterization. More research is needed in this area.

6 Root detection

GPR has been used extensively in the detection of roots. Roots can be detected when the water content in the roots is greater than that of the surrounding soil (Guo et al., 2013), as this creates the needed contrast in permittivity to generate a reflection. This occurs primarily in relatively dry sandy or sandy loam soils, so most studies of GPR root detection have been conducted under these conditions (Liu et al., 2016). GPR is best suited to detect individual roots that are greater than 2.5 mm to 5 mm in diameter (Cui et al., 2011; Wielopolski et al., 2000). Most studies using GPR for individual root detection have thus focused on trees (Guo et al., 2013), making this technique most applicable in orchards and vineyards.

Root detection with GPR is generally used either for detecting individual roots (root mapping) or for estimating root diameter/biomass. Both methods depend on correctly identifying roots in the radargram; methods for root identification are based on either the amplitude or the travel time of the root reflection (Guo et al., 2013). For amplitude methods, the amplitude can be output directly from the raw data or the radargram can be transformed into a pixelated image, and the intensity of the color in different pixels can be assessed (Butnor et al., 2008; Stover et al., 2007). For time-based methods, the travel time to the top of a root reflection can be used to estimate root depth, while the diameter of the root can be approximated by considering the time interval between different portions of the GPR reflection wavelet, which are related to the reflections from the top and bottom of the root (Cui et al., 2011).

For root mapping, GPR has been used successfully to image single roots and to determine whether coarse roots are present in an area (Bassuk et al.,

2011; Cox et al., 2005) or to map the extent of the root zone (Leucci, 2010). Seyfried and Schoebel (2016) used an air-launched GPR system to determine the depth of asparagus roots in a sandy test plot, which could be used to guide automatic harvesting of asparagus. Attempts to use GPR for architectural 3-D root mapping have not been overly successful, as the root systems predicted by GPR data often did not correlate well to the actual root structure as measured using conventional techniques (Zhu et al., 2014; Zenone et al., 2008; Stokes et al., 2002). Limitations with using GPR for mapping the 3D structure of roots are that GPR methods are best suited to detect roots that run perpendicular to the GPR traverse, GPR cannot distinguish between roots from different plants, and vertically inclined roots are difficult for GPR to detect (Zhu et al., 2014).

GPR has been more successful in estimating root biomass than in estimating the diameter of individual roots. Studies of root diameter have focused on analyzing the amplitude or time intervals of reflection events and have been conducted in very controlled circumstances that are ideal for GPR (roots buried in parallel lines in dry sand) (Cui et al., 2011; Hirano et al., 2009; Dannoura et al., 2008). Even under these idealized conditions, correlations between root diameter and GPR signals have been low, and relationships developed in these studies are site-specific (Zhu et al., 2014). Yamase et al (2018) estimated root diameter under natural conditions in the field, but their results did not show a strong correlation between GPR parameters and root diameter. Estimation of root biomass has focused on analyzing the number and amplitude of root reflections in a given area (Butnor et al., 2003) and has more recently included search algorithms in the data processing (Zhu et al., 2014). The success of biomass estimation in different studies varied, but more successful studies used the total root biomass (includes both living and dead roots of all sizes and depths), as GPR is not well suited to distinguishing between these parameters, and time-interval-based methods were relatively more successful (Yamase et al., 2018; Zhu et al., 2014; Cui et al., 2011).

Although processing of GPR data is usually minimal and straightforward, root detection with GPR requires slightly more complex processing than most other agricultural applications. Most studies that measured roots with GPR followed a four-step processing flow (Guo et al., 2013). The first step is to standardize the radargram to have consistent horizontal and vertical scales (Cui et al., 2011; Cox et al., 2005). The second step is to remove background noise, which can be caused by low-frequency energy sources. For root detection, the airwave and groundwave may have a similar travel time as reflections from shallow roots and so may obscure these signals. To remove low-frequency background noise, a bandpass filter can be used (Cui et al., 2011). To remove airwave and groundwave energy, a background subtraction filter is applied (Stover et al., 2007). After background noise is removed, migration is often used to collapse the reflection hyperbolas generated by roots to a point at the hyperbola peak. Kirchoff migration is often used for GPR data (Yao and Qifu, 2012). Migration requires knowledge of the electromagnetic velocity of the soil, so this must be determined before this processing step is undertaken. The final processing step is the Hilbert transformation, which expresses signal phase based on amplitude and reduces the likelihood of misinterpreting the response from a single root as multiple roots (Zheng et al., 2016).

7 Case study: soil water content measurement using ground penetrating radar groundwaves

One of the most straightforward applications of GPR for agricultural applications is the estimation of soil water content. In this case study, 500 MHz and 250 MHz groundwave data were used to estimate the water content in an apple orchard in Eau Claire, Wisconsin, USA. Data were first acquired using WARR surveys to determine the optimal antenna separation distance for common-offset surveying. Then common-offset data were acquired in six traverses across the site (Fig. 12). Examples of the WARR data used for data interpretation and the common-offset data are shown in Fig. 13. The groundwave travel time was calculated using Equation 5, and Topp's equation (Equation 7) was used to estimate the soil water content.

Figure 14 shows the water content estimates acquired using the 500 MHz and 250 MHz antennas, which have sampling depths of approximately 18 cm and 30 cm, respectively (Grote et al., 2010). The average water content is slightly higher for the 500 MHz data than for the 250 MHz, which probably occurs because there was precipitation at the site a few days before data acquisition. Water content measurements acquired in boreholes using gravimetric

Figure 12 Common-offset data were acquired in six traverses across the site.

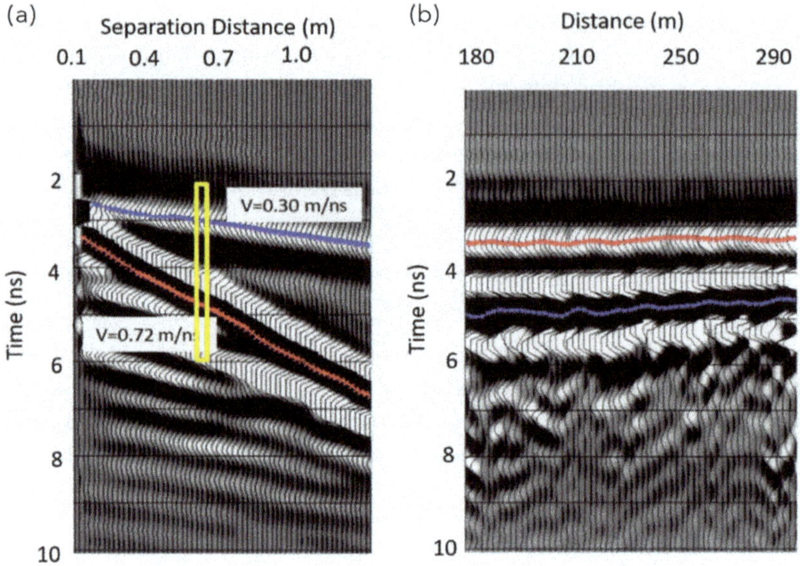

Figure 13 Airwave and groundwave data acquired with 250 MHz antennas. The airwave arrives first, and the groundwave is the second event in each radargram. (a) WARR survey. The yellow box shows the separation distance between antennas at which the common-offset data were acquired. (b) Common-offset data acquired along row 1. Variations in the groundwave travel time indicate changes in soil water content.

sampling at different depths also show wetter soil at shallower depths. Similar patterns of water content are observed between the two frequencies across most of the orchard, although some differences occur. These differences can be analyzed to better understand the vertical distribution of water content over the site.

8 Conclusion and future trends

GPR techniques have much potential for contributing to agricultural site characterization. GPR data can be acquired rapidly over large areas, and since GPR is non-destructive, it is ideal for measuring changes with time. GPR methods are primarily sensitive to changes in soil water content, so can be used to map interfaces with different soil moisture, such as different soil layers or water-filled roots, as well as to determine the water content of different soil layers.

One of the strengths of GPR is its versatility; data can be acquired in several different modes and processed in different ways. Combinations of data acquisition and processing techniques can yield information at different depths and resolutions. Recent advances in instrumentation include the capability

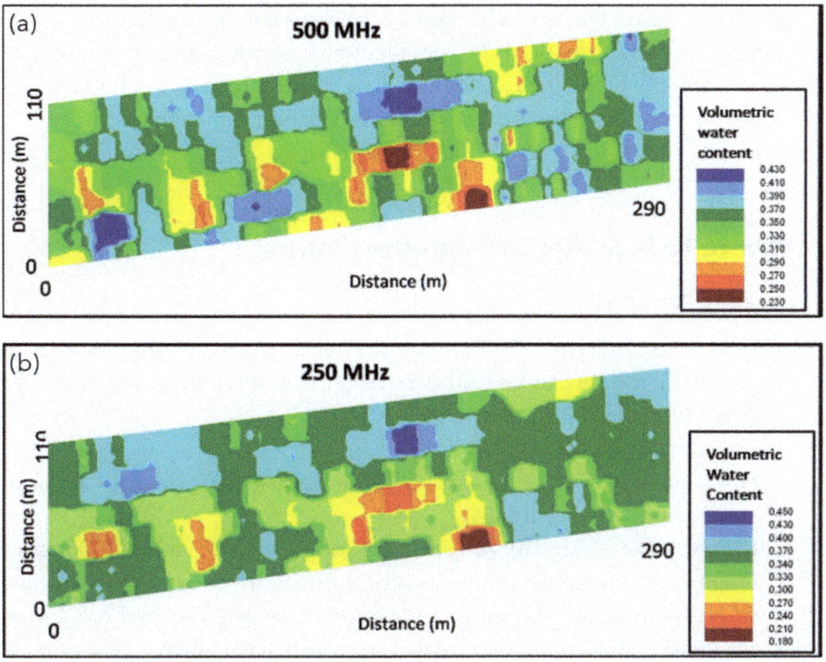

Figure 14 Volumetric water content distribution from (a) 500 MHz and (b) 250 MHz antennas.

of collecting multi-offset data along a traverse (by acquiring data with one transmitter and multiple antennas) and drone-mounted GPR systems.

Although GPR instrumentation has progressed rapidly over the last decade, the main challenge for the future is to develop data processing software that can be used by non-expert users. Most quantitative data processing is complex and cannot be performed by non-geophysicists, and several techniques, such as full waveform inversion, cannot be performed using commercially available software. Development of novice-friendly data interpretation methods and commercially available software for applying these methods will be needed before these techniques can be widely adopted. Similarly, methods that currently require site-specific calibration, such as early time analysis, will require petrophysical relationships that can be applied with reasonable accuracy without time-consuming site-specific calibration or knowledge of other soil properties that are usually unknown.

Future developments that will be most useful to agriculture will probably use ground-coupled methods and advanced data processing techniques. Ground-coupled methods have sampling depths that can penetrate into the root zone, and they are not as sensitive to surface roughness or vegetation as air-launched data are. Advanced data processing and interpretation techniques that can provide information on the electrical conductivity as well as the dielectric permittivity will help GPR become more effective at characterizing soil properties that are affected by parameters other than water content, such as soil density.

9 Where to look for further information

9.1 Further reading

- A good summary of GPR methods and practical suggestions for acquiring GPR data can be found in *Ground Penetrating Radar Theory and Applications* (ed. Harry Jol).
- Review articles on soil moisture:
  - Huisman et al. (2003): Covers basic methods in considerable detail and summarizes many studies.
  - Klotzsche et al. (2018): An update to the Huisman et al. (2003) study. Does not contain as much background information as the Huisman et al. review, but describes advanced data analysis techniques and updated instrumentation that has occurred since 2003.
  - Zajícová and Chuman (2019): Review of both older and newer methods, but is not exclusively focused on soil water content, so contains less information on individual topics.
  - Paz et al. (2017): Reviews groundwater case studies and provides suggestions for GPR usage in groundwater applications.
- Review articles on soil structure/mapping:
  - Zajícová and Chuman (2019): Covers GPR background and basic methods, as well as GPR applications for soil stratigraphy, soil water content, and a shorter section soil salinity and roots.
  - Doolittle et al. (2007): This article provides an introduction and explanation to the 'soil suitability' maps developed by the US Department of Agriculture (USDA). These maps show where (in general terms) conventional GPR techniques are likely to be successful and where more recent data analysis methods (early time analysis, full waveform analysis) might be needed. Maps for most of the United States are available on the NRCS - USDA website.
- Review articles on root detection and biomass:
  - Guo et al. (2013): Comprehensive review of studies for detecting individual roots, root mapping, and biomass estimation. Provides basic GPR background and describes limitations and future directions for GPR-based root detection.

9.2 Key journals/conferences

- Journals that often publish articles related to GPR/geophysical soil characterization:
  - *Vadose Zone Journal.*
  - *Soil Science Society of America Journal.*
  - *Geoderma.*
  - *Journal of Environmental and Engineering Geophysics.*
  - *Journal of Applied Geophysics.*
- SAGEEP annual meeting (Symposium on the Application of Geophysics to Engineering and Environmental Problems) has agricultural geophysics sessions that include GPR.
- EAGE (European Association of Geoscientists and Engineers) has multiple meetings each year, and some of these have agricultural geophysics sessions.
- Society of Exploration Geophysicists (SEG) has presentations on agricultural geophysics during some of their meetings.

10 References

Akinsunmade, A., Tomecka-Suchon, S. and Pysz, P. (2019). Correlation between agrotechnical properties of selected soil types and corresponding GPR response, *Acta Geophys.* 67(6), 1913–1919.

Algeo, J., van Dam, R. L. and Slater, L. (2016). Early-time GPR: A method to monitor spatial variations in soil water content during irrigation in clay soils, *Vadose Zone J.* 15(11), 1–9. https://doi.org/10.2136/vzj2016.03.0026.

Al-Qadi, I. L., Leng, Z., Lahouar, S. and Baek, J. (2010). In-place hot-mix asphalt density estimation using ground-penetrating radar, *Transp. Res. Rec.* 2152(1), 19–27.

André, F., Jonard, M. and Lambot, S. (2015). Non-invasive forest litter characterization using full-wave inversion of microwave radar data, *IEEE Trans. Geosci. Remote Sens.* 53(2), 828–840.

Arcone, S. A., Peapples, P. R. and Liu, L. B. (2003). Propagation of a ground penetrating radar (GPR) pulse in a thin-surface waveguide, *Geophysics* 68(6), 1922–1933. https://doi.org/10.1190/1.1635046.

Ardekani, M. R. (2013). Off- and on-ground GPR techniques for field-scale soil moisture mapping, *Geoderma* 200–201, 55–66. https://doi.org/10.1016/j.geoderma.2013.02.010.

Bassuk, N., Grabosky, J., Mucciardi, A. and Raffel, G. (2011). Ground penetrating radar accurately locates tree roots in two soil media under pavement, *Arboricult. Urban For.* 37(4), 160–166.

Benedetto, A. and Tosti, F. (2013). Inferring bearing ratio of unbound materials from dielectric properties using GPR: The case of runaway safety areas. *Airfield and Highway Pavement, Los Angeles*, 1336–1347.

Bittelli, M. (2011). Measuring soil water content: A review, *hortTechnology* 21(3), 293–300.

Bradford, J. H. (2008). Measuring water content heterogeneity using multifold GPR with reflection tomography, *Vadose Zone J.* 7(1), 184–193. https://doi.org/10.2136/vzj2006.0160.

Busch, S., van der Kruk, J. and Vereecken, H. (2014). Improved characterization of fine texture soils using on ground GPS full-waveform inversion, *IEEE Trans. Geosci. Remote Sensing* 52(7), 3947–3958.

Butnor, J. R., Doolittle, J. A., Johnsen, K. H., Samuelson, L., Stokes, T. and Kress, L. (2003). Utility of ground-penetrating radar as a root biomass survey tool in forest systems. *Soil Sci. Soc. Am. J.* 67, 1607–1615. https://doi.org/10.2136/sssaj2003.1607.

Butnor, J. R., Stover, D. B., Roth, B. E., Johnsen, K. H., Day, F. P. and McInnis, D. (2008). 30 using ground-penetrating radar to estimate tree root mass. Handbook of Agricultural Geophysics, p. 375.

Cassidy, N. J. (2009). Electrical and magnetic properties of rocks, soils and fluids. In: Jol, H. M. (Ed.), *Ground Penetrating Radar Theory and Applications*. Elsevier Science, pp. 41–72. https://doi.org/10.1016/B978-0-444-53348-7.00010-7.

Castro, A. C. M., Meixedo, J. P., Santos, J. M., Góis, J., BentoGonçalves, A., Vieira, A. and Lourenço, L. (2015). On sampling collection procedure effectiveness for forest soil characterization, *Flamma* 6, 98–100.

Collins, M. E. and Doolittle, J. A. (1987). Using ground-penetrating radar to study soil microvariability, *Soil Sci. Soc. Am. J.* 51(2), 491–493.

Collins, M. E., Puckett, W. E., Schellentrager, G. W. and Yust, N. A. (1990). Using GPR for micro-analyses of soils and karst features on the Chiefland Limestone Plain in Florida, *Geoderma* 47(1–2), 159–170. https://doi.org/10.1016/0016-7061(90)90053-C.

Cox, K. D., Scherm, H. and Serman, N. (2005). Ground-penetrating radar to detect and quantify residual root fragments following peach orchard clearing, *hortTechnology* 15(3), 600–607.

Cui, X., Chen, J., Shen, J., Cao, X., Chen, X. and Zhu, X. (2011). Modeling tree root diameter and biomass by ground-penetrating radar, *Sci. China Earth Sci.* 54(5), 711–719. https://doi.org/10.1007/s11430-010-4103-z.

Danjon, F. and Reubens, B. (2008). Assessing and analyzing 3D architecture of woody root systems, a review of methods and applications in tree and soil stability, resource acquisition and allocation, *Plant Soil* 303(1–2), 1–34.

Dannoura, M., Hirano, Y., Igarashi, T., Ishii, M., Aono, K., Yamase, K. and Kanazawa, Y. (2008). Detection of Cryptomeria japonica roots with ground penetrating radar, *Plant Biosyst.* 142(2), 375–380. https://doi.org/10.1080/11263500802150951.

Di Matteo, A., Pettinelli, E. and Slob, E. (2013). Early-time GPR signal attributes to estimate soil dielectric permittivity: A theoretical study, *IEEE Trans. Geosci. Remote Sens.* 51(3), 1643–1654. https://doi.org/10.1109/TGRS.2012.2206817.

Dobson, M. C., Ulaby, F. T., Hallikainen, M. T. and El-Rayes, M. A. (1985). Microwave dielectric behavior of wet soils: II. Dielectric mixing models, *IEEE Trans. Geosci. Remote Sens.* 23, 35–46.

Doolittle, J. and Butnor, J. (2009). Soils, peatlands, and biomonitoring. In: Jol, H. M. (Ed.), *Ground Penetrating Radar Theory and Applications*. Elsevier Science, pp. 179–202. https://doi.org/10.1016/B978-0-444-53348-7.00006-5.

Doolittle, J. A., Minzenmayer, F. E., Waltman, S. W., Benham, E. C., Tuttle, J. W. and Peaslee, S. D. (2007). Ground-penetrating radar soil suitability map of the conterminous United States, *Geoderma* 141(3–4), 416–421. https://doi.org/10.1016/j.geoderma.2007.05.015.

Durand, S. and Slodic, M. (2011). Fully discrete finite element method for Maxwell's equations with nonlinear conductivity, *IMA J. Numer. Anal.* 31(4), 1713–1733. http://doi.org/10.1093/imanum/drr007.

Ernst, J. R., Maurer, H., Green, A. G. and Holliger, K. (2007). Full-waveform inversion of crosshole radar data based on 2-D finite-difference time-domain solutions of Maxwell's equations, *IEEE Trans. Geosci. Remote Sens.* 45(9), 2807–2828.

Escorihuela, M. H., Chanzy, A., Wigneron, J. P. and Kerr, Y. H. (2010). Effective soil moisture sampling depth of l-band radiometry: A case study, *Remote Sens. Environ.* 114(5), 995–1001.

Famiglietti, J. S., Ryu, D., Berg, A. A., Rodell, M. and Jackson, T. J. (2008). Field observations of soil moisture variability across scales, *Water Resour. Res.* 44(1), W01423. doi: 10.1029/2006WR005804.

Ferrara, C., Barone, P. M., Steelman, C. M., Pettinelli, E. and Endres, A. L. (2013). Monitoring shallow soil water content under natural field conditions using the early-time GPR signal technique, *Vadose Zone J.* 12(4). https://doi.org/10.2136/vzj2012.0202.

Freeland, R. S., Yoder, R. E. and Ammons, J. T. (1998). Mapping shallow underground features that influence site-specific agricultural production, *J. Appl. Geophys.* 40(1-3), 19–27.

Galagedara, L. W., Parkin, G. W. and Redman, J. D. (2003). An analysis of the ground-penetrating radar direct ground wave method for soil water content measurement, *Hydrol. Processes* 17(18), 3615–3628.

Galagedara, L. W., Parkin, G. W., Redman, J. D., Von Bertoldi, P. and Endres, A. L. (2005a). Field studies of the GPR ground wave method for estimating soil water content during irrigation and drainage, *J. Hydrol.* 301(1–4), 182–197. https://doi.org/10.1016/j.jhydrol.2004.06.031.

Galagedara, L. W., Redman, J. D., Parkin, G. W., Annan, A. P. and Endres, A. L. (2005b). Numerical modeling of GPR to determine the direct ground wave sampling depth, *Vadose Zone J.* 4(4), 1096–1106. https://doi.org/10.2136/vzj2004.0143.

Garambois, S., Senechal, P. and Perroud, H. (2002). On the use of combined geophysical methods to assess water content and water conductivity of near-surface formations, *J. Hydrol.* 259(1–4), 32–48.

Goodman, D., Piro, S., Nishimura, Y., Schneider, K., Hongo, H., Higashi, N., Steinberg, J. and Damiata, B. (2009). GPR archaeometry. In: *Ground Penetrating Radar Theory and Applications*. Elsevier Science. https://doi.org/10.1016/B978-0-444-53348-7.00015-6.

Greaves, R. J., Lesmes, D. P., Lee, J. M. and Toksöz, M. N. (1996). Velocity variations and water content estimated from multi-offset, ground-penetrating radar, *Geophysics* 61(3), 683–695.

Grote, K., Crist, T. and Nickel, C. (2010). Experimental estimation of the GPR groundwave sampling depth, *Water Resour. Res.* 46(10), W10520. https://doi.org/10.1029/2009wr008403.

Grote, K., Hubbard, S. S., Harvey, J. and Rubin, Y. (2005). Evaluation of infiltration in layered pavements using surface GPR reflection techniques, *J. Appl. Geophys.* 57(2), 129–153.

Grote, K., Hubbard, S. S. and Rubin, Y. (2002). GPR monitoring of volumetric water content in soils applied to highway construction and maintenance, *Leading Edge Explor.* 21(5), 482–504.

Grote, K., Hubbard, S. S. and Rubin, Y. (2003). Field-scale estimation of volumetric water content using GPR groundwave techniques, *Water Resour. Res.* 39(11), 1321–1334.

Guo, L., Chen, J., Cui, X., Fan, B. and Lin, H. (2013). Application of ground penetrating radar for coarse root detection and quantification: A review, *Plant Soil* 362(1–2), 1–23. https://doi.org/10.1007/s11104-012-1455-5.

Hirano, Y., Dannoura, M., Aono, K., Igarashi, T., Ishii, M., Yamase, K., Makita, N. and Kanazawa, Y. (2009). Limiting factors in the detection of tree roots using ground-penetrating radar, *Plant Soil* 319(1–2), 15–24. https://doi.org/10.1007/s11104-008-9845-4.

Huisman, J. A., Hubbard, S. S., Redman, J. D. and Annan, A. P. (2003). Monitoring soil water content with ground-penetrating radar: A review, *Vadose Zone J.* 2(4), 476–491.

Huisman, J. A., Sperl, C., Bouten, W. and Verstraten, J. M. (2001). Soil water content measurements at different scales: Accuracy of time domain reflectometry and ground-penetrating radar, *J. Hydrol. (Amsterdam)* 245(1–4), 48–58. https://doi.org/10.1016/s0022-1694(01)00336-5.

Jacobsen, O. H. and Schjønning, P. (1993). Field evaluation of time domain reflectometry for soil water measurements, *J. Hydrol.* 151(2–4), 159–172.

Jadoon, K. Z., Lambot, S., Scharnagl, B., van der Kruk, J., Slob, E. and Vereecken, H. (2010). Quantifying field-scale surface soil water content from proximal GPR signal inversion in the time domain, *Near Surf. Geophys.* 8(6), 483–491.

Jol, H. (2009). *Ground Penetrating Radar Theory and Applications*. Elsevier Science, p. 524. https://doi.org/10.1016/B978-0-444-53348-7.X0001-4.

Jonard, F., Weihermuller, L., Jadoon, K. Z., Schwank, M., Vereecken, H. and Lambot, S. (2011). Mapping field-scale soil moisture with L-band radiometer and ground-penetrating radar over bare soil, *IEEE Trans. Geosci. Remote Sens.* 49(8), 2863–2875.

Jonard, F., Mahmoudzadeh, M., Roisin, C., Weihermüller, L., André, F., Minet, J., Vereecken, H. and Lambot, S. (2013). Characterization of tillage effects on the spatial variation of soil properties using ground-penetrating radar and electromagnetic induction, *Geoderma* 207–208, 310–322.

Karuss, J. and Berzins, D. (2015). Ground-penetrating radar study of the Cena Bog, Latvia: Linkage of reflections with peat moisture content, *Bull. Geol. Soc. Finl.* 87(2), 87–98.

Kettridge, N., Binley, A., Comas, X., Cassidy, N. J., Baird, A. J., Harris, A., Van Der Kruk, J., Strack, M., Milner, A. M. and Waddington, J. M. (2012). Do peatland microforms move through time? Examining the developmental history of a patterned peatland using ground-penetrating radar, *J. Geophys. Res. Biogeosci.* 117(G3). https://doi.org/10.1029/2011JG001876.

Kettridge, N., Comas, X., Baird, A., Slater, L., Strack, M., Thompson, D., Jol, H. and Binley, A. (2008). Ecohydrologically important subsurface structures in peatlands revealed by ground-penetrating radar and complex conductivity surveys, *J. Geophys. Res. Biogeosci.* 113(G4), 1–15. https://doi.org/10.1029/2008JG000787.

Klotzsche, A., Jonard, F., Looms, M. C., van der Kruk, J. and Huisman, J. A. (2018). Measuring soil water content with ground penetrating radar: A decade of progress, *Vadose Zone J.* 17(1). https://doi.org/10.2136/vzj2018.03.0052.

Kummode, S., Thitimakorn, T. and Kupongsak, S. (2020). Determination of the volumetric soil water content of two soil types using ground penetrating radar, *Environ. Asia* 13(2), 78-87. https://doi.org/10.14456/ea/2020.31.

Lambot, S., Antoine, M., Vanclooster, M. and Slob, E. C. (2006). Effect of soil roughness on the inversion of off-ground monostatic GPR signal for noninvasive quantification of soil properties, *Water Resour. Res.* 42(3), 1–10. https://doi.org/10.1029/2005WR004416.

Lambot, S., Slob, E., Chavarro, D., Lubczynski, M. and Vereecken, H. (2008). Measuring soil surface water content in irrigated areas of southern Tunisia using full-waveform inversion of proximal GPR data, *Near Surf. Geophys.* 6(6), 403–410.

Lambot, S., Slob, E. C., van den Bosch, I., Antoine, M., Gregoire, M. and Vanclooster, M. (2004). Modeling of GPR signal and inversion for identifying the subsurface dielectric properties frequency dependence and effect of soil roughness. In: *Proceedings of the Tenth International Conference Grounds Penetrating Radar* (vol. 2004), GPR, pp. 79–82. https://doi.org/10.1109/ICGPR.2004.179918.

Leucci, G. (2010). The use of three geophysical methods for 3D images of total root volume of soil in urban environments, *Explor. Geophys.* 41(4), 268–278. https://doi.org/10.1071/EG09034.

Li, L., Xia, Y., Liu, S., Zhang, W., Chen, X., Zheng, H., Qiu, H., He, X. and Su, Y. (2015). Modified method for estimating organic carbon density in discontinuous karst soil using ground-penetrating radar and geostatistics, *J. Mt. Sci.* 12(5), 1229–1240. https://doi.org/10.1007/s11629-015-3431-z.

Liu, X., Chen, J., Cui, X., Liu, Q., Cao, X. and Chen, X. (2019). Measurement of soil water content using ground-penetrating radar: A review of current methods, *Int. J. Digit. Earth* 12(1), 95–118. https://doi.org/10.1080/17538947.2017.1412520.

Liu, X., Dong, X. and Leskovar, D. I. (2016). Ground penetrating radar for underground sensing in agriculture: A review, *Int. Agrophys.* 30(4), 533–543. https://doi.org/10.1515/intag-2016-0010.

Lu, Y., Song, W., Lu, J., Wang, X. and Tan, Y. (2017). An examination of soil moisture estimation using ground penetrating radar in dessert steppe, *Water* 9(7), 1–11. https://doi.org/10.3390/w9070521.

Lunt, I. A., Hubbard, S. S. and Rubin, Y. (2005). Soil moisture content estimation using ground-penetrating radar reflection data, *J. Hydrol.* 307(1–4), 254–269.

Minet, J., Bogaert, P., Vanclooster, M. and Lambot, S. (2012). Validation of ground penetrating radar full-waveform inversion for field scale soil moisture mapping, *J. Hydrol.* 424–425, 112–123. https://doi.org/10.1016/j.jhydrol.2011.12.034.

Minet, J., Lambot, S., Slob, E. C. and Vanclooster, M. (2010). Soil surface water content estimation by full-waveform GPR signal inversion in the presence of thin layers, *IEEE Trans. Geosci. Remote Sens.* 48(3), 1138–1150.

Minet, J., Wahyudi, A., Bogaert, P., Vanclooster, M. and Lambot, S. (2011). Mapping shallow soil moisture profiles at the field scale using full-waveform inversion of ground penetrating radar data, *Geoderma* 161, 225–237. https://doi.org/10.1016/j.geoderma.

Mourmeaux, N., Tran, A. P. and Lambot, S. (2014). Soil permittivity and conductivity characterization by full-wave inversion of near-field GPR data. In: *Proceedings of the 15th International Conference on Ground Penetrating Radar*, pp. 497–502.

Muñiz, E., Shaw, R. K., Gimenez, D., Williams, C. A. and Kenny, L. (2016). Use of ground-penetrating radar to determine depth to compacted layer in soils Under pasture. In: Hartemink, A. and Minasny, B. (Eds.), *Digital Soil Morphometrics: Progress in Soil Science*. Cham: Springer. https://doi.org/10.1007/978-3-319-28295-4_26.

Nakashima, Y., Zhou, H. and Sato, M. (2001). Estimation of groundwater level by GPR in an area with multiple ambiguous reflections, *J. Appl. Geophys.* 47(3–4), 241–249.

Novakova, E., Karous, M., Zajíček, A. and Karousova, M. (2013). Evaluation of ground penetrating radar and vertical electrical sounding methods to determine soil horizons and bedrock at the locality Dehtáře, *Soil Water Res.* 8(3), 105–112.

Oimbe, S., Ingle, R. and Awale, R. (2018). Detection of soil water content using continuous wave ground penetrating radar, *JOIV: Int. J. Inform. Visualization* 2(1), 44–50.

Olhoeft, G. R. (2000). Maximizing the information return from ground penetrating radar, *J. Appl. Geophys.* 43(2–4), 175–187.

Pallavi, B., Saito, H. and Kato, M. (2011). On mapping surface moisture content of Japanese andisol using GPR, *IEEE J. Sel. Top. Appl. Earth Obs. Remote Sens.* 4(4), 804–808.

Pan, X., Zhang, J., Huang, P. and Roth, K. (2012). Estimating field-scale soil water dynamics at a heterogeneous site using multi-channel GPR, *Hydrol. Earth Syst. Sci.* 16(11), 4361–4372. https://doi.org/10.5194/hess-16-4361-2012.

Paz, C., Alcalá, F. J., Carvalho, J. M. and Ribeiro, L. (2017). Current uses of ground penetrating radar in groundwater-dependent ecosystems research, *Sci. Total Environ.* 595, 868–885. https://doi.org/10.1016/j.scitotenv.2017.03.210.

Pettinelli, E., Di Matteo, A., Beaubien, S. E., Mattei, E., Lauro, S. E., Galli, A. and Vannaroni, G. (2014). A controlled experiment to investigate the correlation between early-time signal attributes of ground-coupled radar and soil dielectric properties, *J. Appl. Geophys.* 101, 68–76. https://doi.org/10.1016/j.jappgeo.2013.11.012.

Pettinelli, E., Vannaroni, G., Di Pasquo, B., Mattei, E., Di Matteo, A., De Santis, A. and Annan, P. A. (2007). Correlation between near-surface electromagnetic soil parameters and early-time GPR signals: An experimental study, *Geophysics* 72(2), A25–A28. https://doi.org/10.1190/1.2435171.

Plado, J., Sibul, I., Mustasaar, M. and Jõeleht, A. (2011). Ground-penetrating radar study of the Rahivere peat bog, eastern Estonia. *Est. J. Earth Sci.* 60(1), 31. https://doi.org/10.3176/earth.2011.1.03.

Plati, C. and Loizos, A. (2013). Estimation of in-situ density and moisture content in HMA pavements based on GPR trace reflection amplitude using different frequencies, *J. Appl. Geophys.* 97, 3–10.

Redman, J. D., Davis, J. L., Galagedara, L. W. and Parkin, G. W. (2002). Field studies of GPR air-launched surface reflectivity measurements of soil water content. In: *Proceedings of the Ninth Conference on Ground-Penetrating Radar, Proc. SPIE* 4758, 156–161.

Reppert, P. M., Morgan, F. D. and Toksoz, M. N. (2000). Dielectric constant determination using ground-penetrating radar Reflection coefficients, *J. Appl. Geophys.* 43(2–4), 189–197.

Reynolds, J. (2011). *An Introduction to Applied and Environmental Geophysics*. West Sussex: Wiley-Blackwell.

Robinson, D. A. (2004). Measurement of the solid dielectric permittivity of clay minerals and granular samples using a time domain reflectometry immersion method, *Vadose Zone J.* 3(2), 705–713.

Robinson, D. A., Campbell, C. S., Hopmans, J. W., Hornbuckle, B. K., Jones, S. B., Knight, R., Ogden, F., Selker, J. and Wendroth, O. (2008). Soil moisture measurement for ecological and hydrological watershed-scale observatories: A review, *Vadose Zone J.* 7(1), 358–389.

Robinson, D. A., Gardner, C. M. K. and Cooper, J. D. (1999). Measurement of relative permittivity in sandy soils using TDR, capacitance and theta probes: Comparison, including the effects of bulk soil electrical conductivity, *J. Hydrol.* 223(3–4), 198–211.

Roth, C. H., Malicki, M. A. and Plagge, R. (1992). Empirical evaluation of the relationship between soil dielectric constant and volumetric water content as the basis for calibrating soil moisture measurements by TDR, *J. Soil Sci.* 43(1), 1-13.

Roth, K., Schulin, R., Flühler, H. and Attinger, W. (1990). Calibration of time domain reflectometry for water content measurements using a composite dielectric approach, *Water Resour. Res.* 26(10), 2267-2273.

Serbin, G. and Or, D. (2003). Near-surface soil water content measurements using horn antenna radar: Methodology and overview, *Vadose Zone J.* 2(4), 500-510.

Serbin, G. and Or, D. (2005). Ground-penetrating radar measurements of crop and surface water content dynamics, *Remote Sens. Environ.* 96(1), 119-134.

Seyfried, D. and Schoebel, J. (2016). Ground penetrating radar for asparagus detection, *J. Appl. Geophys.* 126, 191-197. https://doi.org/10.1016/j.jappgeo.2016.01.022.

Simeoni, M. A., Galloway, P. D., O'Neil, A. J. and Gilkes, R. J. (2009). A procedure for mapping the depth to the texture contrast horizon of duplex soils in south-Western Australia using ground penetrating radar, GPS and kriging, *Soil Res.* 47(6), 613-621. https://doi.org/10.1071/SR08241.

Slater, L. D. and Reeve, A. (2002). Investigating peatland stratigraphy and hydrogeology using integrated electrical geophysics, *Geophysics* 67(2), 365-378. https://doi.org/10.1190/1.1468597.

Slob, E., Sato, M. and Olhoeft, G. (2010). Surface and borehole ground-penetrating radar developments, *Geophysics* 75(5), 75A103-75A120.

Steelman, C. M. and Endres, A. L. (2010). An examination of direct ground wave soil moisture monitoring over an annual cycle of soil conditions, *Water Resour. Res.* 46(11), W11533. https://doi.org/10.1029/2009wr008815.

Steelman, C. M. and Endres, A. L. (2011). Comparison of petrophysical relationships for soil moisture estimation using GPR ground waves, *Vadose Zone J.* 10(1), 270-285.

Stoffregen, H., Zenker, T. and Wessolek, G. (2002). Accuracy of soil water content measurements using ground penetrating radar: Comparison of ground penetrating radar and lysimeter data, *J. Hydrol.* 267(3-4), 201-206.

Stokes, A., Fourcaud, T., Hruška, J., Čermák, J., Nadezhdina, N., Nadyezhdin, V. and Praus, L. (2002). An evaluation of different methods to investigate root system architecture of urban trees in situ: 1. Ground-penetrating radar, *AUF* 28(1), 1-10.

Stover, D. B., Day, L. F., Butnor, J. R. and Drake, B. G. (2007). Effect of elevated CO_2 on coarse-root biomass in Florida scrub detected by ground-penetrating radar, *Ecology* 88(5), 1328-1334. https://doi.org/10.1890/06-0989.

Strobach, E., Harris, B. D., Dupuis, J. C. and Kepic, A. W. (2014). Time-lapse borehole radar for monitoring rainfall infiltration through podosol horizons in a sandy vadose zone, *Water Resour. Res.* 50(3), 2140-2163. https://doi.org/10.1002/2013wr014331.

Sucre, E. B., Tuttle, J. W. and Fox, T. R. (2011). The use of ground-penetrating radar to accurately estimate soil depth in rocky forest soils, *Forest Sci.* 57, 59-66.

Topp, G. C., Davis, J. L. and Annan, A. P. (1980). Electromagnetic determination of soil water content: Measurements in coaxial transmission lines, *Water Resour. Res.* 16(3), 574-582.

Tran, A. P., Bogaert, P., Wiaux, F., Vanclooster, M. and Lambot, S. (2015). High-resolution space-time quantification of soil moisture along a hillslope using joint analysis of ground penetrating radar and frequency domain reflectometry data, *J. Hydrol.* 523, 252-261. https://doi.org/10.1016/j.jhydrol.2015.01.065.

Tronicke, J. and Hamann, G. (2014). Vertical radar profiling: Combined analysis of travel times, amplitudes, and reflections, *Geophysics* 79(4), H23-H35. https://doi.org/10.1190/geo2013-0428.1.

Turesson, A. (2006). Water content and porosity estimated from ground-penetrating radar and resistivity, *J. Appl. Geophys.* 58(2), 99–111.

van Dam, R. L., van den Berg, E. H., Schaap, M. G., Broekema, L. H. and Schlager, W. (2003). Radar reflections from sedimentary structures in the vadose zone, *Ground Penetrating Radar Sediments* 211(1), 257–273.

van der Kruk, J., Jacob, R. W. and Vereecken, H. (2010). Properties of precipitation induced multi-layer surface waveguides derived from inversion of dispersive TE and TM GPR Data, *Geophysics* 75(4), WA263–WA273. https://doi.org/10.1190/1.3467444.

van Overmeeren, R. A., Sariowan, S. V. and Gehrels, J. C. (1997). Ground penetrating radar for determining volumetric soil water content; results of comparative measurements at two test sites, *J. Hydrol.* 197(1–4), 316–338.

Vischel, T., Pegram, G. G. S., Sinclair, S., Wagner, W. and Bartsch, A. (2008). Comparison of soil moisture fields estimated by catchment modeling and remote sensing: A case study in South Africa, *Hydrol. Earth Syst. Sci.* 12(3), 751–767.

Wang, P., Hu, Z., Zhao, Y. and Li, X. (2016). Experimental study of soil compaction effects on GPR signals, *J. Appl. Geophys.* 126, 128–137.

Weihermüller, L., Huisman, J. A., Lambot, S., Herbst, M. and Vereecken, H. (2007). Mapping the spatial variation of soil water content at the field scale with different ground penetrating radar techniques, *J. Hydrol.* 340(3–4), 205–216. https://doi.org/10.1016/j.jhydrol.2007.04.013.

Weiler, K. W., Steenhuis, T. S., Boll, J. and Kung, K.-J. S. (1998). Comparison of ground penetrating radar and time domain reflectomery as soil water sensors, *Soil Sci. Soc. Am. J.* 62(5), 1237–1239.

Wielopolski, L., Hendrey, G., Daniels, J. J. and McGuigan, M. (2000). Imaging tree root systems in situ. In: Noon, D. A., Stickley, G. F. and Longstaff, D. (Eds.), *Proceedings of the 8th International Conference* Ground *Penetrating* Radar, Gold Coast, Queensland, Australia, 23–26 May 2000.

William, W. and Comas, X. (2016). Estimating methane gas production in peat soils of the Florida Everglades using hydrogeophysical methods, *J. Geophys. Res.* 121(4), 1190–1202. https://doi.org/10.1002/2015JG003246.

Winkelbauer, J., Völkel, J., Leopold, M. and Bernt, N. (2011). Methods of surveying the thickness of humous horizons using ground penetrating radar (GPR): An example from the Garmisch-Partenkirchen area of the Northern Alps, *Eur. J. Forest Res.* 130(5), 799–812.

Wollschläger, U. and Roth, K. (2005). Estimation of temporal changes of volumetric soil water content from ground-penetrating radar reflections, *Subsurface Sens. Tech. Appl.* 6(2), 207–218.

Yamase, K., Tanikawa, T., Dannoura, M., Ohashi, M., Todo, C., Ikeno, H., Aono, K. and Hirano, Y. (2018). Ground-penetrating radar estimates of tree root diameter and distribution under field conditions, *Trees* 32(6), 1657–1668. https://doi.org/10.1007/s00468-018-1741-9.

Yao, Q. and Qifu, W. (2012). Kirchoff migration algorithm for ground penetrating radar data. In: *International Conference on Computer Science and Electronics Engineering* (vols. 396–398). https://doi.org/10.1109/iccsee.2012.256.

Zajícová, K. and Chuman, T. (2019). Application of ground penetrating radar methods in soil studies: A review, *Geoderma* 343, 116–129. https://doi.org/10.1016/j.geoderma.2019.02.024.

Zenone, T., Morelli, G., Teobaldelli, M., Fischanger, F., Matteucci, M., Sordini, M., Armani, A., Ferre, C., Chiti, T. and Seufert, G. (2008). Preliminary use of ground-penetrating radar and electrical resistivity tomography to study tree roots in pine forests and poplar plantations, *Funct. Plant Biol.* 35(10), 1047–1058. https://doi.org/10.1071/FP08062.

Zhang, D., Tang, R., Wu, H., Shao, K. and Li, Z. (2014a). Surface soil water content estimation from thermal remote sensing based on the temporal variation of land surface temperature, *Remote Sens.* 6(4), 3170–3187. http://dx.doi.org.libproxy.mst.edu/10.3390/rs604317.

Zhang, J., Lin, H. and Doolittle, J. (2014b). Soil layering and preferential flow impacts on seasonal changes of GPR signals in two contrasting soils, *Geoderma* 213, 560–569. https://doi.org/10.1016/j.geoderma.2013.08.035.

Zheng, L., Liu, Z., Wang, G. and Zhang, Z. (2016). Research on application of Hilbert transform in radar signal simulation. In: *Proceedings of the 7th International Conference on Environment and Engineering Geophysics & Summit Forum of Chinese Academy of Engineering on Engineering Science and Technology*. https://doi.org/10.2991/iceeg-16.2016.93.

Zhu, S., Huang, C., Su, Y. and Sato, M. (2014). 3D ground penetrating radar to detect tree roots and estimate root biomass in the field, *Remote Sens.* 6(6), 5754–5773. https://doi.org/10.3390/rs6065754.

Chapter 6

Using proximal electromagnetic/electrical resistivity/electrical sensors to assess soil health

Alain Tabbagh, Sorbonne Université, CNRS, EPHE, UMR7619, Métis, 4 place Jussieu 75252 Paris CEDEX 05, France; and Seger Maud and Cousin Isabelle, INRAE, Centre Val de Loire, UR0272 SOLS, 2163 Avenue de la Pomme de Pin, CS40001 Ardon, F-45075 Orléans Cedex 2, France

1 Introduction

'Caring for Soil is Caring for Life' is the title of the mission proposed by the Soil Health and Food Mission Board of the European Union (EU, 2020). Summarizing this Mission objective, the European Soil Strategy for 2030 - 'Reaping the benefits of healthy soils for people, food, nature and climate' - aims at developing concrete actions of the protection, restoration, and sustainable use of soils, in synergy with other EU Green Deal policies, and especially to achieve healthy soils by 2050. Among its concrete actions, the EU is currently developing a new Soil Health Law by 2023, which should help in reaching the aim that '75% of soils are healthy by 2030 and are able to provide essential ecosystem services'. As central components of ecosystem, soils indeed contribute to essential ecosystem services (biomass production, climate regulation, flood control, availability of water, and nutrients for crops), when neither their functions nor their properties, say their essential

http://dx.doi.org/10.19103/AS.2022.0107.13

physicochemical-biological characteristics (Fig. 1), are not degraded. However, up to 60–70% of European soils are actually considered to be unhealthy, and reaching the highly ambitious objectives of the European Soil Strategy necessitates as a first step to precise both the soil health concept and its associated indicators.

In order to raise awareness about the multi-functionality of soil, the soil quality concept is being defined for decades. Karlen et al. (2001) basically expressed it as 'the capacity of a specific kind of soil to function, within natural or managed ecosystem boundaries, to sustain plant and animal productivity, maintain or enhance water and air quality, and support human health and habitation'. To enlarge the soil quality definition, the soil health concept has then been introduced and characterizes 'the capacity of soil to function as a vital living system to sustain biological productivity, promote environmental quality, and maintain plant and animal health' (Doran and Zeiss, 2000). It then refers explicitly to environmental health, say to some biotic components of soils. While the actual definitions of both soil quality and soil health remain controversial, one may however consider that, to some extent, soil quality may be more related to basic inherent soil characteristics (e.g. clay content and stone content), whereas soil health may be more related to some manageable characteristics (e.g. pH and cation exchange capacity), say to soil functioning and agricultural practices.

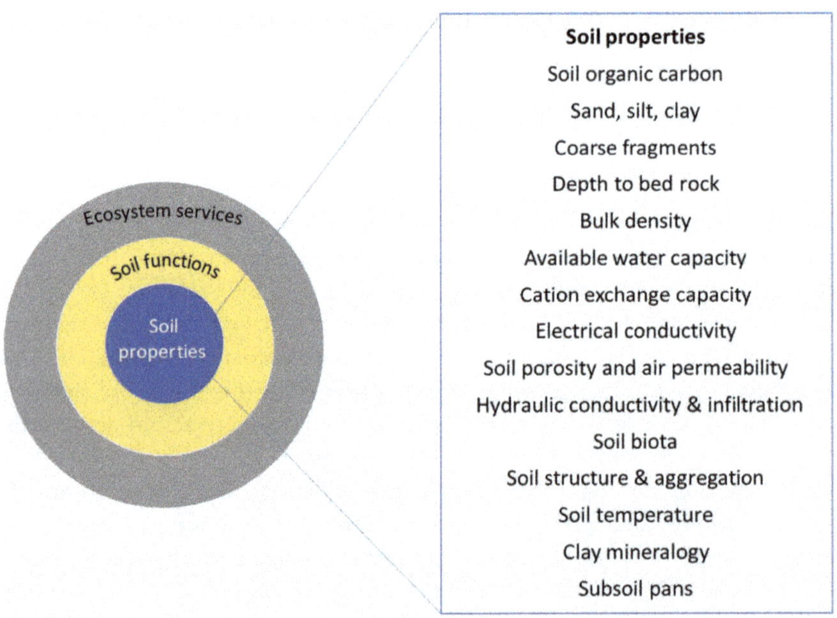

Figure 1 Soil properties involved in the evaluation of soil functions and soil services.

For both soil quality and soil health concepts, complex indicators including several soil characteristics have been defined to take into account the holistic approach associated to these concepts (Rinot et al., 2019), in order to score soil types according to the different functions they fulfill (Karlen et al., 2003), despite the huge variability of soil types and their associated functions (Nortcliff, 2002). For example, Sanchez-Navarro et al. (2015) have developed an integrated index to characterize Mediterranean soils, whereas Obade and Lal (2016a,b) have developed a standardized soil quality index to compare management practices in US agriculture. Soil health and soil quality indexes are developed for a large range of soil use, including cropped soil, pasture (Amorim et al., 2020), forestry (Zornoza et al., 2007), and urban soils (Howard et al., 2016). Basically, soil health/quality indicators mix physical, chemical, and biological characteristics into an integrated index, including inherent and manageable characteristics. For example, Obade and Lal (2016a) developed an index using soil bulk density, available water capacity, pH, electrical conductivity, and C/N ratio, whereas Raiesi (2017) also added the aggregate mean weight diameter, electrical conductivity (EC), microbial biomass, C and N mineralization, urea, alkaline and acid phosphatase, invertase, and arysulfatase. In some cases, more specific soil characteristics can be used, like in the Soil Physical Health approach developed by Amirinejad et al. (2011), who took into account bulk density, saturated hydraulic conductivity, available water capacity, and organic carbon content. However, in most cases, a recurring physicochemical soil characteristic used in the evaluation of soil health indexes is the measurement of the soil electrical conductivity (see, for example, Arnold et al. (2005); Arshad and Martin (2002); Islam et al. (2003); (Ye et al., 2021)); specific sensors have even been developed for a direct measurement of several soil health parameters, including electrical conductivity (Dudala et al., 2020; Liebig et al. 1996; Svoray et al., 2015). The evaluation of soil health status using geophysical tools then remains attractive, while numerous soil characteristics or properties, as listed in Fig. 1, may be evaluated by geophysics.

Table 1 presents a non-exhaustive list of published studies addressing soil characteristics/properties by geophysical tools.

To basically evaluate soil health, each individual soil characteristic can be evaluated, and an integrated index can be calculated (Rinot et al., 2019), either at the local scale (Takoutsing et al., 2016) or over large areas (Svoray et al., 2015; Verma et al., 2008), usually by using data fusion approaches (Grunwald et al., 2015; Veum et al., 2017). The objective of this chapter is then to analyze how some geophysical tools (electromagnetic/electrical), which, basically, provide integrated measurements of soil characteristics, can be used to assess some specific characteristics involved in the evaluation of soil health. Section 2 is dedicated to the presentation of some theoretical elements of subsurface geophysics in order to precisely define the concepts supporting

Table 1 Non-exhaustive reference list studies addressing soil properties by proximal sensing (electrical resistivity or electrical conductivity measurements)

Soil Properties	References
Soil organic carbon	Martinez et al. (2009), De Benedetto et al. (2022)
Sand, silt, clay	Stepien et al. (2015), Ciampalini et al. (2015)
Coarse fragments	Tetegan et al. (2012), Rossi et al. (2013), Rey et al. (2006)
Depth to bed rock	Yamakawa et al. (2012), Bourennane et al. (2017)
Bulk density/compaction	Bertermann and Schwarz (2018), Ren et al. (2022), Kowalczyk et al. (2014)
Electrical conductivity	*Direct measurement*
Soil porosity and air permeability	Fu et al. (2021), Friedman (2005)
Hydaulic conductivity & infiltration	Doussan and Ruy (2009), Clément et al. (2009)
Soil structure & aggregation	Romero-Ruiz et al. (2019), Rossi et al. (2013)
Soil temperature	Besson et al. (2008), Ma et al. (2011)
Clay mineralogy	
Subsoil pans	Jeřábek et al. (2017)
Available water capacity	Abdu et al. (2008), Cousin et al. (2022)
Cation exchange capacity	
Soil biota	Joschko et al. (2010) (earthworm), Amato et al. (2009) (root zone)

electromagnetic and electrical measurements. Section 3 presents different measurement techniques, at both the local and the spatial scale. Section 4 demonstrates the use of the described geophysical tools to evaluate some soil characteristics involved in soil health measurements. Finally, Section 5 is dedicated to a critical evaluation of the potential of electromagnetic, electrical, and impedance measurements for a direct or indirect evaluation of soil health, as well as some future research directions on this topic.

2 Soil physical properties involved in electrical and electromagnetic domains

2.1 Definitions: conductivity, permittivity, and magnetic susceptibility

Electric charges can move in reaction to the application of an electric field, $\vec{E}$. The corresponding electrical current, $\vec{i}$, that is the vector quantity of electric charges crossing the unit surface during the unit of time is proportional to it. One has the Ohm's law, $\vec{i} = \sigma\vec{E}$, where σ is the electrical conductivity, whose unit is Siemens by meter ($S.m^{-1}$). If the medium is not isotropic, the conductivity is no more expressed by a simple scalar value but by a tensor. In geophysics, one often uses the electrical resistivity, which is the inverse of the electrical conductivity ($\rho = \frac{1}{\sigma}$), whose unit is ohm.meter (Ω.m).

If the electric charges cannot move at macroscopic scale, their displacements are limited to microscopic motions, and an electrical polarization, $\vec{P}$, exists in the medium where the location of the barycenter of the positive charges no more coincides with the barycenter of the negative charges. A couple of a positive and of a negative charge at small distance forms an electric dipole. $\vec{P}$ expresses the density of electric dipoles in the unit volume. The corresponding property is the dielectric permittivity, ε, defined by:

$$\varepsilon_0 \vec{E} + \vec{P} = \varepsilon \vec{E} = \varepsilon_0 \varepsilon_r \vec{E},$$

with ε_0 being the free space permittivity. The unit of ε and ε_0 is Farad by meter ($F.m^{-1}$), with $\varepsilon_0 = 8.84 \times 10^{-12}$ $F.m^{-1}$, and ε_r, the relative permittivity, is without unit.

In reaction to the application of a magnetic field, $\vec{H}$, the medium acquires a magnetization, $\vec{M} = \kappa \vec{H}$, which corresponds to the density of magnetic dipoles per unit of volume. κ is named the magnetic susceptibility, which is without unit (but its value depends on the considered system of unit). Another property, the magnetic permeability, μ, can be used, which is related to the susceptibility by $\mu = \mu_0 (1 + \kappa)$, where μ_0 is the vacuum magnetic permeability.

Any variation with time generates a coupling between electric and magnetic fields which are described by Maxwell equations by the end of the nineteenth century. Thus, three different properties σ, ε, and κ must be taken into account when considering electromagnetic phenomena.

However, both the electrical polarization and the magnetization are subject to the relaxation mechanism which corresponds to the fact that a time constant, a delay, τ, is necessary to reach the polarization (respectively the magnetization) final value, $\overrightarrow{P_0}$. If the electric field variation corresponds to a step function, the time variation of the polarization is thus written:

$$\overrightarrow{P(t)} = \overrightarrow{P_0} \left(1 - e^{-\frac{t}{\tau}} \right).$$

Consequently in the frequency domain, ε_r and κ must be considered as complex quantities with a real part and an imaginary part expressing the delay: $\varepsilon_r = \varepsilon_r' - i\varepsilon_r''$ and $\kappa = \kappa' - i\kappa''$ (with $i = \sqrt{-1}$). Due to the complex structure of soils, several relaxation mechanisms can take place, and a possibly large distribution of the time constant values must be considered. The same reasoning applied to the conductivity would lead to $\sigma = \sigma' + i\sigma''$ (with a sign '+' in order to be coherent with the Maxwell-Ampère equation).

2.2 Approximations resulting from considered time and space scales

The behavior of the conduction and polarization phenomena follows the Maxwell-Ampère equation,

$$\nabla \times \vec{H} = \sigma\mu\vec{E} + \varepsilon\mu\frac{\partial \vec{E}}{\partial t}.$$

Its right-hand side contains two terms: the direct current ($\sigma\vec{E}$) and the displacement current ($\varepsilon\frac{\partial \vec{E}}{\partial t}$), which can have significantly different magnitudes. The comparison between them is simpler in the frequency domain, with ω being the angular frequency, where both properties are grouped in one expression: $(\sigma + i\omega\varepsilon)$ where the displacement current term can be seen as the imaginary part of the total current which can be descried by $\sigma = \sigma' + i\sigma''$. The second term depends on the frequency and would be negligible for lower frequencies: for instance, if $\sigma = 0.02$ S.m^{-1}, $\omega = 2\pi \times 10^3$ rd.s^{-1}, and $\varepsilon_r = 500$, one has $\omega\,\varepsilon_0\varepsilon_r = 0.28$ 10^{-4} S.m^{-1} which can be neglected in comparison with σ. This approximation is totally independent of the space scale.

The behavior of the electric field (respectively the magnetic field) follows in the frequency domain the equation:

$$\nabla\vec{E} - \left(i\sigma\mu\omega - \varepsilon\mu\omega^2\right)\vec{E} = 0.$$

In the general case, the three properties take place in the quantity, $\gamma^2 = i\sigma\mu\omega - \varepsilon\mu\omega^2$. If it reduces to: $\gamma^2 = i\sigma\mu\omega$, in the so-called 'induction case', σ and μ only take place. If $\omega = 0$, that is in the absence of any time variation (static or steady case) $\gamma^2 = 0$: only σ takes place through the application of the Ohm's law and there is no more coupling between the electric and the magnetic fields. In the induction case, if L is the characteristic length (size of the instrument, distance to the targets, and thickness of the layers), the main parameter determining the field behaviors is the so-called 'induction number': $IN = \sigma\mu\omega L^2$.

2.3 Specificities of porous multi-phase media, formation factor, electrical double layer on grain surfaces, link with the water content and water mineralization, and spectral dependence of the properties

Soils contain gas (mainly air), fluids (mainly water), and solid particles (minerals and organic matter) - what can be the moving electric charges? The relationship between the resistivity and the temperature which is linear and increasing, $\rho(T)=\rho(T_0)(1+\alpha(T-T_0))$ where $\alpha \cong 0.02$ (Campbell et al., 1948), demonstrates that they are ions and not electrons (with possible exceptions in the presence of sulfides). These ions are free in the water phase or coating the mineral grain surface in a double-negative and double-positive layer where the outer cations can move tangentially at the surface. The first case corresponds to a volume conductivity, σ_v, and the second case a surface conductivity, σ_s. The resultant total conductivity is the sum of both (Rhoades et al., 1976). The relative importance of these two mechanisms depends on the specific surface: the surface conductivity is negligible for the sand fraction but is very important for the clay fraction. If the soil water has a high salt content, the volume conductivity will dominate, but for fresh water the surface conductivity will dominate. In any case, the total conductivity will increase with the water content. The volumes of gas and mineral grains are insulating.

In the absence of a clay fraction, only the volume conductivity acts. If the path of the moving ions would be straight, the formation factor, F, the ratio between the bulk resistivity ρ_b and the water resistivity ρ_w, would be inversely proportional to the volumetric water content θ. In fact, the paths are tortuous and then longer, and the ratio follows the empirical Archie's law:

$$F=\frac{\rho_b}{\rho_w}=\theta^{-m},$$

where m is called the cementation factor which depends on the shape of the solid grains and is comprised between 1.3 and 2.5. For instance, a sandy soil with a 30% content of 20 Ω.m fresh water would have a 96 Ω.m bulk resistivity if m = 1.3. The same soil with a 1 Ω.m brackish water would have a 4.8 Ω.m bulk resistivity. In the presence of a clay fraction, one has: $\sigma_b=\frac{\sigma_w}{F}+\sigma_s$.

In the lower frequency domain considered in this chapter, below 1 MHz, various relaxation phenomena may take place in the electrical polarization: the rotation of the polar water molecule (as in higher frequencies, see Chapter 11) and the limitation of ion motions due either to the shapes and sizes of the pore and connecting throats or to the sizes and shapes of clay platelets. This complex structure induces a large time constant distribution which generates a decrease of the real part of the permittivity as the frequency, f, increases. The slope of the decrease follows a statistical law $\left(\frac{\varepsilon'(f)}{\varepsilon'(f_0)}\right)^{-n}$, if f_0 is taken as the reference, where the exponent, n, depends on the water content. Values of ε' between 100 and 10 000 are usual in soils at 10 kHz frequency (Tabbagh et al., 2021).

3 Measurement techniques

3.1 The electrical method: quadripole principle, profiling, sounding, fixing electrical resistivity tomography, and moving multipole instruments

In the electrical method, also called the resistivity method (Samouëlian et al., 2005), a direct electrical current (DC) is injected in the soil through two metallic stakes, the A and B electrodes. However, the simple measurement of the ratio between the applied voltage and the injected current intensity would only give access to the sum of the two contact resistances. To determine the soil electrical resistivity, the use of four electrodes is necessary. If a current, I, is injected in a homogeneous soil by a point electrode A, the current density vector is radially isotropic and of amplitude $i_r = \frac{I}{2\pi r^2}$ at distance r. The corresponding electric field is thus $E_r = \frac{\rho I}{2\pi r^2}$ and the corresponding potential $V = \frac{\rho I}{2\pi r} + V_0$, V_0 being the potential when $r \rightarrow \infty$. To generate the current flow a $-I$ current must be injected in B and between two other different points M and N at the soil surface one have a voltage difference:

$$\nabla V = V_M - V_N = \frac{\rho I}{2\pi}\left[\frac{1}{MA} - \frac{1}{MB} - \frac{1}{NA} + \frac{1}{NB}\right]$$

The resistivity value, ρ, is thus proportional to the ratio $\frac{\nabla V}{I}$ by a factor, K, resulting from the geometrical configuration of the A, B, M, and N quadripole: $= \frac{2\pi}{\frac{1}{MA} - \frac{1}{MB} - \frac{1}{NA} + \frac{1}{NB}}$. In reality, the soil is never homogeneous, and the above formula allows to express any result by the so-called 'apparent resistivity': the resistivity that would have a homogeneous ground giving the same $\frac{\nabla V}{I}$ with the same electrode arrangement, i.e. the same array. This allows the expression of any measurement result by only one quantity and facilitates the comparison between the results acquired with different arrays or with electromagnetic (EM) methods.

Depending of the objective, an infinity of arrangements of four electrodes over the ground surface can be designed. For soil studies at metric scale, one commonly uses either the Wenner array, where the four electrodes are in-line and equally separated by the distance a; thus, $K = 2\pi a$, or the square array where $K = 10.72a$ (a being the side of the square). The investigation depth depends on the geometry of the electrodes, and it can be approximated by the shortest distance existing between an injection electrode and a voltage electrode.

For mapping the resistivity at ground surface, a fixed-size array can be moved along parallel profiles, but it is also possible to achieve a vertical (one dimension, 1D) investigation, called a sounding, by increasing the array extent while keeping the central position to increase the thickness of the terrain taken into account in the measurement. The electrical method offers thus the possibility to describe the soil resistivity variations in 2D or 3D by combining a lateral displacement and an increase of the inter-electrode distance. This can be achieved either by a series of pre-set fixed electrodes spread over the studied surface or by moving a multipole array. In the former, the measuring instrument successively switches the predefined four electrodes arrays, as illustrated in Fig. 2. This technique is called electrical resistivity tomography (ERT) and can be applied over an in-line arrangement of electrodes, two-dimensional ERT, or over a surface arrangement of electrodes, three-dimensional ERT. The limited depth of investigation required for soil studies facilitates the use of mobile multipole pulled in the field by a quad or an

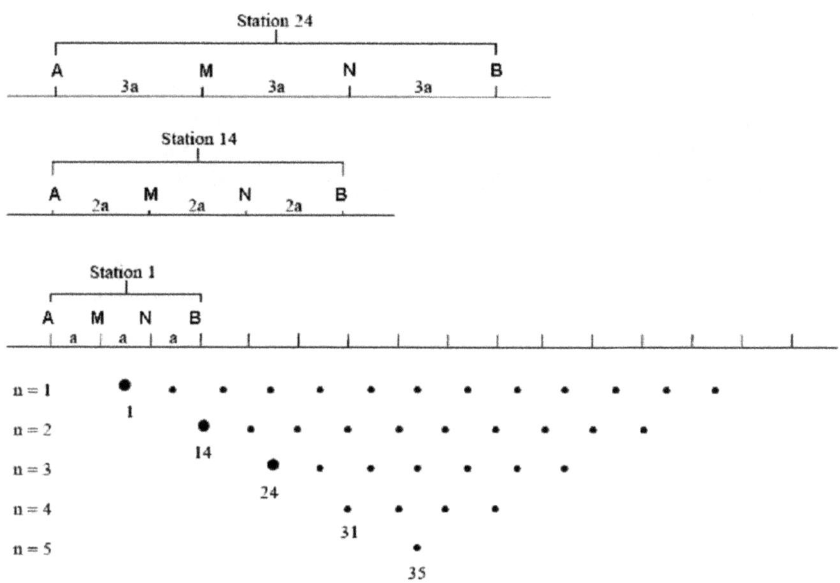

Figure 2 Two-dimensional tomography principle for a line of 16 electrodes. For the first investigation level, $n = 1$, by switching electrodes from the left to the right of the line, 13 successive measurements are achieved with *a* inter-electrode Wenner array by moving 'station 1'. In the 'pseudo-section', the apparent resistivity are affected to the *a* pseudo-depth. Then, for $n = 2$, 10 successive Wenner arrays with 2*a* inter-electrode distance are achieved and affected to the 2*a* pseudo-depth, etc. until $n = 5$ where only one array can exist. After that acquisition, the interpretation of the pseudo-section is performed by a non-linear inversion that is the repetitions of finite difference or finite element numerical methods forward calculations until a satisfying fit with the experimental data.

Figure 3 Mobile multipole (ARP, automatic resistivity profiling, Courtesy of Geocarta, France) pulled by a quad and scheme of electrode arrays. The injection dipole is associated with three measurement dipoles corresponding to approximate 0.56 m, 1 m, and 2 m depths of investigation. Measurements are acquired with 0.2 m step, while continuously moving at 2 m.s^{-1} to 4 m.s^{-1} depending on the terrain surface.

all-terrain vehicle. The galvanic contacts are achieved by spiked wheels as illustrated in Fig. 3 but the number of different array sizes is more limited than with fixed electrodes.

3.2 Electromagnetic induction: coil-coil frequency-domain (frequency domain electromagnetism) instrument principle, and multi-coil and multi-frequency development and separation of the different property contributions

The transmission of a time-varying magnetic field, the primary field, allows the induction of eddy currents in the soil, the magnitude of which depends on the electrical conductivity of the terrain. In turn, these currents generate a secondary magnetic field, H_s, which can be measured. It is thus possible to determine the electrical conductivity without any contact with the ground which facilitates the motion in the field. A convenient solution for soil studies (Corwin and Lesch, 2003) is the use of small coils as transmitter and receiver. They are separated by distances varying from a fraction of meter to several meters, and the relevant frequency range extends from several kHz to one-hundredths of kHz. These characteristic distances and frequency range correspond to a low induction number (*LIN*) in which condition the depth of investigation is controlled by geometrical parameters: the spacing between the transmitter and the receiver coils (inter-coil spacing), their relative orientation and the clearance above the ground surface. As with the resistivity method, the description of the conductivity variation is thus possible in 3D. Moreover, the two other EM properties, magnetic susceptibility and dielectric permittivity, may intervene, and both complicate and enrich the interpretation. The determination of the other properties can be achieved by considering the phase lag between the secondary and the primary fields and measuring at several frequencies - i.e. spectroscopy - allow the determination of two or three of the properties.

Considering numerical values allows illustrating the multi-property capability by comparing two different apparatus: one has a 25 kHz frequency and a 1 m inter-coil spacing ('L = 1 m apparatus') and is held at 0.20 m clearance, and the second has a 10 kHz frequency and a 4 m inter-coil spacing ('L = 4 m apparatus') and is held at 1 m clearance. The three usual coil configurations are considered: horizontal coplanar (HCP) where the plane of the two coils is horizontal (the equivalent magnetic dipoles having vertical axes), vertical coplanar (VCP) where the plane of the two coils is vertical (the equivalent magnetic dipoles having horizontal axes), and perpendicular (PERP) where the transmitter has a horizontal plane and the receiver a vertical one (the equivalent magnetic dipole having a radial orientated axis). In all cases, the results are expressed by ratios H_s/H_p in parts per million (ppm), with $H_p = M/(4\pi L^3)$, M being the transmitter moment.

One observes that whatever the coil configuration, the quadrature components have monotonous and quasilinear dependences, with the conductivity in the range from 1 mSm^{-1} to 500 mSm^{-1}, with around 30 ppm/mSm^{-1} for the L = 1 m apparatus and 150 ppm/mSm^{-1} for the L = 4 m apparatus. The small curvature is more marked in HCP for the L = 4 m apparatus, but the sensitivity is significantly greater.

All the in-phase components are linearly dependent on both the permittivity and the magnetic susceptibility. The sensitivity to the permittivity is greater for the L = 4 m apparatus and slightly higher for the HCP configuration (129 ppm for a change of 1000 in relative permittivity) than for PERP or VCP configurations (96 ppm for a change of 1000 in relative permittivity). For the L = 1 m apparatus, these sensitivities reduce to 60 ppm of HCP and 44 ppm for VCP or PERP. The VCP and PERP sensitivity to the magnetic susceptibility is 4 ppm/1 × 10^{-5} SI, that of HCP is of opposite sign and lower (for this configuration, it is very dependent on the clearance): –2.3 ppm/1 × 10^{-5} SI for the L = 1 m apparatus and –1.4 ppm/1 × 10^{-5} SI for the L = 4 m apparatus. As the responses of both properties linearly add, will it be possible to separate the respective contributions of ε' and κ' to the in-phase response? With only one coil configuration, the answer is no, but when considering both HCP and VCP or HCP and PERP, the answer is yes, because the sensitivities to each property are not identical.

In practice, the part of the frequency is also a determining factor. For most common values of soil electrical and magnetic properties, the following rules of thumb can guide the interpretation: The relevant criterion is obtained by multiplying the frequency by the inter coil distance: if $(f\ L) > 10^6$ Hz.m is respected, the susceptibility response would be a priori negligible, if $(f\ L) < 3.10^4$ Hz.m is respected, the permittivity response would be a priori negligible. Time-domain EM measurements may also be considered as they proved to be invaluable for the investigation of the first several tenths of meters in depth; however, for shallower soil investigation, the identification of the part of the different properties is simpler with the frequency-domain measurements.

In conclusion, the measurement of the electrical conductivity is reliable and easy with electromagnetic induction (EMI) instruments. Contrary to the resistivity method where the signal is proportional to the resistivity, here the signal is increasing with the conductivity, which favors the EMI use in high-conductivity context such as salted areas. The simultaneous determination of the magnetic susceptibility is complemented with a valued information about the redox history of the studied soil, and the determination of the permittivity is a new path, which yet merits to be explored.

4 Field examples

As seen in the preceding sections, mobile electrical and electromagnetic devices measuring geophysical properties are well adapted for non-invasive soil mapping as they allow the collection of lateral and in-depth information sampled with a fine mesh (Corwin and Lesch 2005). Using such maps, three main developments can be considered:

- Establishment of petro-physical relationships to transform geophysical property maps into soil property maps of interest;
- Help in the mapping of soil types using expert and statistical approaches;
- Delineation of homogeneous units by using statistical and geostatistical approaches.

The two examples presented below illustrate the use of maps of electrical resistivity (example of Epoisses) and of electrical conductivity (El Haouareb case) for mapping the soil characteristics, delineating pedological units, or identifying homogeneous areas.

4.1 Epoisses (France, temperate climate), multi-depth electrical resistivity mapping, and interpretation for soil mapping

The Epoisses experimental site is located in France, west of Dijon (Côte-d'Or, France). It is an experimental site managed by INRAE with an area of approximately 120 ha where agro-ecological practices are experimented. The soils are alluvial and consist in clayey horizons based on a mixture of limestone pebbles and fine earth. The thickness of soil is highly variable and fluctuates between 20 cm and 200 cm, which makes the soils of this site spatially heterogeneous.

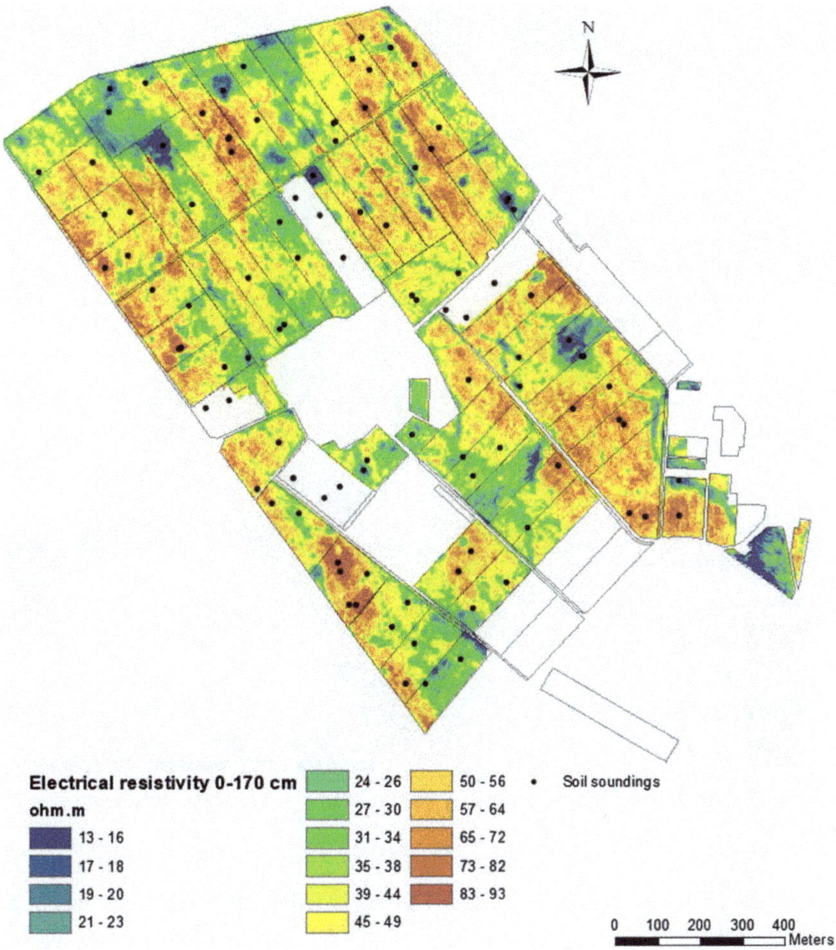

Figure 4 Soil electrical resistivity map generated by the geoelectrical prospecting using the ARP tool with the 170 cm inter-electrode spacing.

An electric prospecting was carried out in 2011 using the Geocarta ARP tool, generating three maps of resistivity corresponding to three investigation thicknesses: approximately 0–50 cm; 0–100 cm; and 0–170 cm. The resistivity map of the 0–170 cm inter-electrode spacing (Fig. 4) was used in the development of a stratified sampling scheme for positioning 105 pedological excavations allowing to record soil vertical profiles. For each of these positions, classical pedological characteristics of the soils were described and analyzed (e.g. horizonation, granulometry, pH, and carbon content) including the soil thickness.

On the basis of the georeferenced pedological observations and the geoelectrical information, several maps were developed. The first map

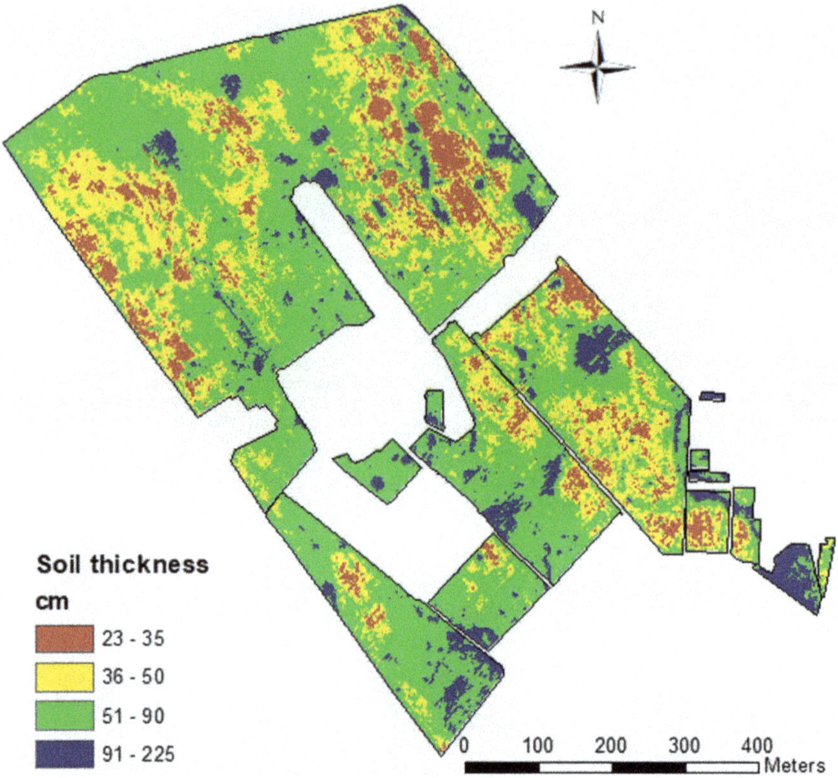

Figure 5 Soil thickness map estimated by kriging with electrical resistivity as an external drift.

(Fig. 5) shows soil thickness kriged to a 4 m grid using electrical resistivity as an external drift.

The very strong statistical links between the electrical resistivity ($R^2 = 0.81$) and the soil thickness allowed the pedological expert to use the contours of the resistivity maps to refine the contours of the soil map, initially conceptualized according to the traditional expert approach (Fig. 6). From this map, soil characteristics used in the evaluation of soil health (e.g. available water content and pH) can be mapped over the whole studied area. Once a soil health index has been calculated, it could be mapped also.

4.2 El Haouareb (Tunisia, arid zone) electromagnetic induction mapping in salted soil conditions (Jouini et al., 2019)

The olive tree plantation was created in 2001 over a 1.5 ha area in a region where the annual rainfall varies between 250 mm and 300 mm, while the potential evapotranspiration equals 1500 mm. The substratum is alluvial. The

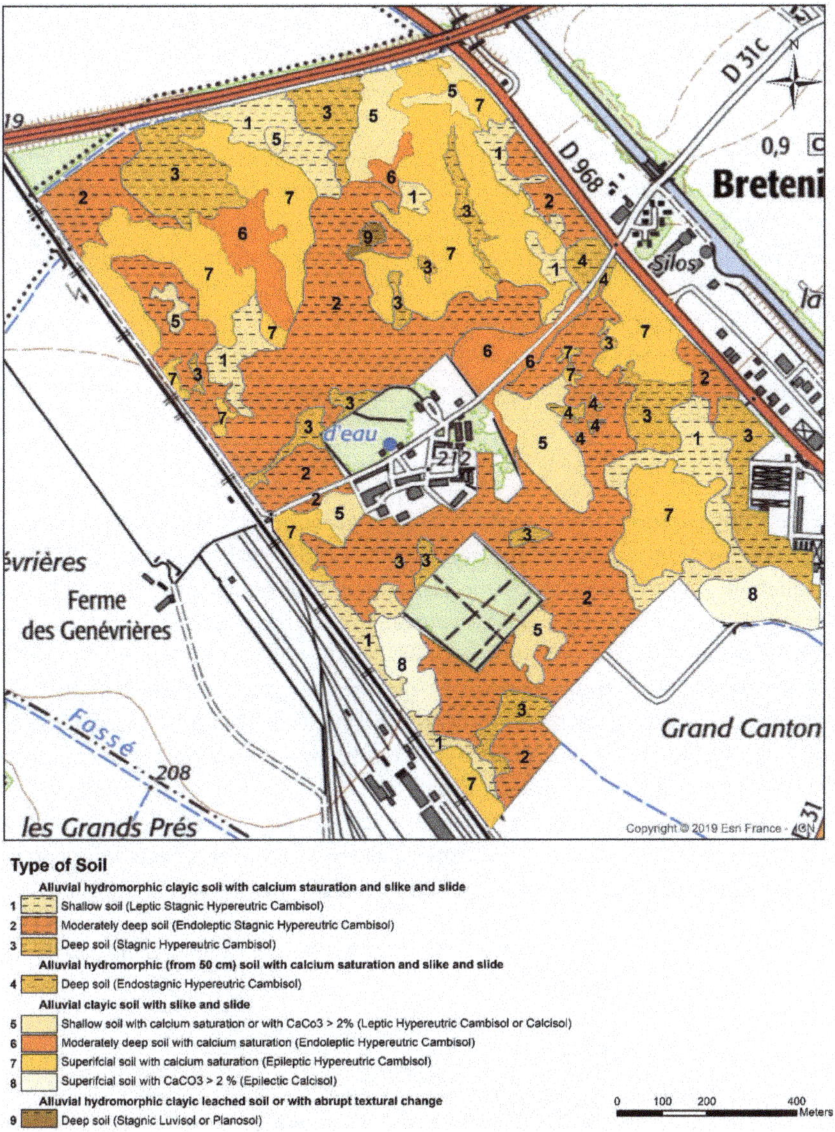

Figure 6 Map of the soil typology based on the interpretation of soil direct observations (soundings and profiles) coupled with the electrical resistivity map.

soil between 0 m and 0.4 m in depth has a significant clay content reaching until 47%. It is crossed by an ancient river bed (oued or wadi) where the coarse sand content is higher. The salted water table is at about 70 m in depth. The distance between trees is 10 m, and in a 1.08 ha (180 × 60 m^2) area of the plot, a fava bean crop is achieved between the tree rows. This portion of the plot is irrigated by a drip-feeding system using water pumped from wells about 80 m deep and having a salt content of 2 g/L (0.23 $S.m^{-1}$).

The electromagnetic survey was achieved in April, 2 weeks after the fava bean harvest using the EM31 (Geonics Ltd, Mississauga, Canada) instrument in both HCP and VCP configurations along parallel profiles 5 m apart. The profiles were orientated perpendicularly to the irrigation lines. Each measurement location was recorded at a 0.3 m step using a global navigation satellite system positioning system. The EM31 is experienced for 40 years and well known where f = 9.8 kHz and L = 3.66 m (Geonics Ltd). The clearance above the ground is h = 0.95 m. Its investigation depths in HCP and VCP configurations are in relevant fit with the depth and thickness of the olive tree root zone.

The objectives of the survey were to assess the global salt content, to map its variations together with that of the soil texture, and to delineate the irrigation impact.

Figure 7 shows the apparent electrical conductivity measured with EM31 in HCP configuration. The range of values corresponds to a moderately saline soil. At the center of the map, the course of the old wadi is observable by lower conductivity values associated with the alluvium of coarser granularity. In the rectangular part where the irrigation system exists, the apparent conductivity is higher than in the northern part and the irrigation lines correspond to higher conductivity strips. One thus observes, on the one hand, that the irrigated part is more conductive and, on the other hand, that the granularity significantly influences the conductivity values. The VCP apparent conductivity map confirms all these observations. However, as can be observed in soundings S1 and S2 (Table 2), the granulometry and the saturated past extract conductivity vertical variations (according to the 1/5 protocol) are more complex than what can be evidenced by only HCP and VCP EM31 measurements.

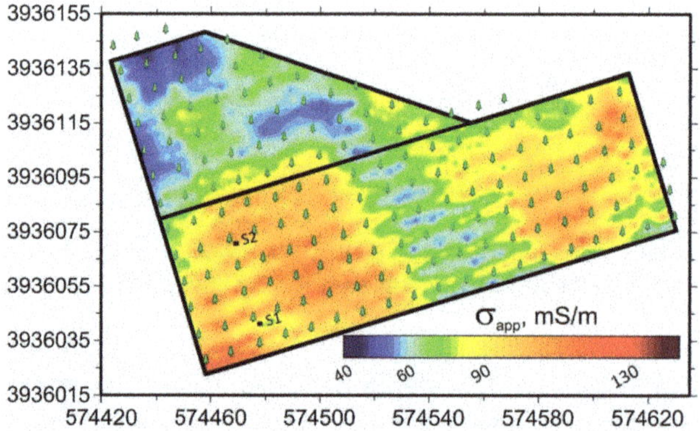

Figure 7 Apparent conductivity map EM31 HCP where are marked each olive tree location. S1 and S2 correspond to the hand auger soundings. The irrigation system is deployed into the rectangular part where the fava bean is cultivated.

Table 2 Granulometry and saturated past extract conductivity variations versus depth with 20 cm intervals

	Sounding 1				Sounding 2			
Depth range (cm)	Clay (%)	Silt (%)	Sand (%)	σ (S/cm)	Clay (%)	Silt (%)	Sand (%)	σ (μS/cm)
0–20	33.5	34.25	30.95	885	44.25	26.5	26.835	500
20–40	36.75	30	29.85	890	23.75	8.5	66.165	720
40–60	35	34.75	27.805	840	39	13	45.635	900
60–80	39.5	27.25	31.615	855	27.75	9	61.555	550
80–100	31.75	15.5	49.91	820	30	12.75	53.13	510
100–120	37	24	35.255	6220	41.75	25.25	30.185	660
120–140	35.5	36.5	26.11	2700	67.5	8.5	19.5	1200
140–160	31	18	46.22	1970	47.5	16.25	33.57	1220
160–180	32.25	15	49.965	1940	19.5	27.5	50.12	970
180–200	24.5	5.5	68.255	1390	27.5	0.5	69.065	680
200–220	40.5	29.25	28.805	1500	28	1.75	68.41	750

Significant differences can be observed between the two soundings, while the apparent conductivity values are comparable in the EM31 HCP map.

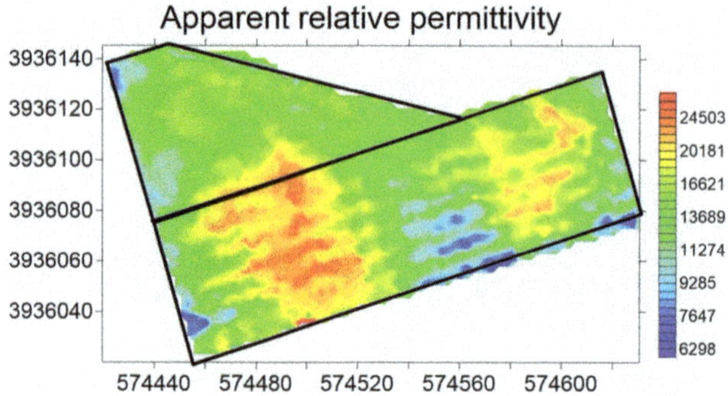

Figure 8 Apparent relative permittivity map EM31 deduced from both HCP and VCP measurements.

Using both HCP and VCP in-phase data, it is possible to draw in Fig. 8 the apparent permittivity map. This second property confirms a higher ion content in the irrigated part of the parcel with the presence of strips in relation with the irrigation lines and the higher leaching down in the course of the old wadi. However, the interpretation of the prospecting needs to associate both knowledge of the irrigation (and then of the soil water content), and salinity, which is definitely a soil characteristic relevant to evaluate soil health in some contexts.

5 The use of electrical and electromagnetic tools to evaluate soil health

As demonstrated in the previous sections, both geoelectrical and EMI prospections are useful to delineate the spatial extension of significant soil properties - here the soil thickness and the ion content. By using transfer functions using independent data, geophysical measurements are indeed used to map individual soil characteristics (Corwin and Scudiero, 2020), like the soil stoniness (Tetegan et al., 2012), the soil clay content, and the soil depth (Bourennane et al., 2017). The latter are inherent soil characteristics, which are supposed to be stable over time, and have been used for decades to characterize the spatial extension of soil types. They are definitely useful to describe soil quality. But, due to their sensitivity to a large number of soil characteristics including manageable ones like pH, soil organic carbon content, or bulk density, geophysical tools are then also relevant to analyze some spatio-temporal evolutions of all the components of soil health, including both inherent and manageable soil characteristics, or even by integrating some temporal soil variables like temperature and/or water content. Examples of the use of geophysical tools to disentangle inherent and manageable soil properties are numerous, and some experiments have specifically been designed to evaluate the effect of management practices on soils, in the developing discipline of agro-geophysics (Garre et al., 2021). An early work by Arnold et al. (2005) has demonstrated the potential of electrical conductivity as an integrated soil characteristic to derive some information on soil health; Howard et al. (2016) have also suggested to associate magnetic and electric data from proximal sensing prospection to evaluate soil health in urbanized areas.

Nevertheless, one important challenge is to deal with the biotic components of soils, which are important to evaluate soil health, but which are usually missing information in the description of the soil functioning, especially because they are strongly dependent on the seasonality. How to deal with the biotic components of soil health? Doran and Zeiss (2000) remind that soil organisms have to be taken into account as indicators of soil health, and Sharma et al. (2010) underline the usefulness of microbial community structure and diversity to evaluate soil quality; Neher (2001), for example, advocate for the use of nematodes in soil health evaluation, due to their high relationship with nitrogen cycling and decomposition, whereas Rudisser et al. (2015) suggest the use of a biological soil quality index which takes into account the micro-arthropod groups as it is a sensitive indicator of land use practices. By looking for a minimum data set of soil characteristics necessary for the evaluation of soil health, Marzaioli et al. (2010) as well as Raiesi (2017) suggest biological parameters (including, for example, microbial biomass, fungal mycelium, and arysulfatase activity) and, among other physicochemical parameters, the use of soil electrical conductivity. A way could be the development of experimental field kits enabling the measurements of several soil characteristics in order to derive an integrated index [see, for example, Liebig et al. (1996) or Ramson et al. (2021)], as suggested by Yang et al. (2020) who militate for the development of integrated sensors dedicated to a simplified evaluation of soil health. Electromagnetic, electrical, and impedance spectroscopy tools could be a gateway toward such sensors. Further research work could be oriented in this direction, say toward the ability of a geophysical measurement to be easily transformed into a soil health indicator, without an intermediate evaluation of soil characteristics or properties.

Another challenge with the evaluation of soil health consists in the scale at which the question is addressed. Electrical tools are especially relevant to further elaborate an evaluation at the field scale, by creating a so-called pedo-electrical model including some relevant soil characteristics; the latter could then be transformed into a soil health map, by using pedotransfer functions designed to transform the electrical signal into a soil health indicator from laboratory measurements of electrical conductivity in controlled conditions. However, both ad hoc measurements strategies and methodologies enabling advanced signal analyses still need to be developed to propose the most relevant methods to evaluate soil health.

6 Conclusion

The EU Mission Board for Soil Health and Food has specified a list of several soil characteristics to be evaluated and mapped over Europe to reach the ambitious objectives of the EU Green Deal related to soil. Soil structure, soil organic carbon, nutrients or salts in excess are considered as essential components of soil health and need to be mapped. Electrical and electromagnetic sensors are sensitive to several soil parameters considered as essential in the evaluation of soil health. By looking at two examples at the local scale, we have emphasized that, on the one hand, these tools are useful to evaluate some individual soil parameters, like the soil thickness or the delineation of salt areas. But on the other hand, the data interpretation remains difficult insofar as lots of soil characteristics have an influence on the signal. Knowing that, by definition, the soil health concept embraces a holistic vision, the ability of geophysical tools to be sensitive to numerous soil parameters should be an opportunity to derive indirect integrated indexes directly linked to soil health, and including bio-physicochemical soil characteristics. Instead of disentangling or disaggregating the signal to recover the value of one specific soil parameter, future studies should focus on the integrated interpretation of geophysical signals - even recorded by multi-sensor systems to be designed and developed - by taking into account the influence of numerous soil parameters, for a direct evaluation and mapping of soil health. Whereas this Graal is not yet within reach, electrical/electromagnetic tools have a key role to play in reaching such objectives.

7 Where to look for further information

For more information about electrical/electromagnetic theory and use, the readers are advised to look at 'Proximal Soil Sensing', by Viscarra Rossel, Raphael A., McBratney, Alex B. and Minasny, Budiman (Eds.), published in 2011.

8 References

Abdu, H., Robinson, D. A., Seyfried, M. and Jones, S. B. 2008. Geophysical imaging of watershed subsurface patterns and prediction of soil texture and water holding capacity. *Water Resources Research* 44(4): W00D18.

Amato, M., Bitella, G., Rossi, R., Gómez, J. A., Lovelli, S. and Gomes, J. J. F. 2009. Multi-electrode 3D resistivity imaging of alfalfa root zone. *European Journal of Agronomy* 31(4): 213-222. https://doi.org/10.1016/j.eja.2009.08.005.

Amirinejad, A. A., Kamble, K., Aggarwal, P., Chakraborty, D., Pradhan, S. and Mittal, R. B. 2011. Assessment and mapping of spatial variation of soil physical health in a farm. *Geoderma* 160(3-4): 292-303.

Amorim, H. C. S., Ashworth, A. J., Moore, P. A., Wienhold, B. J., Savin, M. C., Owens, P. R., Jagadamma, S., Carvalho, T. S. and Xu, S. T. 2020. Soil quality indices following long-term conservation pasture management practices. *Agriculture, Ecosystems and Environment* 301: 10.

Arnold, S. L., Doran, J. W., Schepers, J., Wienhold, B., Ginting, D., Amos, B. and Gomes, S. 2005. Portable probes to measure electrical conductivity and soil quality in the field. *Communications in Soil Science and Plant Analysis* 36(15-16): 2271-2287.

Arshad, M. A. and Martin, S. 2002. Identifying critical limits for soil quality indicators in agro-ecosystems. *Agriculture, Ecosystems and Environment* 88(2): 153-160.

Bertermann, D. and Schwarz, H. 2018. Bulk density and water content-dependent electrical resistivity analyses of different soil classes on a laboratory scale. *Environmental Earth Sciences* 77(16): 570.

Besson, A., Cousin, I., Dorigny, A., Dabas, M. and King, D. 2008. The temperature correction for the electrical reistivity measurements in undisturbed soil samples. *Soil Science* 173(10): 707-720.

Bourennane, H., Hinschberger, F., Chartin, C. and Salvador-Blanes, S. 2017. Spatial filtering of electrical resistivity and slope intensity: Enhancement of spatial estimates of a soil property. *Journal of Applied Geophysics* 138: 210-219.

Campbell, R. B., Bower, C. A. and Richards, L. A. 1948. Change of electrical conductivity with temperature and the relation of osmotic pressure to electrical conductivity and ion concentration for soil extracts. *Soil Sci. Soc. Proc.* 66-69.

Ciampalini, A., André, F., Garfagnoli, F., Grandjean, G., Lambot, S., Chiarantini, L. and Moretti, S. 2015. Improved estimation of soil clay content by the fusion of remote hyperspectral and proximal geophysical sensing. *Journal of Applied Geophysics* 116: 135-145.

Clément, R., Descloitres, M., Günther, T., Ribolzi, O. and Legchenko, A. 2009. Influence of shallow infiltration on time-lapse ERT: Experience of advanced interpretation. *Comptes Rendus Geoscience* 341(10-11): 886-898.

Corwin, D. L. and Lesch, S. M. 2003. Application of soil electrical conductivity to precision agriculture: Theory, principles, and guidelines. *Agronomy Journal* 95(3): 455-471.

Corwin, D. L. and Lesch, S. M. 2005. Apparent soil electrical conductivity measurements in agriculture. *Computers and Electronics in Agriculture* 46(1-3): 11-43.

Corwin, D. L. and Scudiero, E. 2020. Field-scale apparent soil electrical conductivity. *Soil Science Society of America Journal* 84(5): 1405-1441.

Cousin, I., Buis, S., Lagacherie, P., Doussan, C., Le Bas, C. and Guerif, M. 2022. Available water capacity from a multidisciplinary and multiscale viewpoint: A review. *Agronomy for Sustainable Development* 42(3): 1-29.

De Benedetto, D., Barca, E., Castellini, M., Popolizio, S., Lacolla, G. and Stellacci, A. M. 2022. Prediction of soil organic carbon at field scale by regression kriging and multivariate adaptive regression splines using geophysical covariates. *Land* 11(3): 381.

Doran, J. W. and Zeiss, M. R. 2000. Soil health and sustainability: Managing the biotic component of soil quality. *Applied Soil Ecology* 15(1): 3-11.

Doussan, C. and Ruy, S. 2009. Prediction of unsaturated soil hydraulic conductivity with electrical conductivity. *Water Resources Research* 45(10): W10408.

Dudala, S., Dubey, S. K. and Goel, S. 2020. Microfluidic soil nutrient detection system: Integrating nitrite, pH, and electrical conductivity detection. *IEEE Sensors Journal* 20(8): 4504-4511.

EU. 2020. *Caring for Soil is Caring for Life*.

Friedman, S. P. 2005. Soil properties influencing apparent electrical conductivity: A review. *Computers and Electronics in Agriculture* 46(1-3): 45-70.

Fu, Y., Horton, R., Ren, T. and Heitman, J. L. 2021. A general form of Archie's model for estimating bulk soil electrical conductivity. *Journal of Hydrology* 597: 126160.

Garre, S., Hyndman, D., Mary, B. and Werban, U. 2021. Geophysics conquering new territories: The rise of "agrogeophysics." *Vadose Zone Journal* 20(4): 4.

Grunwald, S., Vasques, G. M. and Rivero, R. G. 2015. Fusion of soil and remote sensing data to model soil properties. In: Sparks, D. L. (Ed.), *Advances in Agronomy*. Elsevier Academic Press Inc, San Diego, Vol. 131, pp. 1-109.

Howard, J. L., Orlicki, K. M. and LeTarte, S. M. 2016. Evaluation of some proximal sensing methods for mapping soils in urbanized terrain, Detroit, Michigan, USA. *CATENA* 143: 145-158.

Islam, K., Singh, B. and McBratney, A. 2003. Simultaneous estimation of several soil properties by ultra-violet, visible, and near-infrared reflectance spectroscopy. *Soil Research* 41(6): 1101-1114.

Jeřábek, J., Zumr, D. and Dostál, T. 2017. Identifying the plough pan position on cultivated soils by measurements of electrical resistivity and penetration resistance. *Soil and Tillage Research* 174: 231-240.

Joschko, M., Gebbers, R., Barkusky, D. and Timmer, J. 2010. The apparent electrical conductivity as a surrogate variable for predicting earthworm abundances in tilled soils. *Journal of Plant Nutrition and Soil Science* 173(4): 584-590.

Jouini, W., Hovhannissian, G., Azaiez, H., Charni, W., Hajlaoui, S. and Hachicha, M. 2019. Caractérisation de la Variabilité Spatiale des Propriétés des Sols dans une Parcelle Agricole par des Méthodes Géophysiques: Cartographie Électromagnétique et Tomographie de Résistivité Électrique 2D. Résumés de Colloque Maghrébin de la Géophysique Appliquée - CMGA 8, Hammamet, Tunisie, 21-23 Mars 2019.

Karlen, D. L., Andrews, S. S. and Doran, J. W. 2001. Soil quality: Current concepts and applications. *Advances in Agronomy* 74(74): 1-40.

Karlen, D. L., Ditzler, C. A. and Andrews, S. S. 2003. Soil quality: Why and how? *Geoderma* 114(3-4): 145-156.

Kowalczyk, S., Maślakowski, M. and Tucholka, P. 2014. Determination of the correlation between the electrical resistivity of non-cohesive soils and the degree of compaction. *Journal of Applied Geophysics* 110: 43-50.

Liebig, M. A., Doran, J. W. and Gardner, J. C. 1996. Evaluation of a field test kit for measuring selected soil quality indicators. *Agronomy Journal* 88(4): 683-686.

Ma, R., McBratney, A., Whelan, B., Minasny, B. and Short, M. 2011. Comparing temperature correction models for soil electrical conductivity measurement. *Precision Agriculture* 12(1): 55-66.

Martinez, G., Vanderlinden, K., Ordonez-Fernandez, R. and Muriel, J. L. 2009. Can apparent electrical conductivity improve the spatial characterization of soil organic carbon? *Vadose Zone Journal* 8(3) : 586-593.

Marzaioli, R., D'Ascoli, R., De Pascale, R. A. and Rutigliano, F. A. 2010. Soil quality in a Mediterranean area of Southern Italy as related to different land use types. *Applied Soil Ecology* 44(3): 205-212.

Neher, D. A. 2001. Role of nematodes in soil health and their use as indicators. *Journal of Nematology* 33(4): 161-168.

Nortcliff, S. 2002. Standardisation of soil quality attributes. *Agriculture, Ecosystems and Environment* 88(2): 161-168.

Obade, V. D. and Lal, R. 2016a. Towards a standard technique for soil quality assessment. *Geoderma* 265: 96–102.

Obade, V. D. and Lal, R. 2016b. A standardized soil quality index for diverse field conditions. *Science of the Total Environment* 541: 424–434.

Raiesi, F. 2017. A minimum data set and soil quality index to quantify the effect of land use conversion on soil quality and degradation in native rangelands of upland arid and semiarid regions. *Ecological Indicators* 75: 307–320.

Ramson, S. R. J., Leon-Salas, W. D., Brecheisen, Z., Foster, E. J., Johnston, C. T., Schulze, D. G., Filley, T., Rahimi, R., Soto, M. J. C. V., Bolivar, J. A. L. and Malaga, M. P. 2021. A self-powered, real-time, LoRaWAN IoT-based soil health monitoring system. *IEEE Internet of Things Journal* 8(11): 9278–9293.

Ren, L., D'Hose, T., Borra-Serrano, I., Lootens, P., Hanssens, D., De Smedt, P., Cornelis, W. M. and Ruysschaert, G. 2022. Detecting spatial variability of soil compaction using soil apparent electrical conductivity and maize traits. *Soil Use and Management* 38: 1749–1760.

Rey, E., Jongmans, D., Gotteland, P. and Garambois, S. 2006. Characterisation of soils with stony inclusions using geoelectrical measurements. *Journal of Applied Geophysics* 58(3): 188–201.

Rhoades, J. D., Raats, P. A. C. and Prather, R. J. 1976. Effects of liquid-phase electrical conductivity, water content, and surface conductivity on bulk soil electrical conductivity. *Soil Science Society of America Journal* 40(5): 651–655.

Rinot, O., Levy, G. J., Steinberger, Y., Svoray, T. and Eshel, G. 2019. Soil health assessment: A critical review of current methodologies and a proposed new approach. *Science of the Total Environment* 648: 1484–1491.

Romero-Ruiz, A., Linde, N., Keller, T. and Or, D. 2019. A review ofgeophysical methods for soilstructure characterization.*Reviews of Geophysics* 56: 672–697.

Rossi, R., Amato, M., Pollice, A., Bitella, G., Gomes, J. J., Bochicchio, R. and Baronti, S. 2013. Electrical resistivity tomography to detect the effects of tillage in a soil with a variable rock fragment content. *European Journal of Soil Science* 64(2): 239–248.

Rudisser, J., Tasser, E., Peham, T., Meyer, E. and Tappeiner, U. 2015. The dark side of biodiversity: Spatial application of the biological soil quality indicator (BSQ). *Ecological Indicators* 53: 240–246.

Samouëlian, A., Cousin, I., Tabbagh, A., Bruand, A. and Richard, G. 2005. Electrical resistivity survey in soil science: A review. *Soil and Tillage Research* 83(2): 173–193.

Sanchez-Navarro, A., Gil-Vazquez, J. M., Delgado-Iniesta, M. J., Marin-Sanleandro, P., Blanco-Bernardeau, A. and Ortiz-Silla, R. 2015. Establishing an index and identification of limiting parameters for characterizing soil quality in Mediterranean ecosystems. *CATENA* 131: 35–45.

Sharma, S. K., Ramesh, A., Sharma, M. P., Joshi, O. P., Govaerts, B., Steenwerth, K. L. and Karlen, D. L. 2010. Microbial community structure and diversity as indicators for evaluating soil quality. In: Lichtfouse, E. (Ed.), *Biodiversity, Biofuels, Agroforestry and Conservation Agriculture. Sustainable Agriculture Reviews*. Springer, Dordrecht, pp. 317–358.

Stępień, M., Samborski, S., Gozdowski, D., Dobers, E. S., Chormański, J. and Szatyłowicz, J. 2015. Assessment of soil texture class on agricultural fields using ECa, Amber NDVI, and topographic properties. *Journal of Plant Nutrition and Soil Science* 178(3): 523–536.

Svoray, T., Hassid, I., Atkinson, P. M., Moebius-Clune, B. N. and van Es, H. M. 2015. Mapping soil health over large agriculturally important areas. *Soil Science Society of America Journal* 79(5): 1420–1434.

Tabbagh, A., Rejiba, F., Finco, C., Schamper, C., Souffaché, B., Camerlynck, C., Thiesson, J., Jougnot, D. and Maineult, A. 2021. The case for considering polarization in the interpretation of electrical and electromagnetic measurements in the 3 kHz to 3 MHz frequency range. *Surveys in Geophysics* 42(2): 377–397.

Takoutsing, B., Weber, J., Aynekulu, E., Martin, J. A. R., Shepherd, K., Sila, A., Tchoundjeu, Z. and Diby, L. 2016. Assessment of soil health indicators for sustainable production of maize in smallholder farming systems in the highlands of Cameroon. *Geoderma* 276: 64–73.

Tetegan, M., Pasquier, C., Besson, A., Nicoullaud, B., Bouthier, A., Bourennane, H., Desbourdes, C., King, D. and Cousin, I. 2012. Field-scale estimation of the volume percentage of rock fragments in stony soils by electrical resistivity. *CATENA* 92: 67–74.

Verma, V. K., Setia, R. K., Sharma, P. K. and Singh, H. 2008. Geoinformatics as a tool for the assessment of the impact of ground water quality for irrigation on soil health. *Journal of the Indian Society of Remote Sensing* 36(3): 273–281.

Veum, K. S., Sudduth, K. A., Kremer, R. J. and Kitchen, N. R. 2017. Sensor data fusion for soil health assessment. *Geoderma* 305: 53–61.

Yamakawa, Y., Kosugi, K., Masaoka, N., Sumida, J., Tani, M. and Mizuyama, T. 2012. Combined geophysical methods for detecting soil thickness distribution on a weathered granitic hillslope. *Geomorphology* 145–146: 56–69.

Yang, T., Siddique, K. H. M. and Liu, K. 2020. Cropping systems in agriculture and their impact on soil health – A review. *Global Ecology and Conservation* 23: e01118.

Ye, R. Z., Parajuli, B., Szogi, A. A., Sigua, G. C. and Ducey, T. F. 2021. Soil health assessment after 40 years of conservation and conventional tillage management in Southeastern Coastal Plain soils. *Soil Science Society of America Journal* 85(4): 1214–1225.

Zornoza, R., Mataix-Solera, J., Guerrero, C., Arcenegui, V., Mayoral, A. M., Morales, J. and Mataix-Beneyto, J. 2007. Soil properties under natural forest in the Alicante Province of Spain. *Geoderma* 142(3–4): 334–341.

Chapter 7

Using ground-penetrating radar to map agricultural subsurface drainage systems for economic and environmental benefit

Barry Allred, USDA-ARS – Soil Drainage Research Unit, USA; and Triven Koganti, Aarhus University, Denmark

1 Introduction

The widespread adoption of subsurface drainage practices to remove excess soil water has enabled the Midwest USA to become one of the most productive agricultural regions in the world. A 1985 economic survey showed that several states within this region (Illinois, Indiana, Iowa, Ohio, Minnesota, Michigan, Missouri, and Wisconsin) had by that year approximately 12.5 million

http://dx.doi.org/10.19103/AS.2022.0107.21
Published by Burleigh Dodds Science Publishing Limited, 2023.

ha containing subsurface drainage systems, with cropland accounting for the vast majority of areas having buried drainage pipes (Pavelis, 1987). Since 1985, substantial amounts of additional agricultural drainage pipes have been installed in the Midwest USA. These systems are typically comprised of 10 cm diameter older clay tile or newer corrugated plastic tubing (CPT) drain line networks buried 0.6–1.5 m beneath the surface. The transition from clay tile to CPT drainage pipe occurred in the 1960s. Farmers within this region, as well as in other parts of the world where agricultural drainage practices are common, often need to repair drain lines that are not functioning properly or, in order to increase crop yields, install new drain lines between the older ones to improve soil water removal efficiency. For system repairs or efficiency improvements, determining the locations of preexisting drain lines is required, and yet, usually, a map of the original subsurface drainage system installation is no longer available.

Furthermore, subsurface drainage practices can release substantial amounts of nitrate (NO_3^-) and phosphate (PO_4^{3-}) from farm fields into adjacent waterways (Sims et al., 1998; Zucker and Brown, 1998). This farm field subsurface drainage discharge of NO_3^- and PO_4^{3-} in turn causes degradation of surface water bodies at the local, regional, and national scales. Risk assessment of this environmental hazard, from a farm field perspective, calls for knowledge of the drainage pipe network, including the extent of coverage and drain line spacing distance.

Regardless of whether the need is economic or environmental, finding drain lines with a handheld tile probe is time consuming and extremely tedious, and, if not performed carefully, can damage buried pipes (Fig. 1a). Using heavy trenching equipment is generally effective but always causes considerable pipe damage requiring costly repairs (Fig. 1b). Subsurface drainage system patterns can be complex (e.g. rectangular, herringbone, and random), further

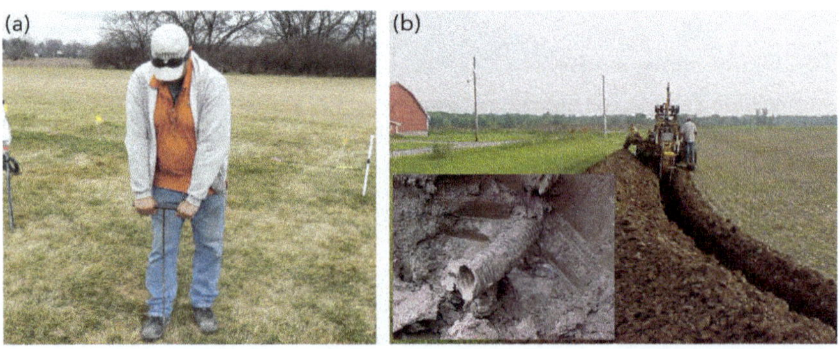

Figure 1 Present methods for finding drainage pipes: (a) handheld tile probe and (b) heavy trenching equipment, with the inset photo showing an example of the drainage pipe damage caused during this excavation operation.

Published by Burleigh Dodds Science Publishing Limited, 2023.

hampering efforts to map drain lines using traditional tile probe or trenching detection methods.

A survey of drainage installation/repair contractors in Ohio, USA, conducted by Allred et al. (2004) found that for half the respondents, at least 60% of their subsurface drainage installation projects required locating preexisting buried drainage pipe. In addition, 60% of the respondents spent 2 days or more per month trying to find drainage pipes; 52% were moderately to completely dissatisfied with the present methods of locating drainage pipe, and 69% were sure that there would be an economic benefit to their business by having a better method for mapping drain lines. Consequently, there is a crucial necessity for effective, efficient, and nondestructive methods to locate buried drainage pipes. Ground-penetrating radar (GPR) can potentially provide a viable means for mapping agricultural subsurface drainage systems. With the exception of Chow and Rees (1989), who successfully demonstrated the use of GPR to locate agricultural drainage pipes in the Maritime Provinces of Canada, no other research on this topic had been done prior to that reported in this chapter; therefore, the additional investigations conducted over the past 20 years were certainly warranted.

2 Comparison of proximal soil-sensing methods for drainage pipe detection

Allred et al. (2004) tested the drainage pipe detection potential of four proximal soil-sensing methods: magnetometry, electromagnetic induction, resistivity, and GPR. Magnetometry was evaluated with a Geometrics, Inc. (San Jose, CA, USA) G-858 cesium gradiometer, electromagnetic induction using a Geophex, Ltd. (Raleigh, NC, USA) GEM-2 ground conductivity meter, resistivity by a Geometrics, Inc. OhmMapper TR1 capacitively coupled dipole–dipole towed electrode array, and GPR via a Sensors & Software, Inc. (Mississauga, Ontario, Canada) Noggin system having 250 MHz center frequency antennas. These four proximal soil-sensing (i.e. near-surface geophysics) methods were evaluated on two test plots in Columbus, OH, USA, one specifically designed for this geophysics research, and both contained functional subsurface drainage systems. Of the four proximal soil-sensing methods tested, only one, GPR, was determined to have the capability of detecting drainage pipes.

A GPR system operates with a transmitting antenna first sending a radar signal into the subsurface. Portions of this signal reflect off buried features, such as soil layer interfaces and infrastructure objects (e.g. drainage pipes), and are then directed back toward the surface where the returning signal is recorded by a receiving antenna (Fig. 2). Two electrical properties, electrical conductivity and dielectric constant, largely govern the GPR subsurface response. Electrical conductivity determines radar signal penetration depth. Where the subsurface

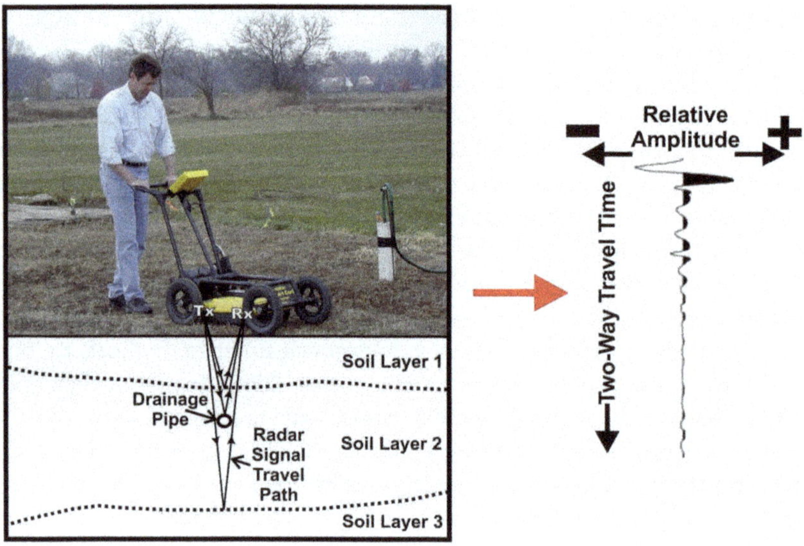

Figure 2 GPR principle of operation.

electrical conductivity is high, there is greater attenuation of the radar signal, while with lower subsurface electrical conductivity, there is less signal attenuation. Soils with high clay content and/or wetter conditions tend to have greater electrical conductivity values.

Subsurface contrasts in dielectric constant (i.e. relative dielectric permittivity) produce radar signal reflections. The greater the dielectric constant contrast, the greater the reflected signal amplitude (and energy). Consider, for example, the interface between two soil layers. If the two soil layers have very different dielectric constant values, then there will be a strong radar reflection at the interface. If the two soil layers have just a small difference in dielectric constant, then the reflection will be weak. Finally, if the two soil layers have no difference in dielectric constant, then there will be no radar signal reflected from the interface. Consequently, for GPR to successfully detect a buried feature, a significant dielectric constant contrast needs to be present. One last note on dielectric constant, besides determining whether there is a radar signal reflection from a buried feature, the dielectric constant also governs radar signal velocity in the subsurface. (Increased dielectric constant reduces radar signal velocity and vice versa.)

The record of returning radar signals obtained at the receiving antenna is a radar signal trace that charts the reflected radar signal amplitudes versus the two-way travel times (Fig. 2). The two-way travel time is the total time it takes for a radar signal to complete its path from the transmitting antenna downward to a subsurface feature and then back upward to the receiving antenna. A signal trace is obtained at each measurement point along a GPR transect. Filling in the positive amplitude side of the signal traces, and then placing the signal traces side by side in the order they were obtained along the transect, is essentially the process whereby a GPR profile is produced. The GPR profile depicts how GPR 'views' the subsurface directly beneath the transect line. The horizontal axis on a profile represents the distance along the transect, while the vertical axis is a two-way travel time, which can be converted to depth, if subsurface radar velocity information is available. If there is a grid of GPR transects spaced fairly close together or a single GPR transect having a very tight serpentine or spiral pattern, then the GPR profile information can be used to produce a GPR map representing the reflected radar energy over the survey area from a particular two-way travel time or depth interval.

After determining that GPR was the best proximal soil-sensing method for finding drainage pipes, Allred et al. (2004) further evaluated this technology at nine additional test plots in Ohio, USA. A GPR system with 250 MHz antennas was employed at all of the added locations. Overall, GPR was found to work relatively well at these nine sites detecting clay tile and CPT drainage pipes down to depths of around 1 m within a variety of different soil materials from clay to sandy loam.

Representative GPR results from one of these sites located in northwest Ohio, USA, are shown in Fig. 3. GPR profiles are provided in Fig. 3a,b. The Fig. 3a east–west profile depicts the GPR responses, banded, mostly horizontal, linear features on the east and west sides of the profile, which are obtained when the measurement transect is directly over the top and along the trend of a drain line. This east–west profile, in the center, also shows the GPR response, a horizontally expanded reflection hyperbola (upside-down U-shaped feature) that is obtained when the measurement transect crosses the buried drain line at an angle of 30–60°. The Fig. 3b north–south profile depicts the GPR response, tight (narrow) reflection hyperbolas (first three from the south and the two furthest north), occurring when the measurement transect is perpendicular to the drain line trend. Note, however, that the fourth reflection hyperbola from the south in Fig. 3b, unlike the others, is somewhat expanded due to the measurement transect crossing the drain line at a 60° angle.

The site GPR map shown in Fig. 3c characterizes the reflected radar energy from the depth interval between 0.9 m and 1.4 m, where the drain lines are buried. Lighter shading in Fig. 3c indicates greater reflected radar energy, while darker shades indicate less. The subsurface drainage system pattern is clearly defined by the lighter shaded linear features in Fig. 3c. Based on the GPR map, the interpreted drain line pattern (dashed black lines) is depicted in Fig. 3d. This interpreted map also provides locations for the Fig. 3a,b GPR profile measurement transects (blue and red lines).

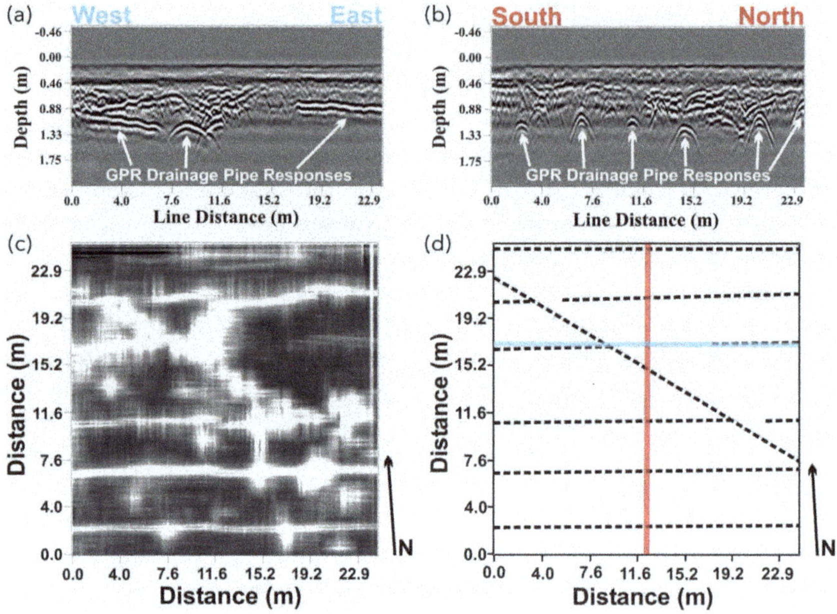

Figure 3 Representative GPR results from a test plot located in northwest Ohio, USA: (a) east-west GPR profile; (b) north-south GPR profile; (c) GPR map; and (d) interpreted drain line map (dashed black lines) with GPR profile measurement transect locations also provided (blue and red lines).

3 Factors potentially impacting ground-penetrating radar drainage pipe detection

Allred et al. (2005) investigated some of the factors potentially affecting GPR drainage pipe detection. Almost all of this research was conducted at the same two Columbus, OH, USA, test plots used in the Allred et al. (2004) study. Particularly, much of the investigation reported by Allred et al. (2005) was carried out at a specially constructed test plot containing a combined clay tile and CPT drainage system. The drainage system for this test plot was designed not only for free drainage but also for its capacity to raise and lower a shallow water table to positions between the surface and the drain line depth. The main categories of GPR drainage pipe detection factors studied by Allred et al. (2005) included equipment and survey setup, site conditions, and computer processing.

3.1 Equipment and survey setup

3.1.1 Antenna frequency

The radar pulse emitted by the transmitting antenna has a rather wide bandwidth of frequencies, but most of the pulse energy is concentrated in frequencies near a center frequency. Consequently, a GPR antenna is defined by its specified center frequency. Choosing the right antenna is one of the most critical decisions for a GPR survey. Antennas with a high center frequency are better able to isolate smaller buried objects but have less signal penetration. Conversely, antennas with a low center frequency have good signal penetration but are less able to isolate the smaller objects. This results in a tradeoff between high- and low-frequency antennas regarding the depth of investigation and resolution of subsurface features. Therefore, employing the best antenna, based on center frequency, will depend on the size and depth of the GPR target, in this case farm field drainage pipes.

Allred et al. (2005) tested GPR antennas with center frequencies of 100, 200, 250, and 500 MHz, with representative results presented in Fig. 4. As shown by the profiles from measurement transects perpendicular to the drain lines (first column GPR profiles in Fig. 4), the GPR drainage pipe reflection hyperbola responses were most clearly distinguishable using 250 MHz antennas. Furthermore, the GPR profiles from measurement transects over the top and along the trend of a drain line (second column GPR profiles in Fig. 4) likewise show the GPR drainage pipe banded, mostly horizontal, linear response was also most clearly distinguishable using 250 MHz antennas. These results indicated that 250 MHz antennas were superior for mapping farm field subsurface drainage systems, providing the best combination of radar signal penetration depth and drainage pipe resolution. Consequently, all continuing research reported in Sections 3, 4, 5, 6, and 8 of this book chapter employed a GPR system with 250 MHz antennas.

3.1.2 Signal trace stacking and station interval

At each measurement point along a transect, a GPR system can record a number of signal traces and average them together in a process called 'stacking' to produce one signal trace in which noise is reduced. The stacking number refers to the number of signal traces recorded and used in this process at a transect measurement point. Station interval refers to the uniform distance between measurement points on a GPR transect. An increase in the signal trace stacking numbers, along with a station interval decrease, will reduce noise and provide more detail in GPR profiles and maps, but adversely increase the time needed in the field for data acquisition. Consequently, there are tradeoffs with these equipment parameters that need to be considered in any GPR survey.

Allred et al. (2005) evaluated signal trace stacking values of 4, 8, 16, and 32, along with station intervals of 2.5 cm, 5 cm, and 10 cm. Results of this investigation determined that drainage pipes could be readily detected by the lowest stacking value (4) and highest station interval (10 cm). This finding implies, within reasonable limits, that smaller signal trace stacking values and larger

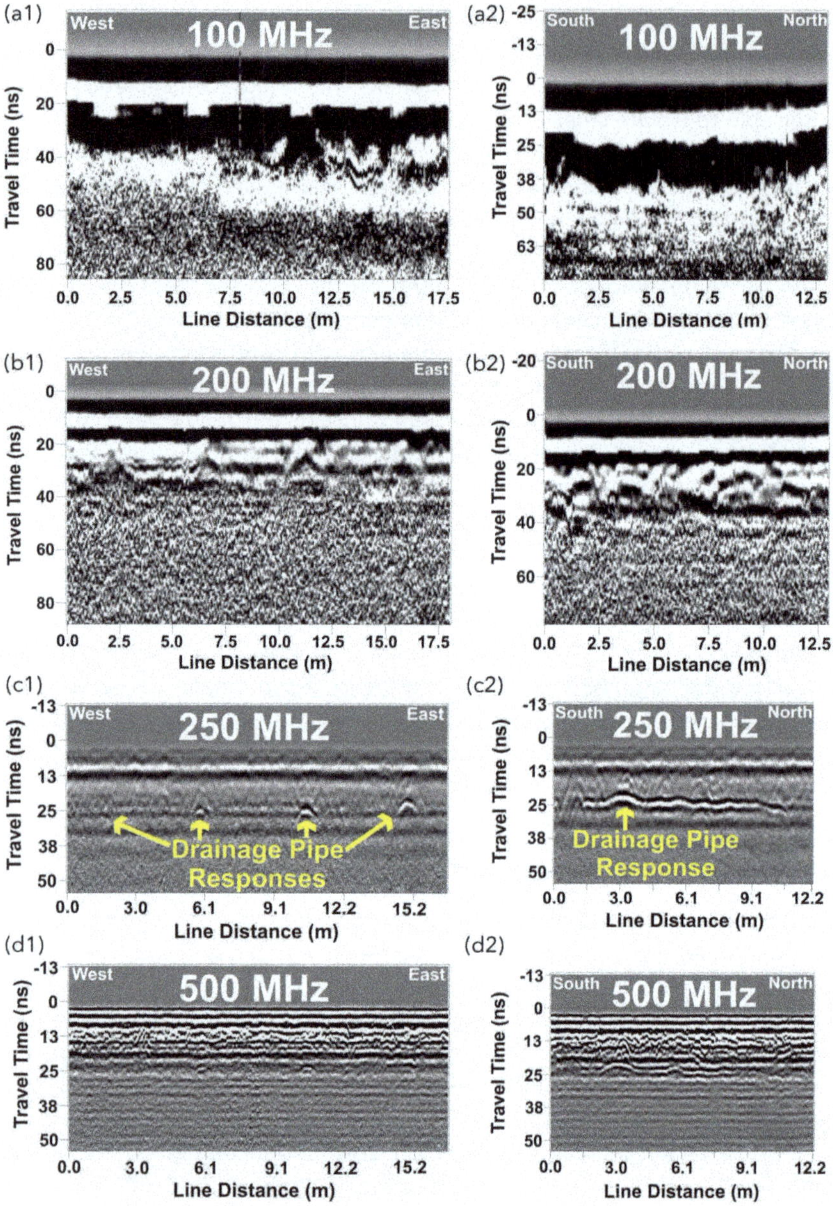

Figure 4 GPR profiles exhibiting antenna frequency effects. First column GPR profiles were obtained from measurement transects oriented perpendicular to drain lines. Second column GPR profiles were obtained from measurement transects oriented over the top and along the trend of a drain line. Results shown include: (a1) and (a2) 100-MHz center frequency antennas; (b1) and (b2) 200-MHz center frequency antennas; (c1) and (c2) 250-MHz center frequency antennas; and (d1) and (d2) 500-MHz center frequency antennas.

Published by Burleigh Dodds Science Publishing Limited, 2023.

station intervals still allow drain lines to be located while maintaining the efficiency at which the GPR data is collected.

3.1.3 Measurement transect orientation (unidirectional versus bidirectional)

The GPR measurement transect orientation needs to be considered for any field survey, especially in instances where linear features (i.e. drainage pipes) are being mapped. If the general directional trend of the drain lines is unknown before conducting the survey, and a unidirectional orientation is chosen for a set of parallel GPR transects, which just happens to align directly with the drainage pipes, then it is possible that many of the drain lines will go undetected. Essentially, if GPR measurement transects and drain pipes have the same orientation, the only way GPR can detect a drain line is if the measurement transect is directly over the top and along the trend of the drainage pipe. For this scenario, the problem can be easily resolved with two sets of parallel GPR transects oriented perpendicular to one another, where it would be ensured that at least one of the GPR measurement transect sets is not directly aligned with the drain lines.

Allred et al. (2005), using GPR data collected at the test plot shown in Fig. 3, demonstrated the importance of employing two sets of parallel GPR transects oriented perpendicular to one another, where there is prior uncertainty of the drain line directional trend. Two sets of parallel GPR transects were collected at this site, one oriented east-west and the other oriented north-south, with a spacing distance of 1.5 m between adjacent parallel transects. The north-south set of GPR transects was oriented perpendicular to the drain lines (with the exception of one drain line trending northwest-southeast), and, by itself, this set detected all of the drainage pipes. Conversely, the east-west set of GPR transects was oriented parallel to the drain lines (with the exception of one drain line trending northwest-southeast) and, by itself, failed to detect most of the drainage pipes. The best subsurface drainage system mapping results for this northwest Ohio, USA, site were obtained by combining the two sets of GPR transects (Fig. 3c). Consequently, unless the drain line directional trend is known with confidence beforehand, the best GPR survey approach is to employ two sets of parallel measurement transects oriented perpendicular to one another.

3.2 Site conditions

3.2.1 Clay tile versus corrugated plastic tubing drainage pipe

Allred et al. (2005) found that the material used to fabricate drainage pipes, clay tiles, or corrugated plastic tubing (CPT) did not affect the GPR drainage pipe response. The GPR drainage pipe response is predominantly determined by the dielectric constant of the soil material surrounding the pipe and the dielectric constant of the air or water inside the pipe. Explanations for the pipe material not affecting the GPR response include 250 MHz antenna center frequency radar signal wavelengths being much greater than pipe wall thickness (210–340 mm versus 1–13 mm) and similar dielectric constant values for clay tiles and CPT (5 and 2.4, respectively).

3.2.2 Soil type

Allred et al. (2005) noted that GPR profiles from finer grained, higher clay content soils had a more uniform appearance that allowed drainage pipe responses, if obtained, to be clearly distinguished. The GPR profiles for courser grained, sandier soils were often found to have a chaotic, jumbled look that sometimes obscured drainage pipe responses. However, on the whole, courser grained soils proved to be a much better environment for mapping subsurface drainage systems than finer grained soils, because finer grained, high clay content soils, due to their greater electrical conductivity, reduced radar signal penetration depth, which in some cases prevented drainage pipe detection.

3.2.3 Shallow hydrology

Allred et al. (2005) studied the effects of shallow hydrology on GPR mapping of subsurface drainage systems using the specially designed and constructed test plot in Columbus, OH, USA, that had the capability for free drainage or water table positioning at any level between the ground surface and the drainage pipe depth. Shallow hydrologic conditions with saturated soil surrounding a water-filled drainage pipe (i.e. water table well above drain line) produced the poorest GPR drainage pipe detection results. Shallow hydrologic conditions with saturated or unsaturated soil surrounding at least partially air-filled drainage pipe (i.e. water table below top of drain line) produced the best GPR drainage pipe detection results, especially if the ground surface is frozen. With wet soil surrounding a water-filled pipe, the dielectric constant contrast was such that the drainage pipe radar signal reflections were rather weak, while with wet soil surrounding at least a partially air-filled pipe, the dielectric constant contrast was such that strong drainage pipe radar signal reflections were produced. Frozen soil, such as occurs near the ground surface in winter months, can be an advantage to GPR, because frozen soils have very low electrical conductivity, which reduces radar signal attenuation.

3.3 Computer processing

Allred et al. (2005) considered the importance of various computer processing steps with respect to GPR drainage mapping. The application of a signal saturation correction filter to remove low-frequency noise, along with a spreading and exponential compensation gain function to amplify radar signal, were the computer processing steps found to be most helpful for enhancement of the drainage pipe response exhibited within the GPR profiles. The GPR maps showing subsurface drainage pipe system patterns required additional computer processing, which included signal trace enveloping to convert all signal traces to positive monopulse records and two-dimensional (2D) migration to collapse hyperbolic responses to point features. In some cases, GPR maps required further processing with a high-frequency noise filter and/or a spatial background subtraction filter that suppressed horizontal features.

4 Ground-penetrating radar assessment of drainage pipe conditions and associated functionality implications

Allred and Redman (2010) carried out computer modeling and field research to document the GPR response to air-, water-, and soil-filled drainage pipes, with implications for determination of drain line functionality. During and shortly after a large rainfall event, farm field drain lines are often completely water filled. Several days later, once most of the excess soil water has been removed, drain lines are likely to be mostly air filled. Although fairly rare, drain lines can become partially or completely clogged with soil that enters the drain lines via joints between clay tile pipe segments or CPT perforations. Computer modeling with gprMax (Warren et al., 2016) provided initial indications of the GPR response to air-, water-, and soil-filled drainage pipes. The modeled GPR responses to air-, water-, and soil-filled drainage pipes were then confirmed by 250 MHz antenna GPR surveys carried out at a Columbus, OH, USA, test plot specially designed for this study (Allred and Redman, 2010).

Some of the research results are depicted in Fig. 5 with GPR profiles representative of measurement transects perpendicular to a drain line, and in Fig. 6 with GPR profiles representative of measurement transects over the top and along the trend of a drain line. Model-simulated profiles perpendicular to a drain line are provided in Fig. 5a–e. The modeled GPR response for an air-filled pipe (Fig. 5a), an air/soil-filled pipe (top one-third air, bottom two-third soil, Fig. 5b), and a water/soil-filled pipe (top one-third water, bottom two-third soil, Fig. 5d) is that of a single reflection hyperbola. The modeled GPR response to a water-filled pipe (Fig. 5c) is that of dual reflection hyperbolas, one directly

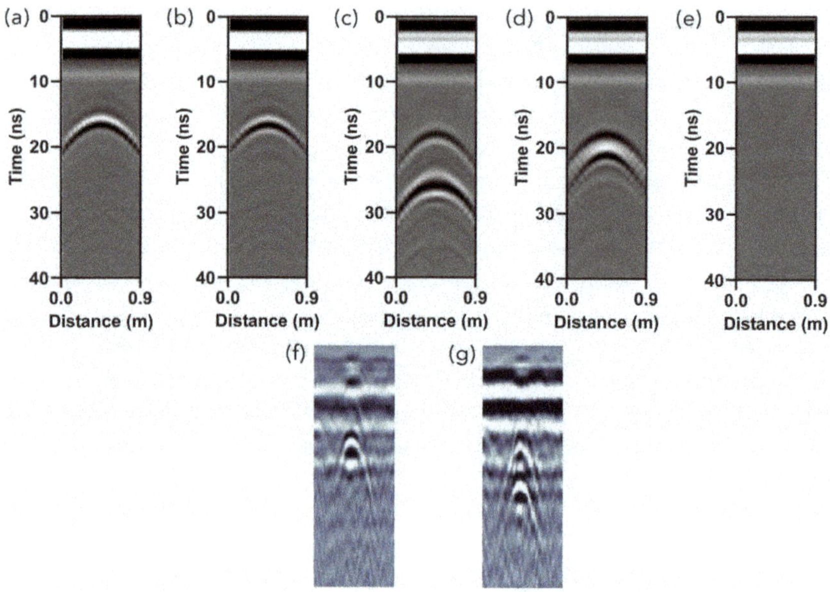

Figure 5 Computer-simulated and actual test plot GPR profiles perpendicular to the drain line: (a) simulation for air-filled pipe; (b) simulation of pipe that is one-third air filled at the top and two-third soil filled at the bottom; (c) simulation for water-filled pipe; (d) simulation of pipe that is one-third water filled at the top and two-third soil filled at the bottom; (e) simulation for soil-filled pipe; (f) the actual test plot air-filled pipe GPR response; and (g) actual test plot water-filled pipe GPR response.

above the other. The model simulation for a completely soil-filled pipe (Fig. 5e) indicates that no GPR reflection hyperbola response is produced.

Radar signals can be reflected off both the top and the bottom of a drainage pipe. Within a water-filled pipe, the radar signal velocity is low, and the reflections off the top and bottom of pipe remain separated in the signal trace record, thereby potentially producing a dual reflection hyperbola in GPR profiles. Within air-filled, air/soil-filled, and water-/soil-filled pipes, the radar signal velocity is relatively high, causing the reflections off the top and bottom of the pipe to become merged in a signal trace record, thereby producing a single GPR profile reflection hyperbola. A refection hyperbola response is not produced with a completely soil-filled pipe, because the dielectric constants of the materials inside and outside the pipe are likely the same, so without the presence of some dielectric constant contrast, drainage pipe reflections cannot occur. GPR profiles from actual test plot data collected along a measurement transect perpendicular to a drain line (Fig. 5f,g) confirm the computer simulation results. A single reflection hyperbola is shown in Fig. 5f for an air-filled pipe (compare to Fig. 5a), and a dual reflection hyperbola is shown in Fig. 5g for a water-filled pipe (compare to Fig. 5c).

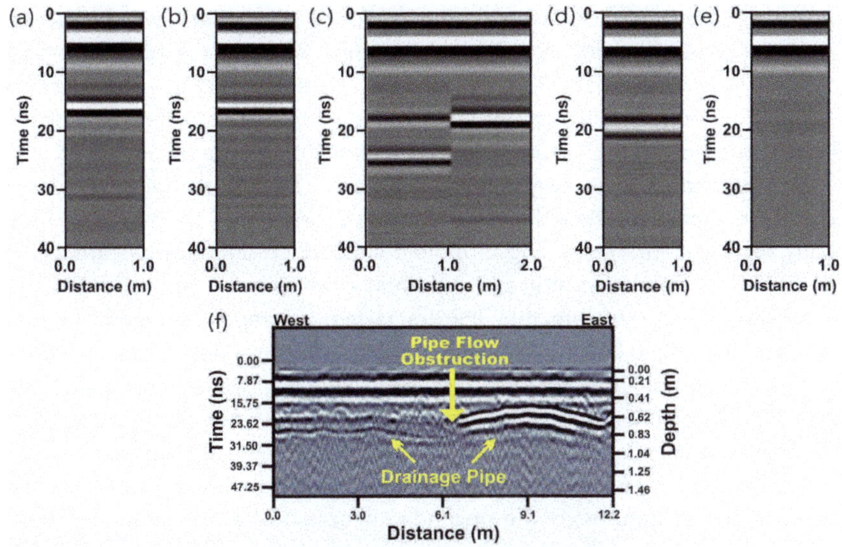

Figure 6 Computer-simulated and actual test plot GPR profiles over the top and along the trend of a drain line: (a) simulation for air-filled pipe; (b) simulation of pipe that is one-third air filled at the top and two-third soil filled at the bottom; (c) simulation for water-filled pipe on the left side that abruptly changes to an air-filled pipe on the right side; (d) simulation of pipe that is one-third water filled at the top and two-third soil filled at the bottom; (e) simulation for soil-filled pipe; and (f) actual test plot GPR response to water-filled pipe on the west side that abruptly changes to an air-filled pipe on the east side.

Model-simulated GPR profiles over the top and along the trend of a drain line are provided in Figs 6a–e. A banded, mostly horizontal, linear feature of some form is the typical drainage pipe response depicted in a GPR profile obtained from a measurement transect directly over the top and along the trend of a drain line. This modeled GPR profile banded, horizontal, linear response is fairly strong for an air-filled pipe (Fig. 6a and right side of Fig. 6c), an air-/soil-filled pipe (top one-third air, bottom two-third soil, Fig. 6b), and a water-/soil-filled pipe (top one-third water, bottom two-third soil, Fig. 6d) but somewhat weak for a water-filled pipe (left side of Fig. 6c). Where the pipe is completely soil filled, there is again, for the reason previously provided, no GPR drainage pipe response. A GPR profile from actual test plot data collected over the top and along the trend of a drain line (Fig. 6f) confirms the computer simulation results. Given the same scenario modeled for Fig. 6c, the Fig. 6f profile from test plot GPR data shows a similar response to a drain line that abruptly changes from a water-filled pipe on the west side (represented by a weak banded, horizontal linear feature) to an air-filled pipe on the east side (represented by a strong banded, horizontal, linear feature).

Differences in the GPR response to air- and water-filled drainage pipes can be employed to provide insight into whether a drain line is functioning properly, particularly in terms of determining the presence of a water flow obstruction. A drainage pipe obstruction that blocks the flow of water will prevent proper soil drainage in field areas adjacent to the drain line segment upgradient of the pipe obstruction. A localized obstruction could be due to a crushed, completely severed, or clogged pipe present at some point along a drain line.

As mentioned previously, several days after a large rainfall event, once most of the excess soil water has been removed, drain lines are very likely to be mostly air filled in a functioning farm field subsurface drainage system. Given this scenario, several days after a large rainfall event with a properly functioning subsurface drainage system, the GPR drainage pipe responses across the farm should be indicative of air-filled pipes (single reflection hyperbolas or strong banded, horizontal linear features). However, due the presence of a localized water flow obstruction, the drain line segment upgradient of the obstruction can remain water filled for days after a large rainfall event. Therefore, water-filled GPR pipe responses (dual reflection hyperbolas and weak banded, horizontal linear features) in the upgradient section of a drain line can reveal and help isolate a water flow obstruction. A good example proving that GPR can determine the location of a drainage pipe obstruction is provided in Fig. 6f. One complicating matter with this approach for isolating pipe obstructions is the observation that dual reflection hyperbolas are not always found for water-filled pipes depicted in GPR profiles obtained from measurement transects perpendicular to drain lines. Furthermore, this approach does not take into account the fairly rare occurrence of drainage pipes that are partially or completely soil filled.

5 Effects of ground-penetrating radar antenna orientation relative to drain line directional trends

GPR systems are typically employed with the antennas oriented perpendicular to the direction of travel along a measurement transect. However, many of these GPR systems provide the flexibility of repositioning the antennas so that they are oriented parallel to the direction of travel along a measurement transect. Allred (2013) employed a GPR system with the capability of positioning 250-MHz antennas either perpendicular or parallel to the measurement transect in order to investigate GPR drainage pipe response effects associated with antenna orientation relative to drain line trends.

The same specially constructed test plot used in the study by Allred and Redman (2010) was employed in the research carried out by Allred (2013). For this test plot, four east–west drain lines were installed in addition to two east–west drain lines that were already in place. The GPR data in this study were collected under moderately dry and very wet soil conditions. The test plot under moderately dry conditions had an average near-surface soil volumetric water content of 0.18, and none of the drain lines contained any water. The test plot under very wet but somewhat drained conditions had a near-surface average soil volumetric water content of 0.43, and the drain lines were mostly empty of water on the west side of the test plot, while mostly full of water on the east side of the test plot. Figure 7 shows test plot GPR maps under different field wetness conditions and with GPR antennas either parallel or perpendicular to the drain lines. Each of the GPR maps was produced from a parallel set of north-south GPR transects spaced 1.5 m apart.

The overall study results provided in Fig. 7 indicate that there are substantial differences in the strength of the GPR drainage pipe response for an antenna orientation perpendicular to a drain line versus an antenna orientation parallel to a drain line. Under the moderately dry soil conditions, a GPR antenna orientation perpendicular to the drain line provided the best GPR drainage pipe response, as indicated by drain lines that are more clearly depicted in Fig. 7b than Fig. 7a. Conversely, under very wet soil conditions, a GPR antenna orientation parallel to a drain line provided the best GPR drainage pipe response, as indicated by drain lines that are more clearly depicted in Fig. 7c than Fig. 7d. These findings imply that on-site assessment of field wetness conditions and knowledge of drain line directional trends can guide the GPR system antenna setup and formulation of a GPR survey plan to optimize GPR drainage pipe detection and assessment capabilities.

6 Integration of ground-penetrating radar with real-time kinematic global navigation satellite system technology

Producing the Figs 3c 7 GPR maps depicting drain line patterns involved collecting GPR data with either one set of parallel measurement transects (Fig. 7) or two sets of parallel measurement transects oriented perpendicular to one another (Fig. 3c). In both cases, the spacing distance between adjacent transects was 1.5 m. Collecting GPR data in this manner on small test plots having surface areas of several hundreds, or even a couple thousand, square meters is certainly feasible. However, this kind of intense GPR data collection strategy would be impractical at the farm field scale (i.e. 10 s of hectares), due to the time and effort required to collect and process such large amounts of GPR data.

Allred et al. (2018a) developed a more efficient approach integrating GPR with real-time kinematic (RTK) global navigation satellite system (GNSS) technology in which data were collected along a limited number of select measurement transects to provide insight on drain line patterns in small farm fields. This approach was tested at three small farm field sites - two near Beltsville, MD, USA and one near Columbus, OH, USA. The 3 field sites ranged

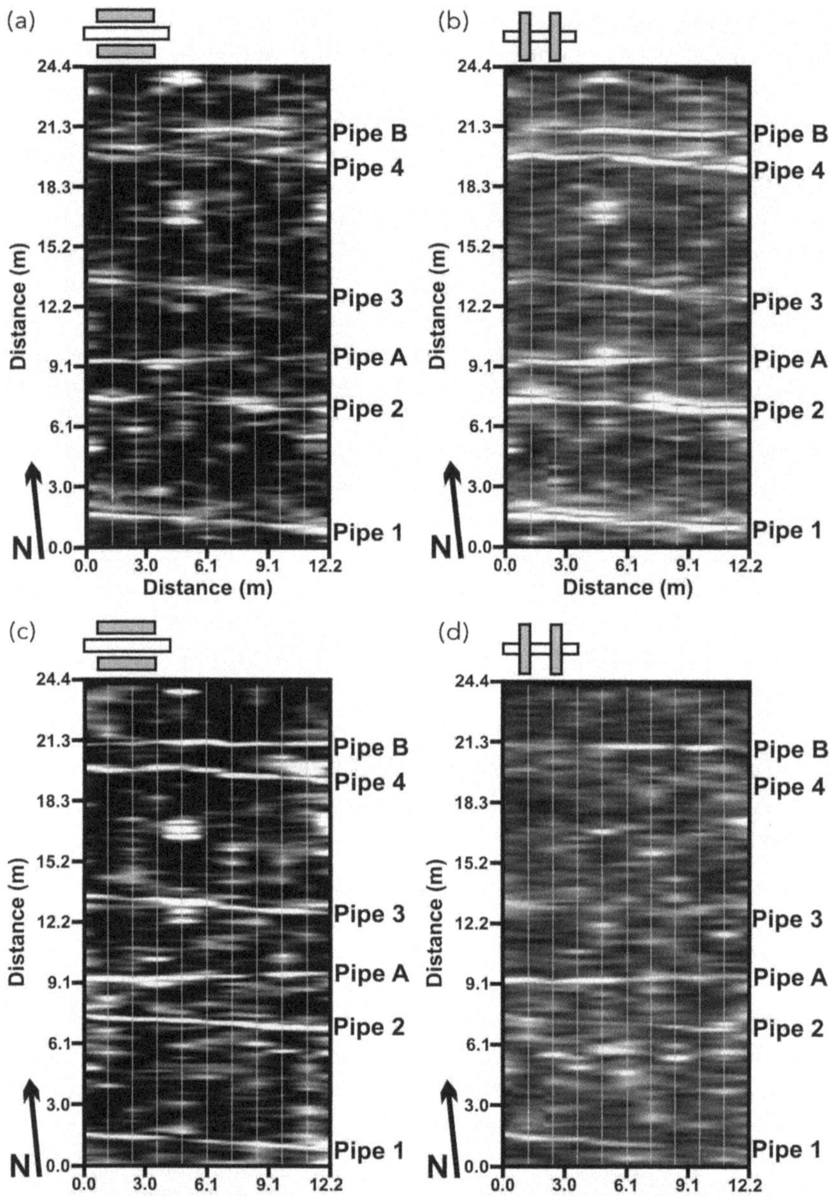

Figure 7 GPR map results from a depth interval of 0.46–0.76 m: (a) moderately dry conditions with antennas parallel to drain lines; (b) moderately dry conditions with antennas perpendicular to drain lines; (c) very wet conditions with antennas parallel to drain lines; and (d) very wet conditions with antennas perpendicular to drain lines. The icon on the top right above for each GPR map indicates the data collection approach for each map regarding orientation of antennas (gray bars) relative to drain lines (white bar).

in size from 6 ha to 13 ha. Results from one of the sites near Beltsville, MD, USA, highlighting this GPR drainage mapping approach are provided in Fig. 8.

Current GPR systems are equipped with digital video loggers (DVLs) capable of real-time display of a GPR profile as data is collected along a measurement transect. At any point along a measurement transect where the DVL GPR profile indicates a possible drainage pipe response (i.e. reflection hyperbola or banded, horizontal, linear feature), a spiral or serpentine measurement path can then be employed in the vicinity to confirm or repudiate drain line presence (Fig. 8a). If drain line presence is confirmed by more than one DVL GPR profile drainage pipe response along the localized spiral or serpentine path (Fig. 8a), the integration of GPR with RTK/GNSS technology will not only provide accurate and precise drainage pipe latitude/longitude coordinates but also the directional trend of the drain line at this location (blue line in Fig. 8a). After the localized spiral or serpentine path is completed, data

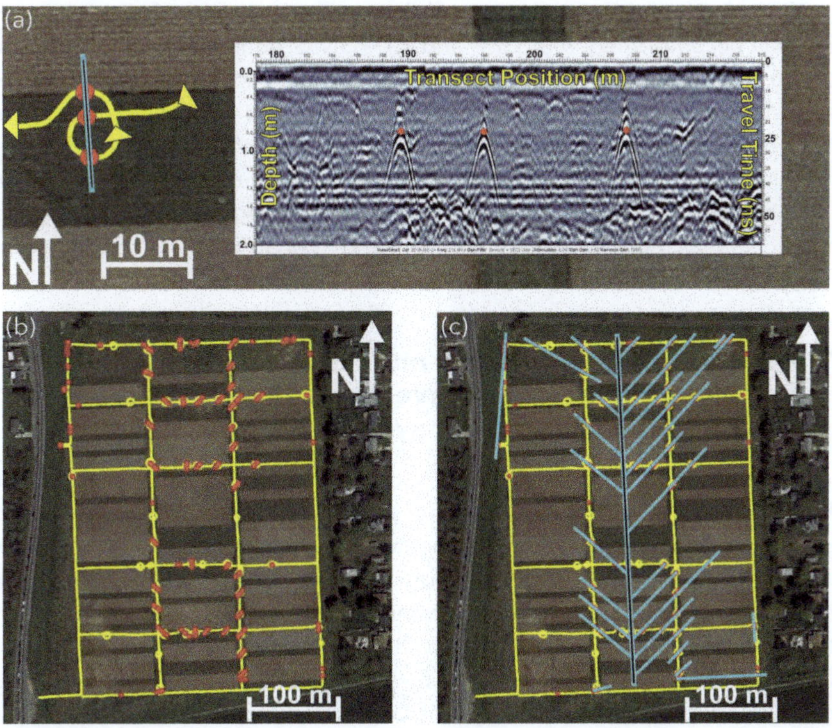

Figure 8 Results from a field site near Beltsville, MD, USA, highlighting an approach integrating GPR with RTK/GNSS technology to map a subsurface drainage system: (a) aerial map of localized spiral measurement path (yellow line) with drain line locations (red dots) detected by a DVL GPR profile similar to inset image; (b) aerial map depicting GPR measurement transects (yellow lines) and drainage pipe locations (red dots); and (c) aerial map with interpreted drain lines (blue lines).

collection continues along the overall measurement transect until another potential DVL GPR profile drainage pipe response is encountered.

If there is no prior knowledge of the drain line directional trend, or the subsurface drainage system pattern is expected to be complex, then the selected measurement transects need to be bidirectional - for example, transects oriented both east-west and north-south or transects oriented both southwest-northeast and southeast-northwest. Just such sets of select measurement transects, essentially east-west and north-south, are depicted with yellow lines on the Fig. 8b aerial image map. Where GPR drainage pipe responses were encountered along a transect, a spiral GPR path was deployed to confirm the presence of the drain line and determine its trend at that location (marked with red dots on Fig. 8b). As shown in Fig. 8c, blue lines were drawn through drainage pipe red dot locations that were in-line with one another, thereby providing evidence of a subsurface drainage system having a herringbone pattern.

GPR data collection took around 5 h at the 11 ha site shown in Fig. 8, and was accomplished with GPR and RTK/GNSS equipment mounted on a cart pushed by hand (Allred et al., 2018a). GPR data collection time could be reduced somewhat by mounting the GPR system and RTK/GNSS equipment on a sled pulled by an all-terrain vehicle (ATV). Further efficiency improvements might have been obtained with a multichannel GPR system described in Section 7 of this book chapter. Using the GPR drainage mapping approach highlighted in Fig. 8, and by mounting the GPR system and RTK/GNSS equipment on a sled pulled by an ATV, it is possible to cover up a 25 ha field area in a single day.

7 Drainage mapping with a multichannel, stepped-frequency, continuous-wave three-dimensional ground-penetrating radar system

Research reported in prior sections (Sections 2, 3, 4, 5, and 6) of this book chapter employed a GPR system with a single transmitting/receiving antenna pair having a specified center frequency (see Fig. 2). Koganti et al. (2020) evaluated a multichannel, stepped-frequency, continuous-wave 3D GPR system for mapping subsurface drainage systems at 12 field sites in Denmark. For this study, a GeoScope Mk IV 3D-Radar with DXG1820 Antenna Array (3D-Radar AS, Trondheim, Norway) was integrated with RTK/GNSS technology and towed by an ATV (Fig. 9a). Operationally, this GPR system collected data in a series of frequency steps (Fig. 9b). The total frequency range for the system was 60 MHz to 3000 MHz. Twenty-one bow-tie monopole antennas were employed by the system to provide 20 transmitting/receiving antenna pairs (i.e. channels), enabling 3D data collection along a 1.5 m swath in any travel direction of GPR measurement.

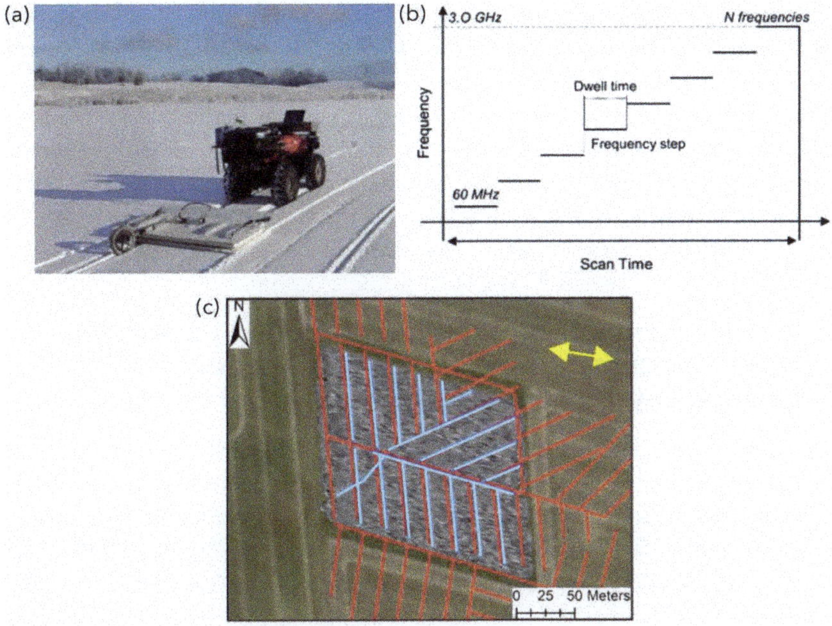

Figure 9 Multichannel, stepped-frequency, continuous-wave 3D GPR system: (a) photo of the system in operation; (b) GPR data is collected in a series of frequency steps; and (c) results from a site near Faardrup, Denmark, with blue lines representing drainage pipes detected by GPR, red lines depicting known drainage pipe locations, and the yellow line with arrowheads at each end indicating the GPR travel direction.

This multichannel, stepped-frequency, continuous-wave 3D GPR system proved successful for mapping agricultural subsurface drainage systems (i.e. 50% of drain lines detected) at 5 of the 12 study sites in Denmark (Koganti et al., 2020). Generally, the best drainage mapping results were found at field sites where the bulk soil electrical conductivity, from the surface down to a depth of 1.5 m, measured via a Dualem, Inc. (Milton, Ontario, Canada) DUALEM 21S ground conductivity meter, was <25 mS/m, which in turn allowed for better radar signal penetration. One example of subsurface drainage mapping success is shown in Fig. 9c for a field site near Faardrup, Denmark. Soil electrical conductivity at the Faardrup, Denmark, site averaged 21.3 mS/m. As shown in Fig. 9c, the multichannel, stepped-frequency, continuous-wave 3D GPR system mapped most of the known subsurface drainage systems (compare blue and red lines within gray shaded GPR survey area).

One advantage to using a multichannel, stepped-frequency, continuous-wave 3D GPR system for mapping subsurface drainage systems is the ability during data processing to choose the best radar frequency that provides optimal drainage pipe resolution and signal penetration depth. Another advantage is that this GPR system integrated with RTK/GNSS technology collects three-dimensional data in a 1.5 m swath along a direction of measurement. Consequently, with a single pass over a drain line, this system not only obtains the drain line location but also determines the drain line directional trend (localized spiral of serpentine measurement paths are not needed). The disadvantages of using a multichannel, stepped-frequency, continuous-wave 3D GPR system for drainage mapping are that it is expensive, field setup requires greater effort, and data processing is more complex as compared to a single-channel GPR system with antennas having a specified center frequency.

8 Complementary employment of ground-penetrating radar and unmanned aerial vehicle imagery for Drainage System Characterization

GPR, even when combining integration with RTK GNSS technology and mounting equipment on a sled pulled by an ATV, is still somewhat inefficient for mapping subsurface drainage systems in areas larger than 25 ha. However, recent research (Tlapáková et al., 2017; Allred et al., 2018b, 2020, 2021; Freeland et al., 2019; Williamson et al., 2019; Kratt et al., 2020) found that visible, multispectral (MS), and thermal infrared (TIR) imagery obtained by unmanned aerial vehicles (UAVs) has proven capability for determining drain line patterns in large farm fields. A UAV flight plan covering an 80 ha farm field can easily be completed in less than an hour, so data collection with this technology is quite efficient time wise.

After a rainfall event, the greatest water removal and drying occurs first in the soils directly above a drain line. A dry soil surface reflects a greater amount of visible and near-infrared wavelength radiation than a wet soil surface. Subsequently, given bare ground conditions outside the growing season, lighter shaded linear features exhibiting increased reflected radiation that are depicted on UAV-visible and MS imagery may be indicative of drain lines. (Note: UAV-visible cameras provide imagery in a broad bandwidth of wavelengths from 400 nm to 700 nm, while a MS camera produces separate images in narrow bands of blue, green, red, red edge, and near-infrared wavelengths.) Additionally, soil water content differences over drain lines versus between drain lines can result in emitted thermal radiation differences that are detected by a UAV TIR camera, with the apparent radiant temperature image generated then employed to map buried drainage pipes. Crop establishment and health during the growing season are often better over the drain lines versus between drain lines. Therefore, spatial variations in crop establishment/health shown on UAV-visible, MS, and TIR imagery sometimes depict agricultural subsurface drainage system patterns.

Allred et al. (2020) conducted UAV surveys with visible-color (VIS-C), MS, and TIR cameras at 29 agricultural field sites in the Midwest USA, and overall results showed that VIS-C, MS, and TIR imagery detected at least some of the drainage pipes present at 48%, 59%, and 69% of the sites, respectively. Three key findings were extracted from the overall results obtained by Allred et al (2020): (1) although TIR generally worked best, there were sites where either VIS-C or MS proved more effective than TIR for mapping subsurface drainage systems (Fig. 10); (2) timing of UAV surveys relative to recent rainfall

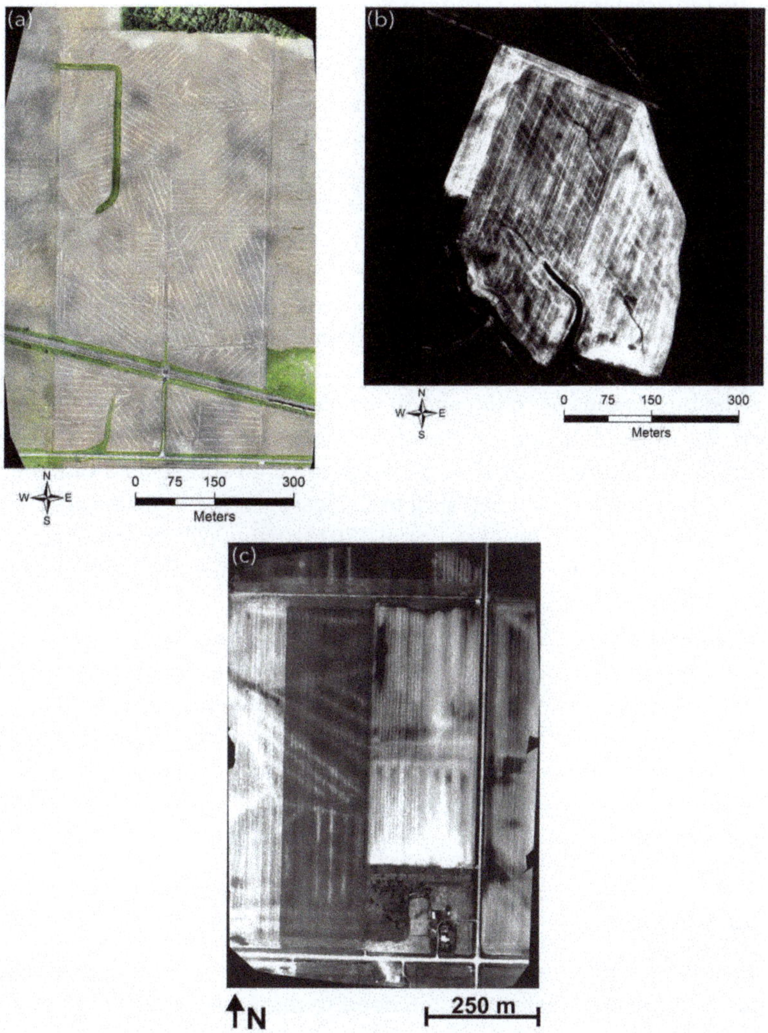

Figure 10 UAV survey results depicting subsurface drainage system patterns: (a) VIS-C image from Seneca County, OH, USA; (b) MS normalized difference vegetation index image (combines red and near-infrared wavelength data) for Ross County, OH, USA, farm field with fully established wheat cover; and (c) TIR image from Clay County, IA, USA.

can sometimes have an important impact on drainage pipe detection results; (3) linear features representing drain lines and farm field operations are both commonly depicted in farm field UAV imagery and can be confused with one another. However, Allred et al. (2020) developed UAV imagery guidelines to distinguish linear features representing drain lines from those representing farm field operations based on knowledge of subsurface drainage system installation and farm field operations. Allred et al. (2021) demonstrated that the UAV TIR drainage pipe response can change during the day. This study determined that the soil surface apparent radiant temperature directly over drain lines is greater than between the drain lines from late morning until late afternoon, while conversely, at sunrise or sunset, the soil surface apparent radiant temperature over drain lines is frequently less than between drain lines. Allred et al. (2021) further established that there could be times of the day, especially early morning, when the soil surface apparent radiant temperature over the drain lines is the same as between the drain lines, and when this is the case, there is no UAV TIR drainage pipe response.

Although UAV imagery can be more efficient than GPR for mapping subsurface drainage systems in large farm fields, GPR can still provide complementary benefits, especially when used in conjunction with UAV surveys. GPR is the best means of confirming drain lines interpreted from UAV imagery. Sometimes, even after application of the developed guidelines, uncertainty remains as to whether a UAV image linear feature represents a drain line or is due to farm field operations, and, in this case, GPR can provide a solution to the problem. Studies by Allred et al. (2018b) and Williamson et al. (2019) both employed GPR to confirm the location of drain lines interpreted from UAV imagery.

GPR also supplies information on subsurface drainage system infrastructure characteristics that cannot be obtained from UAV imagery. In particular, GPR can determine the drain line depth and gradient. Koganti et al. (2021) found field locations where GPR was successful in finding drainage pipes, but UAV imagery was not. Consequently, the ideal approach for mapping subsurface drainage systems is the complementary utilization of both GPR measurements and UAV imagery.

9 Conclusion

Agricultural subsurface drainage systems remove excess soil water, thereby improving crop yields. Farmers often need to repair or retrofit these drainage systems, but in order to do so, maps of the preexisting drain lines are required. Environmental risk assessment regarding farm field discharge of nitrate/phosphate via subsurface drainage can be improved through knowledge of the installed drainage pipe network, including the extent of coverage and

Published by Burleigh Dodds Science Publishing Limited, 2023.

drain line spacing distance. The present methods for finding drainage pipes involving handheld tile probes or trenching equipment are time consuming and tedious and cause pipe damage. Regardless of whether the need is economic or environmental, there is a strong demand for effective, efficient, and nondestructive methods to map buried drainage pipes.

GPR may provide part of the solution to this problem. This book chapter focused on summarizing the research conducted over the past 20 years on mapping subsurface drainage systems with GPR. When evaluated with three other proximal soil-sensing methods (magnetometry, electromagnetic induction, and resistivity), GPR was the only one exhibiting capability for drainage pipe detection. Factors impacting GPR drainage pipe detection were then assessed (i.e. equipment and survey setup, site conditions, and computer processing). Investigations that followed focused on GPR drainage pipe response effects due to pipe conditions (air, water, and/or soil filled) and antenna orientation relative to the drain line directional trend. Highlighted next were GPR drainage mapping efficiency improvements through integration with RTK GNSS technology and the use of a multichannel, stepped-frequency, continuous-wave 3D GPR system. Discussed last was the complementary employment of GPR and UAV imagery for drainage system characterization. The research described in this chapter indicates that without a doubt, GPR can play an important role with respect to mapping subsurface drainage systems across agricultural landscapes.

10 Future trends in research

Further research combining the use of GPR and UAV imagery for drainage mapping is definitely warranted. Future investigations should focus on developing site condition (e.g. soil type, time of day/year, wetness/dryness, and surface characteristics) guidelines for drainage mapping with UAV imagery. GPR can be an important component of this research by providing information on shallow hydrologic conditions and confirmation of drain line locations.

11 Where to look for further information

The following books provide a good overview of GPR:

- Conyers, L. B. 2004. *Ground-Penetrating Radar for Archaeology*. AltaMira Press, Walnut Creek, California, 203 pages.
- Jol, H. M. (Editor). 2009. *Ground Penetrating Radar: Theory and Applications*. Elsevier Science, Amsterdam, Netherlands, 524 pages.

The following book provides a good overview of agriculture subsurface drainage practices.

- Smedema, L. K., Vlotman, W. F. and Rycroft. D. W. 2004. *Modern Land Drainage: Planning, Design and Management of Agricultural Drainage Systems*. A. A. Balkema Publishers, Leiden, Netherlands, 446 pages.

Key research in this area can be found at the following scientific organizations:

- American Society of Agricultural and Biological Engineers (https://www.asabe.org).
- Environmental and Engineering Geophysical Society (http://www.eegs.org).
- Society of Exploration Geophysicists (http://www.seg.org).

12 References

Allred, B. J., Fausey, N. R., Peters, L. Jr., Chen, C., Daniels, J. J. and Youn, H. 2004. Detection of buried agricultural drainage pipe with geophysical methods. *Appl. Eng. Agric.* 20(3), 307–318. https://doi.org/10.13031/2013.16067.

Allred, B. J., Daniels, J. J., Fausey, N. R., Chen, C., Peters, L. Jr. and Youn, H. 2005. Important considerations for locating buried agricultural drainage pipe using ground penetrating radar. *Appl. Eng. Agric.* 21(1), 71–87. https://doi.org/10.13031/2013.17915.

Allred, B. J. and Redman, J. D. 2010. Location of agricultural drainage pipes and assessment of agricultural drainage pipe conditions using ground penetrating radar. *J. Environ. Eng. Geophys.* 15(3), 119–134. https://doi.org/10.2113/JEEG15.3.119.

Allred, B. J. 2013. A GPR agricultural drainage pipe detection case study: effects of antenna orientation relative to drainage pipe directional trend. *J. Environ. Eng. Geophys.* 18(1), 55–69. https://doi.org/10.2113/JEEG18.1.55.

Allred, B., Wishart, D., Martinez, L., Schomberg, H., Mirsky, S., Meyers, G., Elliott, J. and Charyton, C. 2018a. Delineation of agricultural drainage pipe patterns using ground penetrating radar integrated with a real-time kinematic global navigation satellite system. *Agriculture* 8(11), e167. https://doi.org/10.3390/agriculture8110167.

Allred, B., Eash, N., Freeland, R., Martinez, L. and Wishart, D. 2018b. Effective and efficient agricultural drainage pipe mapping with UAS thermal infrared imagery: a case study. *Agric. Water Manag.* 197, 132–137. https://doi.org/10.1016/j.agwat.2017.11.011.

Allred, B., Martinez, L., Fessehazion, M. K., Rouse, G., Williamson, T. N., Wishart, D., Koganti, T., Freeland, R., Eash, N., Batschelet, A. and Featheringill, R. 2020. Overall results and key findings on the use of UAV visible-color, multispectral, and thermal infrared imagery to map agricultural drainage pipes. *Agric. Water Manag.* 232, 1–19. https://doi.org/10.1016/j.agwat.2020.106036.

Allred, B., Martinez, L., Fessehazion, M. K., Rouse, G., Koganti, T., Freeland, R., Eash, N., Wishart, D. and Featheringill, R. 2021. Time of day impact on mapping agricultural subsurface drainage systems with UAV thermal infrared imagery. *Agric. Water Manag.* 256, 107071. https://doi.org/10.1016/j.agwat.2021.107071.

Chow, T. L. and Rees, H. W. 1989. Identification of subsurface drain locations with ground-penetrating radar. *Can. J. Soil Sci.* 69(2), 223–234. https://doi.org/10.4141/cjss89-023.

Freeland, R., Allred, B., Eash, N., Martinez, L. and Wishart, D. 2019. Agricultural drainage tile surveying using an unmanned aircraft vehicle paired with real-time kinematic positioning - a case study. *Comput. Electron. Agric.* 165, e104946. http://doi.org/10.1016/j.compag.2019.104946.

Koganti, T., Van De Vijver, E., Allred, B. J., Greve, M. H., Ringgaard, J. and Iversen, B. V. 2020. Mapping of agricultural subsurface drainage systems using a frequency-domain ground penetrating radar and evaluating its performance using a single-frequency multi-receiver electromagnetic induction instrument. *Sensors (Basel)* 20(14), 1-26, e3922. https://doi.org/10.3390/s20143922.

Koganti, T., Ghane, E., Martinez, L. R., Iversen, B. V. and Allred, B. J. 2021. Mapping of agricultural subsurface drainage systems using unmanned aerial vehicle imagery and ground penetrating radar. *Sensors (Basel)* 21(8), 1–31, 2800. https://doi.org/10.3390/s21082800.

Kratt, C. B., Woo, D. K., Johnson, K. N., Haagsma, M., Kumar, P., Selker, J. and Tyler, S. 2020. Field trials to detect drainage pipe networks using thermal and RGB data from unmanned aircraft. *Agr. Water Manag.* 229, e105895. https://doi.org/10.1016/j.agwat.2019.105895.

Pavelis, G. A. 1987. Economic survey of farm drainage. In: Pavelis, G. A. (Ed.), *Farm Drainage in the United States: History, Status, and Prospects, Miscellaneous Publication Number 1455*. U.S. Dept. of Agriculture and Economic Research Service, Washington, DC. pp. 110–136.

Sims, J. T., Simard, R. R. and Joern, B. C. 1998. Phosphorus loss in agricultural drainage: historical perspective and current research. *J. Environ. Qual.* 27(2), 277–293. https://doi.org/10.2134/jeq1998.00472425002700020006x.

Tlapáková, L., Žaloudík, J. and Kolejka, J. 2017. Thematic survey of subsurface drainage systems in the Czech Republic. *J. Maps* 13(2), 55–65//https://doi.org/10.1080/17445647.2016.1259129.

Warren, C., Giannopoulos, A. and Giannakis, I. 2016. gprMax: open source software to simulate electromagnetic wave propagation for ground penetrating radar. *Comput. Phys. Commun.* 209, 163–170. http://doi.org/10.1016/j.cpc.2016.08.020.

Williamson, T. N., Dobrowolski, E. G., Meyer, S. M., Frey, J. W. and Allred, B. J. 2019. Delineation of tile-drain networks using thermal and multispectral imagery – implications for water quantity and quality differences from paired edge-of-field sites. *J. Soil Water Conserv.* 74(1), 1–11. https://doi.org/10.2489/jswc.74.1.1.

Zucker, L. A. and Brown, L. C. 1998. *Agricultural Drainage: Water Quality Impacts and Subsurface Drainage Studies in the Midwest. Ohio State University Extension Bulletin 871*. Ohio State University Press Extension, Columbus, OH. pp. 40.

Part 3

Advances in sensor data analytics

Chapter 8

Advances in machine vision technologies for the measurement of soil texture, structure and topography

Jean-Marc Gilliot, AgroParisTech Paris Saclay University, France; and Ophélie Sauzet, University of Applied Sciences of Western Switzerland, The Geneva Institute of Technology, Architecture and Landscape (HEPIA), Soils and Substrates Group, Institute Land-Nature-Environment (inTNE Institute), Switzerland

1 Introduction

Although soils are essential for crop production and more generally for life on the Earth, soil resources are under great pressure, and sustainable soil management (SSM) is an important element of sustainable agriculture (FAO, 2015b). In 2015, which was declared as the International Year of Soils during the 68th UN General Assembly, FAO member countries updated the World Soil Charter, originally adopted in 1981, to institutionalize SSM (FAO, 2015a) through 13 principles set out in this guide and more specifically the third principle that defines SSM as:

> Soil management is sustainable if the supporting, provisioning, regulating, and cultural services provided by soil are maintained or enhanced without significantly impairing either the soil functions that enable those services or biodiversity. The balance between the supporting and provisioning services for plant production and the regulating services the soil provides for water quality and availability and for atmospheric greenhouse gas composition is a particular concern.

http://dx.doi.org/10.19103/AS.2022.0107.15

According to the FAO 'Voluntary Guidelines for Sustainable Soil Management' (FAO, 2017), the main ecosystem services provided by soils concerned by this previous definition are supporting services (primary production); provisioning services (food supply); regulating services (water supply); and cultural services (aesthetic). Figure 1 shows the major soil characteristics that are associated with SSM.

Soil can be described as a dispersed system composed of three phases: solid (soil matrix), liquid (water in the soil) and gaseous (air in the soil); the solid matrix is composed of particles of varying composition and shape, including mineral grains and organic matter which can be linked together in aggregates (Hillel, 2003). The spatial arrangement of solid soil constituents into aggregates separated by voids or pores defines the soil structure (Schlüter et al., 2020). Biotic and abiotic factors are responsible for the arrangement of aggregates and pores; in the case of agricultural soils, agricultural practices such as

Figure 1 Soil characteristics associated to sustainable soil management (SSM). Inspired by FAO (2017).

ploughing, various tillage operations and compaction due to the passage of farm machinery are also involved. The particles of the solid phase are generally classified according to their size; the larger ones are visible to the naked eye, while the smaller ones can only be observed under a microscope. Soil texture is the proportion of different particle size ranges that make up a given soil, and it is the intrinsic attribute of the soil that is most often used to characterize its physical composition; three major size classes, called textural fractions, are named in order of increasing size: clay, silt and sand (Hillel, 2003). Vertically, soils are organized according to differentiated strata called horizons, which constitute a homogeneous volume. A vertical succession of horizons constitutes a soil profile, and these horizons are also distributed laterally at the landscape scale (Baize et al., 2013). On the surface, variations in soil height compared to the average ground level define the microtopography of the soil, also called soil surface roughness (SSR). These small variations are usually not recorded when creating topographic maps.

The main ecosystem functions of soil, such as water storage and transport, nutrient cycling and carbon storage, which are essential for plant growth and the maintenance of biodiversity, depend largely on soil structure and texture (Bünemann et al., 2018; Schlüter et al., 2020). SSR is involved in several important processes that can have a significant impact on crop production, including run-off, soil loss, water storage and infiltration or soil deposition (Vermang et al., 2015). The result of these processes is to determine water stagnation time on the soil surface and therefore its possible infiltration to be available to the crop roots, as well as the depletion of the cultivated soil horizon in nutrients due to erosion. Soil workability, which refers to the ease of tillage, also depends on the soil's interrelated characteristics of texture (FAO, 2015b). For these reasons, it is important to measure these soil parameters to determine the agronomic value of an agricultural plot or of an area within a plot in the case of precision agriculture. Within the framework of SSM, the adoption of no-till (NTi) practices instead of conventional tillage (CTi) certainly has impacts on the soil structure, and studies are underway to evaluate them (Malobane et al., 2019).

Classical methods generally used to estimate parameters of texture, structure and SSR of soils are often *ex situ* physicochemical laboratory analyses, performed on soil samples, for quantitative and precise results. For soil structure, the most common method is to measure bulk density and deduce porosity from the dry weight of undisturbed soil cores of a given volume (Schlüter et al., 2020). To measure textural fractions, mechanical analysis is used to determine the particle size distribution of soil samples (Hillel, 2003) dispersing the soil in an aqueous suspension. The separation of the particles into size groups is done by passing the suspension through graduated sieves; for very fine particles, the sedimentation method is generally used; for instance, the pipette method (Robinson, 1922) is considered the standard method.

Optical sizing, x-ray absorption and laser diffraction methods are also used to measure textural fractions (Vitton and Sadler, 1997; Fisher et al., 2017). In some cases, the reference measurement can only be performed *in situ*, as for the measurement of soil roughness using the chain method (Saleh, 1993), which consists of measuring the length of a chain according to the roughness of the ground, which gives an indication of the soil tortuosity.

Classical methods are laborious, time consuming and often expensive; several proximal sensing methods, which measure in close proximity to or in contact with the soil, can be used as an alternative (Viscarra Rossel and McBratney, 1998; Viscarra Rossel et al., 2011) (Fig. 2). The measurement is in some cases made *in situ*, with or without sensor in the field, or *ex situ*, on a horizon soil sample for later analysis in the laboratory. A sensor measures

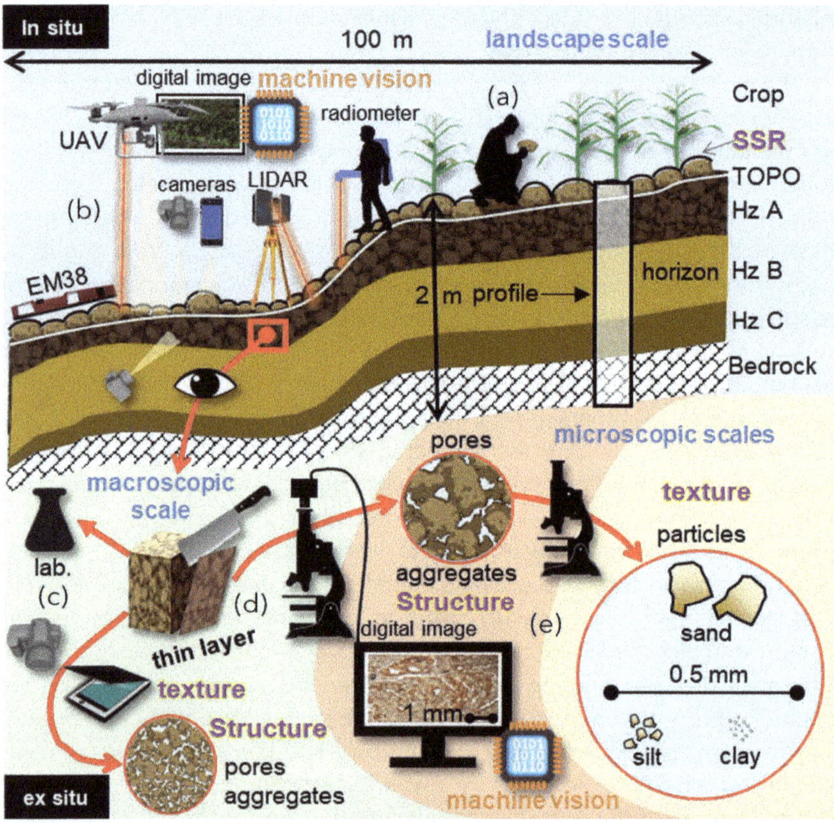

Figure 2 Proximal soil sensing technologies for the measurement of soil texture, structure and topography (SSR) with *in situ* and *ex situ* optical sensors. (a) 'By feel' visual or tactile estimation of the soil; (b) *in situ* sensors; (c) laboratory analyses on disturbed soil samples; (d) resin-impregnated block of soil is cut with a diamond saw to produce a thin layer; (d) microscopic observations on a thin layer.

the interaction of a signal with the components of the soil to analyse some of its properties. If the sensor has an artificial source to generate this signal, it is said to be active; it is passive if it uses a natural signal such as the solar radiation. If the measurement of the property is based on a physical process in the soil, the sensor is said to be direct, whereas when the property is inferred from the measure using a so-called pedotransfer function, then the sensor is indirect (Viscarra Rossel and Adamchuk, 2013). The same sensor, depending on the targeted soil property, can be used in direct or indirect mode; thus, a field radiometer used to measure soil surface properties can give a direct measurement (surface reflectance) or an indirect one (organic matter content). A pedotransfer rule can be a linear equation, established by regression on reference soil samples, predicting organic matter from reflectance, for example. Some simplified methods only require a visual or tactile estimation of the soil. Eyes and fingers are then the sensors of measurement; thus, several guides to 'texture by feel' exist to manually estimate the texture class of a soil corresponding to the soil textural triangle (Thien, 1979). Drop-shatter tests are visual soil structure assessment methods for which a spade of soil is dropped from a height of 1 m, and then a score is calculated from the size distribution and shape of the fallen aggregates (Ball et al., 2007; Emmet-Booth et al., 2016). For *in situ* measurements, geophysical sensors are also commonly used to map soil properties in agricultural plots; they generally carry out continuous measurements, and coupled with a GNSS (Global Navigation Satellite System) receiver, they allow to produce maps and are often referred to as 'on-the-go' sensors (Adamchuk et al., 2004). The sensors of soil electrical resistivity (or conductivity) are among the most used; they acquire a measure on a volume of the soil of the order of one or 2 m deep, and this measure is related to the thickness of the horizons, to their textural composition and to the water content, which is related to the pore spaces and thus to the structure. Domsch and Giebel (2004) obtained a good relationship between soil electrical conductivity in vertical dipole mode (1.5 m deep) and clay and silt content.

However, while general relationships can be found, measurement is most often 'indirect'. Machine vision sensing methods can provide more 'direct' and better measurement; for this reason, image analysis techniques have received great attention and shown potential as proximal sensors for soil analyses (Fig. 2).

In this chapter, we present recent advances in machine vision technologies for the measurement of soil texture, structure and topography with a focus on microtopography by three-dimensional (3D) analysis at a macroscopic scale and the analysis of the structure from soil thin layers at a microscopic scale.

2 Basic principles

2.1 Machine vision for the measurement of soil texture, structure and topography

According to Smith et al. (2021), machine vision (MV) can be defined as 'the practical realization of image understanding, or more specifically computer vision (CV) techniques, to help solve practical industrial problems involving a significant visual component'. Since the beginning of the 1960s, research in MV has attempted to bring the recognition capabilities of machines closer to those of human vision, with moderate success. But more recently, the combination of MV and machine learning methods has allowed significant progress towards this goal. Variations of texture, structure and SSR are sometimes visible in digital images of soil acquired at macroscopic or microscopic scales; many authors have then developed methods to quantify these elements using MV techniques. The elements to be analysed are in some cases directly visible in the images – for example, in the case of soil structure at the microscopic scale, aggregates and pores appear with different colours or image textures (Fig. 28) and can be separated by image segmentation/ classification techniques and their respective surfaces estimated to characterize the structure of the studied soil (Sauzet et al., 2017). In other cases, the element detection in the images is indirect – for example for the measurement of texture, studies have shown significant correlations between clay and silt content and soil colour (Konen et al., 2003) or lightness (Krishna Murti and Satyanarayana, 1971); at a macroscopic scale, soil roughness can be estimated by quantifying shadows in an image of the soil surface (García Moreno et al., 2008). Finally, in some cases, the image undergoes a change of representation or dimension, to make the elements detectable – for example for the measurement of SSR, from a 2D image, or more precisely from a series of images, a 3D model of the soil surface is created by a photogrammetric method, and it is from this 3D model that the soil roughness is estimated (Gilliot et al., 2012).

Data are acquired *in situ* (Fig. 2) or *ex situ* on soil samples returned to the laboratory. Optical *in situ* sensors can be imaging sensors (camera) or not; they are then generally spectroradiometers (spectral analysis), colour sensors (nix, 2022) or LIDAR (Light Detection And Ranging). In-field spectral proximal sensing of soils by reflectance spectroscopy in the visible (VIS), near-infrared (NIR), shortwave infrared (SWIR), mid-infrared (MIR), x-ray fluorescence (XRF) and gamma-ray ranges was used to predict the texture of soils (Viscarra Rossel et al., 2006; van Egmond et al., 2018; Pham et al., 2021; Naimi et al., 2022). The most common sensors are red, green and blue wavelength (RGB) cameras, smartphone or unmanned aerial vehicle (UAV) cameras. Azadnia et al. (2022)

proposed a portable smartphone-based machine vision system, in which texture images were taken from 20 cm, 40 cm and 60 cm heights and were analysed using a neural network. Konen et al. (2003) used a Minolta CR-310 chroma meter. In the field, the measurements are most of the time carried out towards the soil surface, but some authors have proposed methods using images acquired on a soil profile perpendicular to horizons. Thus, Aitkenhead et al. (2016) estimated the structure, texture and drainage of soil horizons from smartphone and digital cameras; Pham et al. (2021) with portable XRF and VIS-NIR data; Aliah Baharom et al. (2015) have experimented in a paddy field a real-time soil sensor composed of two spectrophotometers mounted on a tractor with gage wheels on both sides allowing to bring out the soil at a certain depth to measure its spectrum. SSR measurements are most often made *in situ* from a series of images from RGB cameras or laser measurements with LIDAR sensors; these techniques will be developed in Section 2.2 of the chapter. Regarding the use of *ex situ* sensors, the samples are usually prepared before measurement; most often, they are sifted to eliminate the coarse elements and then dried (Sudarsan et al., 2018), either in the open air or in an oven; because the water content influences the perception of the colour, the more the soil is rich in water the darker it will appear. In the case of structure analysis of undisturbed samples, special sampling techniques may be necessary. Soil blocks are carved out of one horizon, and then at the laboratory, resin-impregnated block of soil is cut with a diamond saw to produce a thin layer, which can be studied under the microscope (Sauzet et al., 2017). Optical sensors that are most often used in the laboratory to measure soil texture and structure are: (i) RGB cameras; (ii) spectrometers with the same spectral ranges seen for the field devices (VIS, NIR, MIR, SWIR and XRF) – some can be used both in the field and in the laboratory; a 'contact probe' adapter with an integrated light source is often used to take measurements in contact with the soil sample (Coblinski et al., 2020); (iii) microscopes – Sudarsan et al. (2018) used a very compact-size microscope inserted in a Teflon holder which was equipped with seven LEDs for standardized lighting conditions; they characterized soil particle sizes using wavelet analysis of soil microscope images, and regression relationship showed a prediction capability of 88% for textural fractions; Sauzet et al. (2017) used an optical microscope Leica DM5500B combined with a digital camera, with an automated stage, to quantify porosity and illuvial clay in large soil thin sections (11 cm × 5 cm) with a 5.3 μm resolution; 576 pictures were automatically taken in 20 min in order to build a final mosaic of about 22 000 x 9000 pixels, covering the whole section; (iv) table scanners – de Oliveira Morais et al. (2020) analysed soil samples collected from different regions of Brazil; samples were transferred to a Petri dish and digitized as images using a table scanner, images were correlated to texture results determined by the standard pipette method, using multivariate linear regressions, and particle size was estimated with correlations above 90%; and (v) x-ray tomography has also been used to produce 3D data of soil (Joschko et al., 1991). Norouzi et al. (2021) introduced a new analytical model that links the particle size distribution and soil reflectance in NIR and SWIR bands of the electromagnetic spectrum, which suggests many future applications for texture prediction from optical measurements.

Data processing methods that were used by authors depend first of all on the type of raw data that were produced by the sensor (Fig. 3).

Although the processing chains greatly depend on the application, there are often several key steps: (i) preprocessing, (ii) feature extraction, (iii) feature classification or quantitative prediction, (iv) postprocessing and (v) object extraction (Fig. 3). For preprocessing, a colorimetric calibration is often performed to correct the effects of ambient lighting on colour, especially for *in situ* measurements, so Konen et al. (2003) used a calibration plate to calibrate soil colour measurements (Minolta CR-310 chroma meter). Geometric preprocessing is also often performed, for example by merging images that were acquired side by side (mosaicking), which allows covering a larger area of study than the sensor field allows, as Sauzet et al. (2017) do on thin-layer microscopic images. Spatial interpolation methods are widely used to generalize point measurements on a continuous surface, usually an image; inverse distance-weighted calculation is the simplest, as done by Gilliot et al. (2017) to produce a digital surface model (DSM) of soil height using a photogrammetric approach. The parameter that can be extracted from the data, for further processing, can be simply the RGB colour as provided by the sensor (Chung et al., 2010), or the colour recoded in another colour system such as Munsell (Konen et al., 2003), or hue, saturation, value (Barman and Choudhury, 2020), or an index combining several wavelengths (Sauzet et al., 2017). Sudarsan et al. (2018) developed a continuous wavelet transform algorithm to characterize soil particle sizes from digital images captured with a microscope; wavelets are a general way to analyse multiresolution images. Chung et al. (2010) used parameters derived from the image histogram.

Previously extracted parameters were generally processed by two main approaches: (i) classifications, to identify constituent elements such as soil aggregates and pores and (ii) prediction model, using a regression calculation between a soil parameter, for example the clay content and the image parameter; this model is then used to predict the clay content from the image.

Supervised or unsupervised classification algorithm was used, and Azadnia et al. (2022) developed a convolutional neural network to predict soil texture from images. Coblinski et al. (2020) applied a Cubist regression algorithm to soil spectroscopy data and standard soil particle size analysis and achieved high-accuracy prediction for both clay and sand content. For classification approaches, postprocessing is often applied to clean the classified image, which corresponds most often to the elimination of isolated pixels as well as

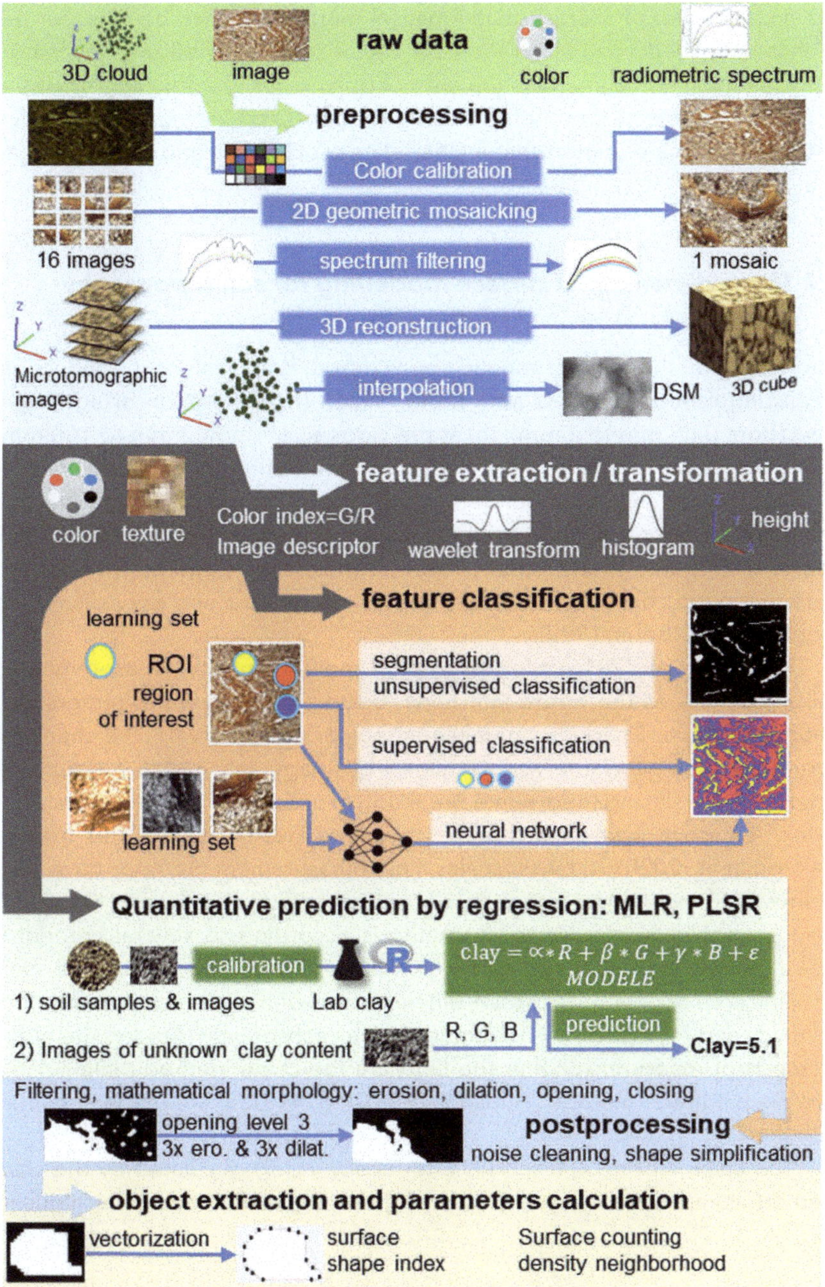

Figure 3 The main classical steps of computer vision processing for soil.

the simplification of shapes; mathematical morphology can be used for this. Such a simplified image can then be vectorized, which allows obtaining a representation of the image in the form of objects, for example aggregates vs. pores; these objects can then be easily characterized individually according to their surface, perimeter, shape (shape index), distance and density (Sauzet et al., 2017).

2.2 Three-dimensional surface modelling for soil topography analysis

Depending on the scale of variation considered, four levels of roughness can be distinguished (Römkens and Wang, 1986): (i) at the grain or aggregate level (from 0.05 mm to 2 mm); (ii) at the clods scale (from 2 mm to 100 mm); (iii) at tillage operations scale (from 100 mm to 300 mm); and (iv) variations in ground-level elevation. Random roughness corresponds to the first two levels, while oriented roughness corresponds to the third (Borselli, 1999) and is mainly due to disturbance from agricultural tillage as well as from water drainage channels which results in an anisotropic and periodic type of roughness (Darboux, 1999).

Fajardo et al. (2016) proposed a CV method for the measurement of soil roughness at the aggregate level. Soil aggregates were immersed in water, and digital images were taken at regular time intervals. The change in aggregate area over time was evaluated by image analyses to measure the kinetics of the slaking (aggregate breakdown) process and produce a slaking index. A mobile app named Slakes was also developed for this method (Fajardo et al., 2020). At larger scales, field measurement methods have been proposed to monitor soil roughness in cultivated plots. Jester and Klik (2005) have proposed to classify these methods according to several criteria: (i) measured with or without contact with the soil, (ii) the measurement is made on a line (2D) or on a surface (3D), (iii) measured with millimetre or centimetre accuracy and (iv) measured depending on the sensor type. Among the oldest methods of measuring SSR in the field are the chain method (Saleh, 1993) and the pin meter method (Hudson, 1993) (Fig. 4); both are made in contact with the soil.

The chain index (CI) (Eq. 1) is a tortuosity index, calculated by the chain method, comprised between 0 and 100; the higher the index, the higher is SSR.

$$\mathrm{CI} = \left(1 - \frac{L_0}{L_1}\right) \times 100 \tag{1}$$

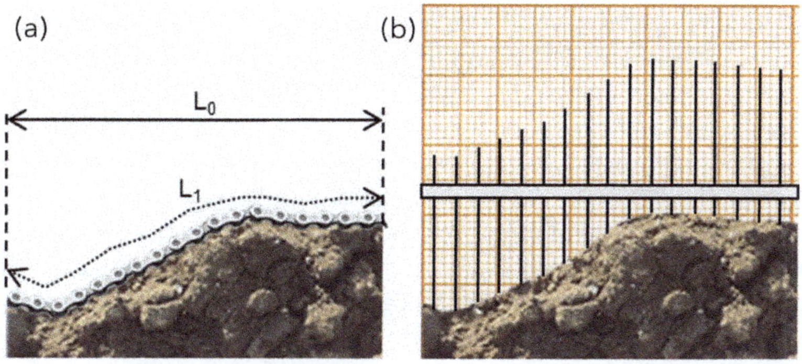

Figure 4 SSR measurement methods with contact to the soil. (a) Chain method - a chain of known length L_1 is placed along a line following the roughness of the soil clods and the horizontal distance L_0 is measured; the ratio of these two lengths is correlated to soil roughness. (b) Pin meter method - a horizontal ruler contains moving pins that move vertically according to the clod height; the height of each pin is finally measured to create the profile.

Boiffin (1984) proposed the roughness index (I_R), also calculated on a profile (Eq. 2)

$$I_R = \frac{(L_1 - L_0)}{L_0} \tag{2}$$

Indeed, in most of the SSR estimation methods, the first step is the measurement of the height in 2D or 3D, which constitutes a model of the soil surface; then, in the second step an index is generally calculated to summarize its roughness state. Many methods have been proposed by authors to calculate roughness indices. The surface chain index (SCI) (Eq. 3) (Kamphorst et al., 2000) is the equivalent of CI but on a surface.

$$SCI = \left(1 - \frac{S_0}{S_1}\right) \times 100 \tag{3}$$

where S_1 is the soil surface and S_0 is the flat surface.

The surface tortuosity index (STI) (Gilliot et al., 2017) (Eq. 4) is the equivalent of I_R but on a surface.

$$STI = \frac{(S_1 - S_0)}{S_0} \tag{4}$$

where S_1 is the soil surface and S_0 is the flat surface.

Other methods of index calculation include standard deviation of height (Guillobez and Arnaud, 1998), correlation length, a combination of the above two (Zribi and Dechambre, 2003), semivariogram analysis (Darboux et al., 2002; Dalla Rosa et al., 2012) and fractal dimension analysis (Zhu and Stein, 2002; Vidal Vázquez et al., 2005; Bretar et al., 2013).

Several authors have proposed a 2D-image analysis method for estimating soil roughness by quantifying shadows under fixed sunlight conditions (García Moreno et al., 2008). Results were significantly correlated to the pin meter ones. But most soil microtopography measurement techniques are based on a 3D model of the scene to be analysed. Sensors are used to measure depth information, which is the basis for 3D reconstruction calculations. This is also known as a depth map or depth image. Depth is estimated at each point and corresponds to the distance between the point and the sensor. Depth estimation is most often based on the analysis of a wave propagating between the object and the sensor, which may be light (visible or infrared), laser, microwaves or ultrasound; optical sensors are the most common. Three measurement principles are generally used by optical sensors for depth estimation (Vázquez-Arellano et al., 2016): triangulation, time of flight (TOF) and interferometry (Fig. 5). For the triangulation method, the measured point forms a triangle with two sensors offset by a distance *b*; e.g. in photogrammetric methods, the depth can then be calculated from *b* and the focal length of the camera.

For the TOF method, the travel time of the signal is measured after reflection on the object, and knowing the speed of light, time is transformed into depth. This is the method used by terrestrial laser scanner (TLS), also known as terrestrial LIDAR. The interferometric method consists of splitting a light beam into two - one beam is projected onto the soil and the other onto a reference surface, and then the two reflected beams are recombined together before being analysed by the sensor. Beam phase shift is correlated to depth. This is the most accurate method, but the most used method is the photogrammetric one by triangulation, because it can be done with a simple digital camera or even a smartphone. The shape and position of objects can be determined

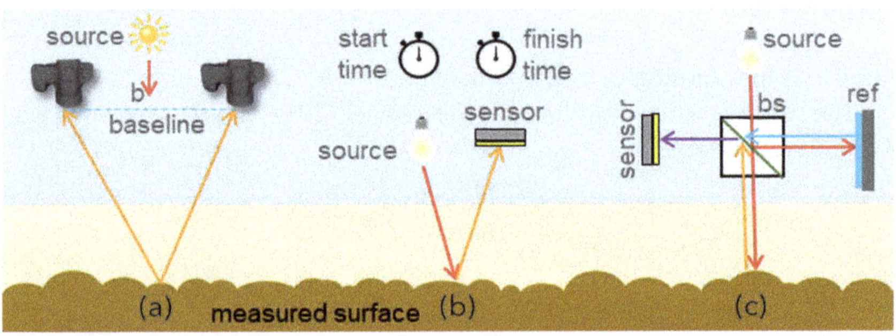

Figure 5 Optical sensor methods for depth estimation. (a) Triangulation from two camera positions; (b) time-of-flight and (c) interferometry. bs is a beam separator and ref is a reference surface (mirror). Source: Adapted from: Vázquez-Arellano et al. (2016).

from photographs; this is the purpose of photogrammetric methods (Kraus and Waldhäusl, 1998). Extracting 3D information of a scene from digital images taken from different positions that give different points of view of the scene is also called stereo vision. An important step of stereo vision algorithms is the correspondence problem, i.e. to find for each point detected in an image, the corresponding point (of the same object) in another image (Lazaros et al., 2008), also called stereo matching or stereo correspondence. The quality of the 3D reconstruction depends greatly on the number of correspondence points found between the images, which implies a very high spatial overlap between the different neighbouring images (Fig. 6).

The creation of a 3D soil model from photographic images to capture the microtopography of the soil surface is relatively simpler than for other landscape elements such as a building (Fig. 7). Indeed, due to the limited height of clods and stones above the average surface of an agricultural soil, a limited number of photographs can capture the 3D structure of the soil microtopography, whereas for a taller structure, such as a building, the camera field of view does not usually cover the whole height of the subject, so it is necessary to multiply the series of photographic acquisitions along the *Z*-axis.

At the scale of a small plot of the order of 1 m², in an agricultural plot with bare soil, the photographs will most often be taken by an operator, at human height (Fig. 8), using a digital camera or a smartphone. The different orientations of the soil microtopography are usually captured by turning around the plot when taking pictures (Fig. 8b). To cover a larger soil area, a pole or a photographic mast can be used. A platform of several meters high composed of a couple of stereo

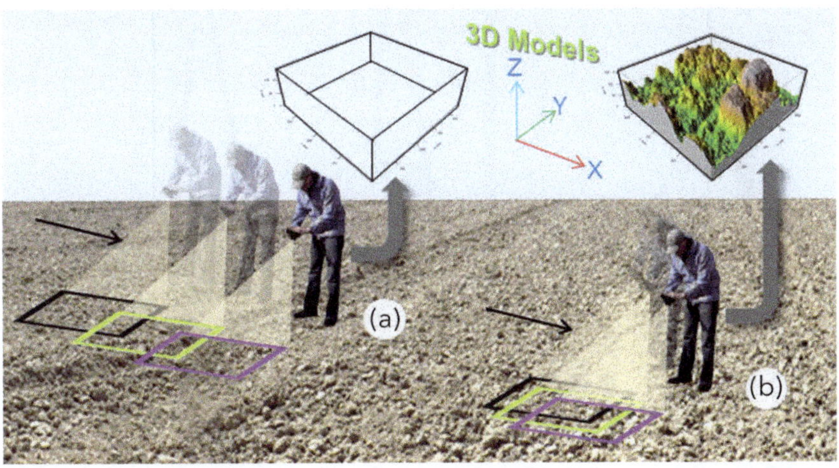

Figure 6 Importance of the percentage of spatial overlap between neighbouring photos. (a) A low overlap rate leads to no 3D reconstruction and (b) a high rate leads to a good 3D model.

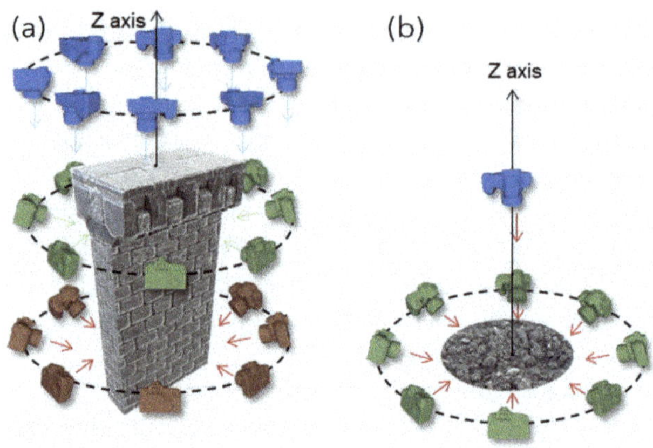

Figure 7 3D photogrammetric acquisition of objects from photographs. (a) Case of a building and (b) case of an agricultural soil surface.

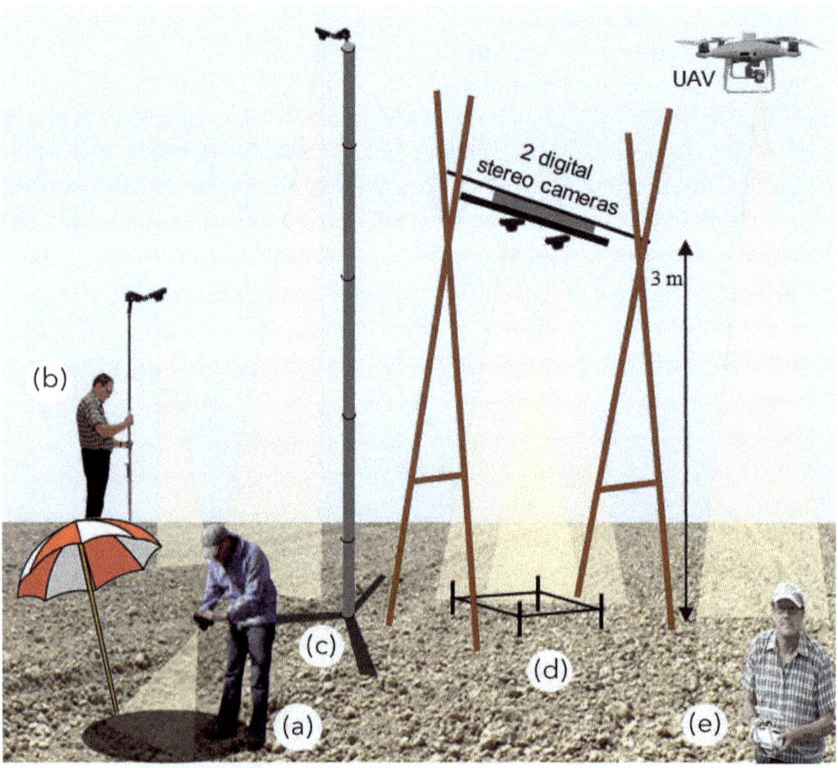

Figure 8 Acquisition configurations of multi-view image series for 3D photogrammetric reconstructions. (a) At human height, (b) with a pole, (c) with a photographic mast, (d) a platform composed of a couple of stereo cameras and (e) UAV.

cameras can also be used as described by Zribi et al. (2000); this configuration is particularly suitable for monitoring the same plot over time (soil erosion study for example). In case of variable light conditions with changing shadows, it may be preferable to take the pictures in the shade of a sun umbrella (Fig. 8). Indeed, the changes of shadow between successive images could prevent the recognition of soil clods by the software or be interpreted as false deformations. The advent of drones or UAVs has revolutionized the shooting of photographs in the field for scientific monitoring. An automatic flight plan can be programmed into a UAV to acquire a large series of images with a high spatial overlap rate allowing 3D reconstruction. UAVs can cover a larger area than previous methods. The most suitable UAV models for this type of measurement are multi-rotor rotating-wing UAVs, which can fly low and slow; several recent models can thus fly at very low altitudes (a few metres) and can therefore also be considered proxy detection sensors. Some models, such as the Dji Phantom 4 RTK (Dji, 2022), are equipped with an RTK GNSS receiver that allows geolocation of the produced models with a centimetric accuracy, which is of great interest to easily compare the same plots during different measurements over time.

In the case of two photographs of the same soil plot, taken in the same direction (Fig. 9), i.e. with their optical axes parallel, from positions 1 and 2 separated by a distance *b* (baseline), one can consider that a point P is projected onto the image planes IP1 and IP2, on points P_1 and P_2, respectively. P_1 and P_2 are called homologous points, i.e. the position of the same object (point P) in image 1 and image 2, respectively. The intersection of the equipolar plane with the two image planes (IP_1 and IP_2) defines the equipolar lines (EL) in both images. Finally, disparity (*d*) is defined as the distance between P_1 and P_2 along the EL line (Dhond and Aggarwal, 1989). In the case of the configuration shown in Fig. 9, the depth Z_p, which gives the distance between the camera and the p-point in the real world, can be expressed as a function of the disparity, the focal length of the camera and the baseline (Eq. 5).

$$Z_p = \frac{b \times f}{d} \tag{5}$$

The production of depth maps, estimating Z_i at each point *i* of the analysed space, is an important step in the photogrammetric calculation.

The first step in the stereo vision process is stereo matching, i.e. matching points representing the same feature in the different images (P_1 and P_2 in Fig. 8); these points are called key points. After key point localization in the images, feature descriptors are calculated for each point. These descriptors are sort of 'identity cards' of image pieces – those with close values correspond to the same point in space. Feature descriptor matching allows key points to be matched with each other, and this is the basis of stereo matching. The calculated

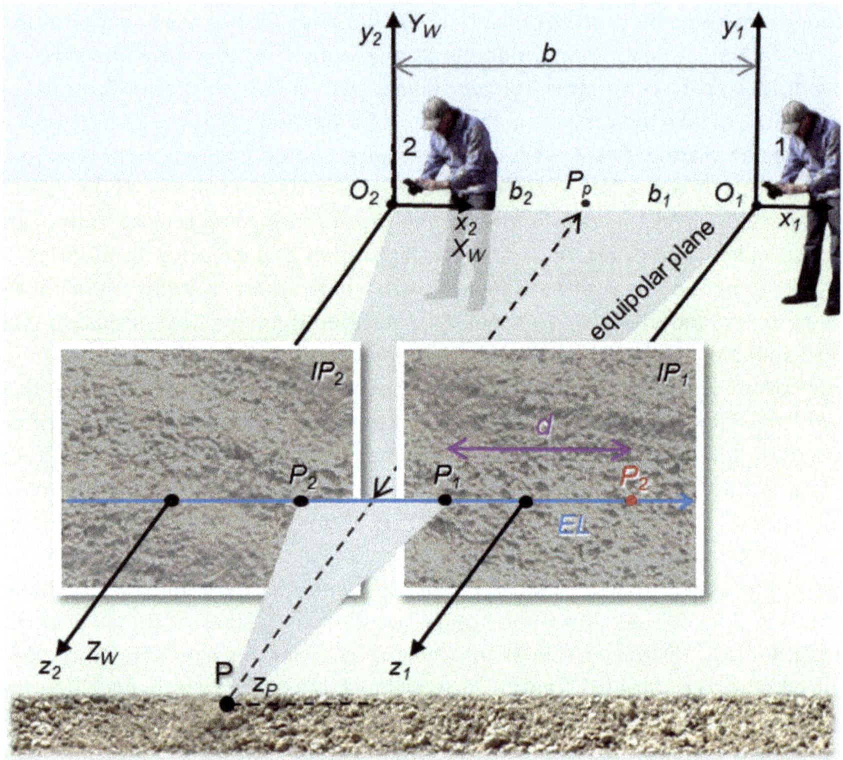

Figure 9 Depth information (Z_p) calculation, from two photo positions O_1 and O_2, P is a soil clod position, P_1 and P_2 are projections of P in image planes IP_1 and IP_2, respectively, EL is the equipolar line, *d* is the disparity (distance between P_1 and P_2) and *b* is the baseline (distance between O_1 and O_2).

descriptors must be invariant to geometric transformations and illumination variation, since the different images can be taken at different orientations and scales; only then can the matching between the points work. The SIFT (scale-invariant feature transform) descriptor is one of the most used; it was proposed by Lowe (2004). Figure 10 shows one of the SIFT algorithm results obtained with the VisualSFM software (Wu, 2022), with two images taken at different angles and scales; 207 key points are well matched between the two images despite differences in geometry.

The SIFT descriptor is based on scale-space filtering (Witkin, 1983); indeed, the recognition of a given object is often only relevant at certain scales. The scale-space concept attempts to integrate a multi-scale object representation on digital images. Scale space is separated into octaves (Fig. 11); the first octave corresponds to the image at the original size; for the second octave, the image is spatially subsampled by a factor of 2; each octave's image is subsampled in relation to the previous octave. Within an octave, the input image I(*x*,*y*) is

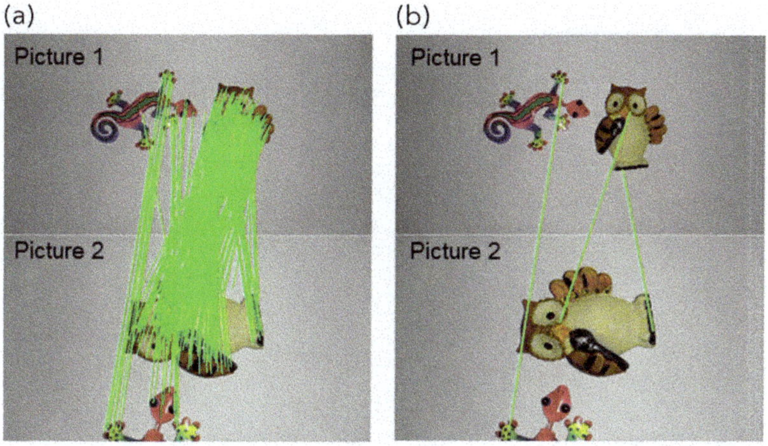

Figure 10 Point matching with the SIFT algorithm between two pictures of the same scene consisting of two figurines (an owl and a gecko). The green lines show key point matching between the two images. Photo number 2 was shot with a 90° rotation compared to photo number 1 and with a different scale; (a) the 207 total matches and (b) three random matches among the 207.

modified by a convolution operator $G(x,y,\sigma)$ of a growing size σ, corresponding to a Gaussian filtering (Eq. 6).

$$L(x,y,\sigma) = I(x,y) * G(x,y,\sigma) \tag{6}$$

where * is the convolution operation.

The difference of Gaussian (DOG) is then computed, which is the difference between two nearby scales (Lowe, 2004); maxima values in the DOG are key point candidates.

Some parameters defining the geometry of the projection (Fig. 9) must also be found to do the 3D reconstruction. Two groups of parameters are needed (Verhoeven et al., 2012) : (i) inner camera orientation and (ii) six exterior parameters. Among the most important parameters are the focal length, the geographical position and the 3D orientation of the camera. According to the structure from motion (SfM) theorem (Ullman, 1979), the 3D structure of objects can then be found from their projected positions. Photogrammetry software is sometimes referred to as an SfM system. The main steps present in most software (Fig. 12) are: (i) photo alignment, which leads to the determination of camera positions and the creation of a sparse 3D point cloud; (ii) point cloud densification which will lead to a much denser cloud which is closer to reality; (iii) 3D cloud can be exported in a format such as the Standford format (.ply) to be modified with software such as Meshlab or a programming language such as Python; (iv) 3D vector Mesh model and raster DSM creation - these

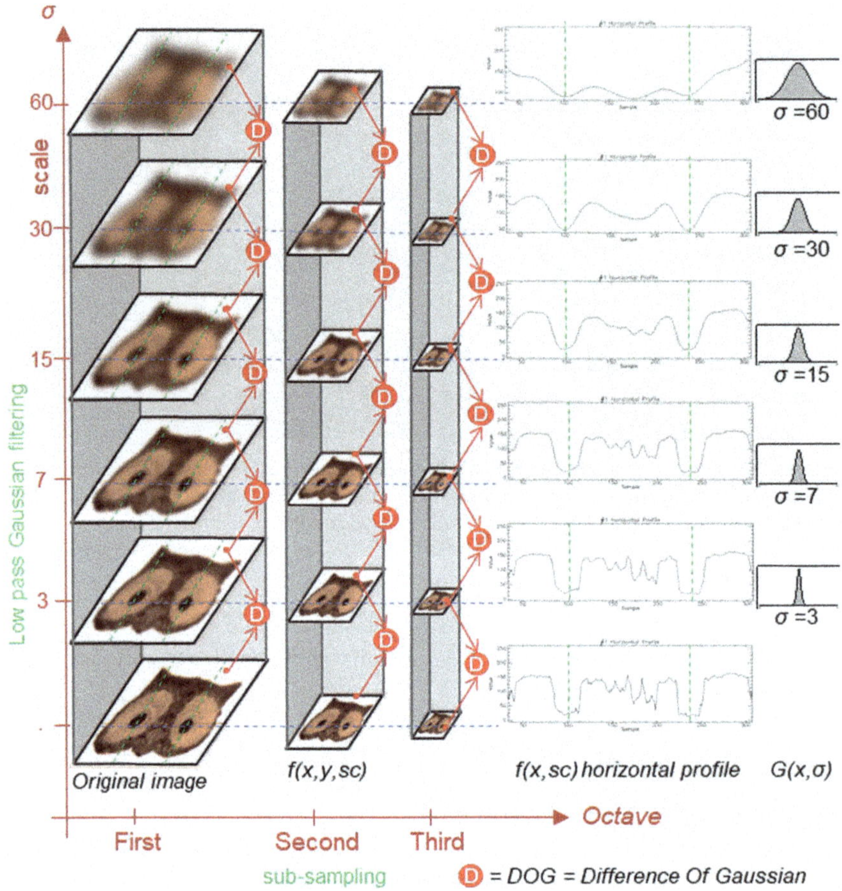

Figure 11 The scale-space filtering in the SIFT algorithm.

models contain the 3D information on the height of the clods and stones above the ground; and (v) finally the mosaic image to merge all photos into one final image. Texture images can also be calculated for the textured rendering of 3D models to give a visual impression close to the real scene. It may happen that the starting point of the photogrammetric processing chain is not a series of photos but a video; in this case, software such as ffmpeg (Ffmpeg, 2022) can convert a video into a series of individual images.

There are several software packages available for 3D reconstruction from a set of overlapping images; they can be classified according to several criteria: (i) commercial or freeware; (ii) with a graphical user interface (GUI) or in command line; (iii) online or stand-alone; (iv) using GPUs (graphical processing unit) to speed up calculations or not; (v) some software only performs part of the processing chain and in this case must be complemented by a second one; (vi)

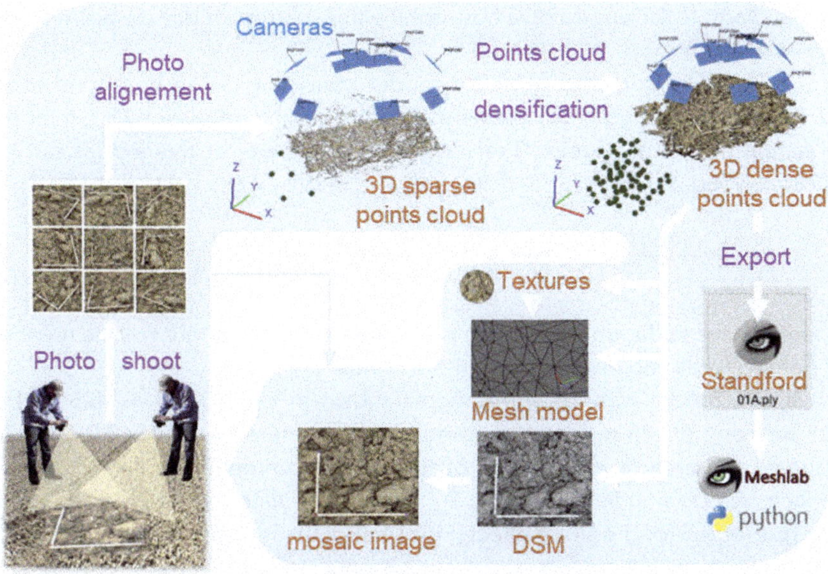

Figure 12 Standard workflow of a photogrammetric software.

depending on the supported operating systems, Windows, Linux or Mac OS/X; (vii) the ability to do distributed computing between several machines over the network (clustering); (viii) be able to call up calculations from scripts to automate processing runs. Metashape (formerly Photoscan) from Agisoft (Agisoft, 2022) is one of the most widely used commercial software packages, known for the quality of its reconstructions and its ease of use, thanks to its intuitive user interface. The Bundler package (Snavely, 2010) is a free command-line tool SfM system, which produces a sparse cloud and must be complemented by a densification tool like PVMS2 (Furukawa and Ponce, 2008) and can be easily integrated into scripts. VisualSFM (Wu, 2022) is also a free tool that can be used with a GUI or command line and supports the use of GPUs. Meshroom from AliceVison (AliceVision, 2022) is a free, open-source 3D reconstruction software based on the AliceVision framework. It has a mature user interface and allows the reconstruction process to be modelled as a pipeline of nodes. Colmap (Colmap, 2022) is an SfM reconstruction pipeline with a graphical and command-line interface; it runs under Windows, Linux and Mac OS/X.

Before shooting in real conditions, for example in open fields on agricultural soils to estimate roughness indices, it can be useful to evaluate the accuracy of the photogrammetric chain that will be used. For this purpose, a 3D artificial model of known geometry can be made to estimate the accuracy of the method for determining the position, height and area of objects in a scene. Models shown in Fig. 13 are composed of several polystyrene hemispheres of known diameter arranged on a horizontal plane, corresponding to a simplified model of clods.

The profile extracted from the model consisting of 16 hemispheres of 50 mm diameter (Fig. 13d) shows estimated heights of about 25.1 mm, for an actual height of 25 mm. The theoretical surface S_t of this model can be calculated (Eq. 7).

$$S_t = \frac{n4\text{p}\,r^2}{2} + S_p - n\text{p}\,r^2 = n\text{p}\,r^2 + S_p \tag{7}$$

where S_t is the surface area, S_p is the plane area of the rectangle, *n* is the number of hemispheres and *r* is the hemisphere radius.

In the model with 16 hemispheres shown in Fig. 13d, S_p is 425 mm × 325 mm or 138 125 mm²; the theoretical area is therefore 169 540.9 mm². To estimate the area from the DSM of the photogrammetric model, if *S* is the surface of a pixel in flat condition, S_{cor} can be calculated as the surface of the pixel corrected for the slope effect (Fig. 14) (Eq. 8).

$$S_{cor} = \frac{S}{\cos\alpha} \tag{8}$$

where S_{cor} is the pixel surface corrected of the slope effect, *S* is the pixel surface at the DSM resolution and *a* is the slope.

The total area of the DSM considering the microtopography is then the sum of all the S_{cor} value of the pixels. In the previous example, the result is (calculated by using the QGIS platform available at: https://www.qgis.org/en/site/) an area of 169 526.29 mm², i.e. a difference with the theoretical surface of 14.61 mm², or an error of 0.009%. The areas estimated by photogrammetry can therefore potentially be very accurate. Gilliot et al. (2017) in a previous study obtained, with a series of 12 similar artificial models with a varying number of hemispheres, a good correlation between theoretical surfaces and photogrammetric surfaces, with an R^2 of 0.99.

In the case of field measurements, the base plane of the measured area is not necessarily perfectly horizontal; this systematic bias can lead to an error in the estimated soil roughness. This inclination, which may be due to a local slope of the field, can be considered a local trend of the height function. Detrending methods have been proposed by several authors (Taconet and Ciarletti, 2007; Vázquez et al., 2009; Marzahn et al., 2012; Bretar et al., 2013; Snapir et al., 2014); a method often used is to find the best fit plane by regression between the *z* elevation and the (x,y) coordinates, which can be done with the 'lm: Fitting Linear Models' R function (Fig. 15).

Figure 16 shows an example of the precision that can be obtained for a DSM created from images acquired by UAV. These data were produced as part

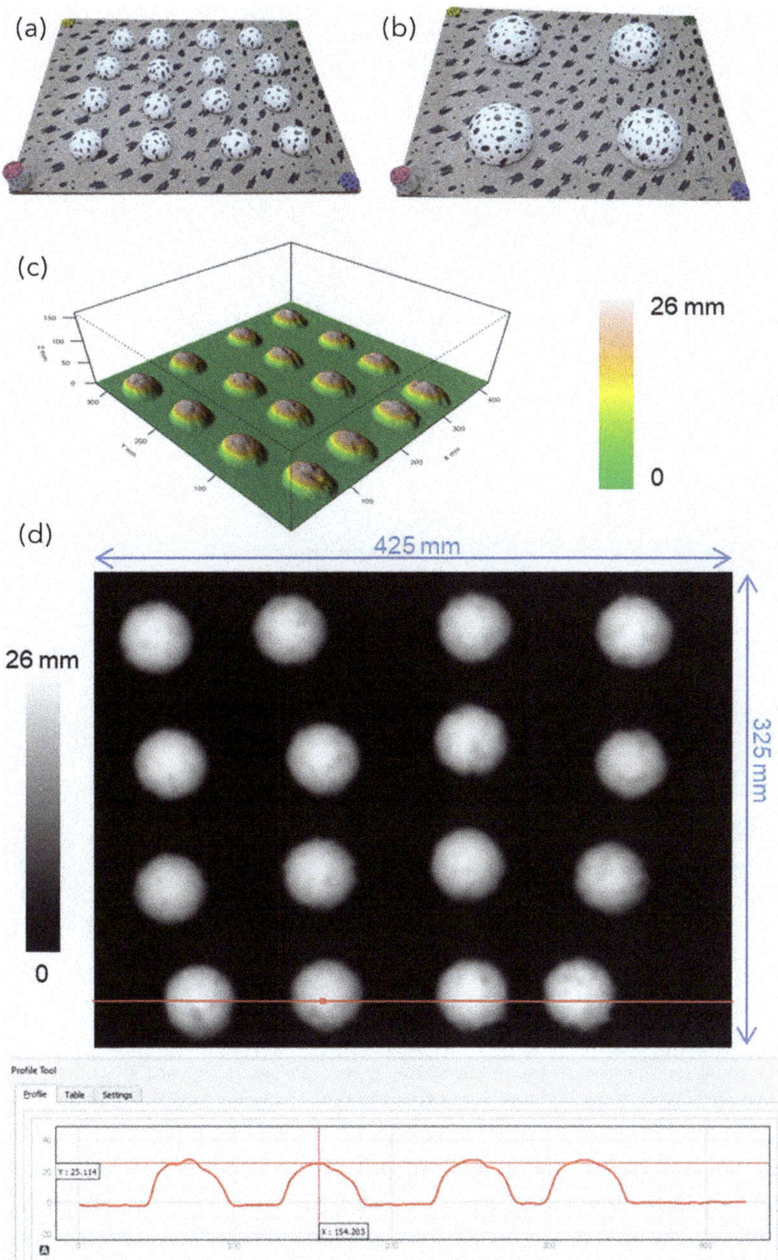

Figure 13 Artificial models composed of polystyrene hemispheres. (a) Sixteen hemispheres (50 mm diameter), (b) 4 hemispheres, (c) 3D model of a 1 mm resolution, (d) DSM of (a) with a *Z* profile (red line).

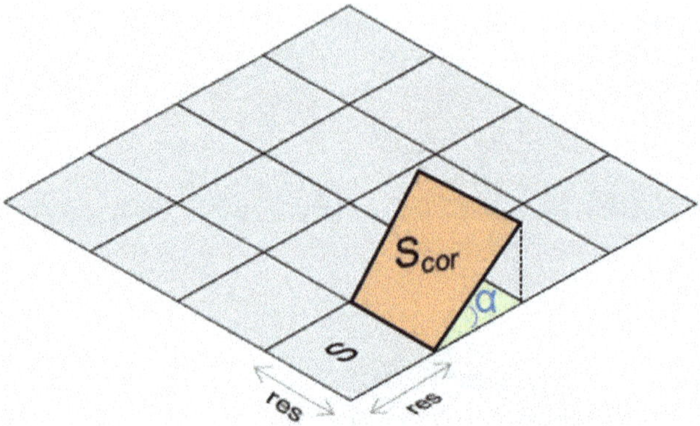

Figure 14 Correction of the area *S* of a DSM pixel according to the effect of the slope *α*; res is the pixel resolution, $S = res^2$ and S_{cor} is the corrected surface.

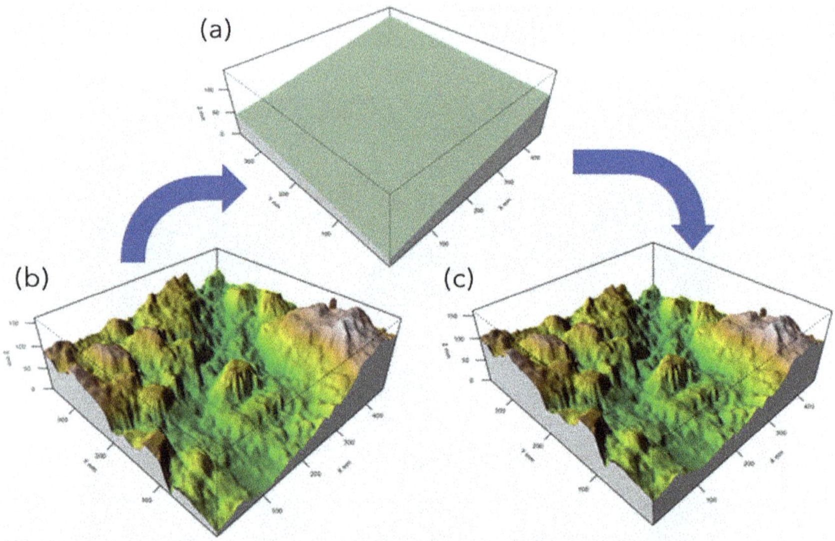

Figure 15 Surface detrending by the best fit plane method. (a) Original model with a visible trend, (b) the best-fit plane found by regression between *z* and (*x,y*) with the lm R function and (c) corrected model by subtracting the trend plane from the original model.

of the AgroParisTech project 'ToUHR-DRONAé' (Ultra-High-Resolution Remote Sensing by UAV for the adaptation to Agroecological practices); the UAV used was a Dji Phantom 4 RTK (P4-RTK) with an 'SDK' remote controller, compatible with the GS-PRO (Ground Station pro) software, which has the capacity to plan automatic flights at a very low height. At 5 m height above ground level (AGL), the images obtained with the 20 MPixel camera of the P4-RTK had a resolution of 1.5 mm and a 3 mm resolution for the DSM. A visual comparison of the DSM and

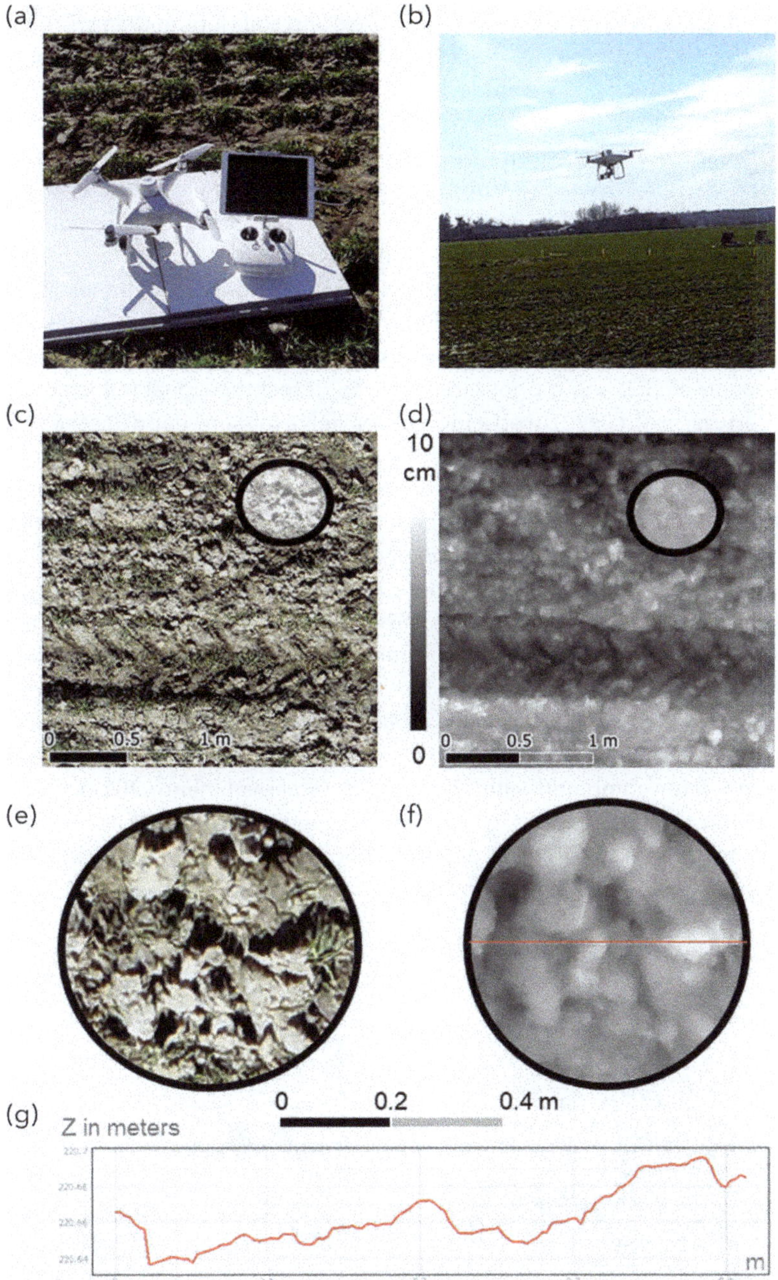

Figure 16 SSR model creation from UAV images. (a) Dji Phantom 4 RTK UAV with its optional SDK controller, (b) P4-RTK in flight at 5 m AGL flight on February 28, 2022, in the Paris region, France, (c) extract from the othomosaic, (d) extract from the DSM, (e) zoom in the mosaic and (f) zoom in the DSM and (g) a profile (red line).

the corresponding image shows small stones and soil clods in detail. The profile plotted on the model shows that the DSM has captured variations in Z of a few millimetres. At such a low altitude, there is a great risk that the images acquired by the UAV during its movements are blurred; this is called 'motion blur'. To avoid this, below 16 m height AGL, the P4-RTK uses a special flight mode called 'Hover & Capture', in which it stops to take each picture, which ensures that there is no blur, but this greatly increases the flight time required for a mission.

An alternative to the photogrammetric approach for acquiring data to create 3D models of soil microtopography is to use distance measurements using a laser beam. Since the end of the 1990s, several authors have developed non-contact profile meters using optical systems. Römkens et al. (1986) used a pulsating IR LED and a photo transistor to detect the signal reflected by the soil. Bertuzzi et al. (1990) used a low-power HeNe (helium–neon) laser as a laser source and a matrix-array detector; the height profile was done at a constant increment of 2 mm and automatically recorded on a computer. Gilliot et al. (2012) designed a profile meter using a Leica Disto Pro 4a laser distance meter (Fig. 17) and a dedicated application on a windows pocket terminal.

Fanigliulo et al. (2020) instrument is a Leica Disto laser telemeter which moves on a horizontal aluminium rail fixed on two tripods, by means of a step-by-step electric motor (Fig. 18). The rail is 4 m long, and measurements are taken every centimetre.

Marinello et al. (2015) used a Microsoft Kinect depth camera for the characterization of soil microtopography; the depth sensor consists of an IR laser projector combined with IR and RGB cameras (Zhang, 2012). The laser passes through a diffraction grating, which generates a set of IR dots on the scene. As the geometry of the generated dots is known, a dot detected in the IR image is associated with its position in the projector, and then a 3D model is reconstructed using triangulation (Fig. 19).

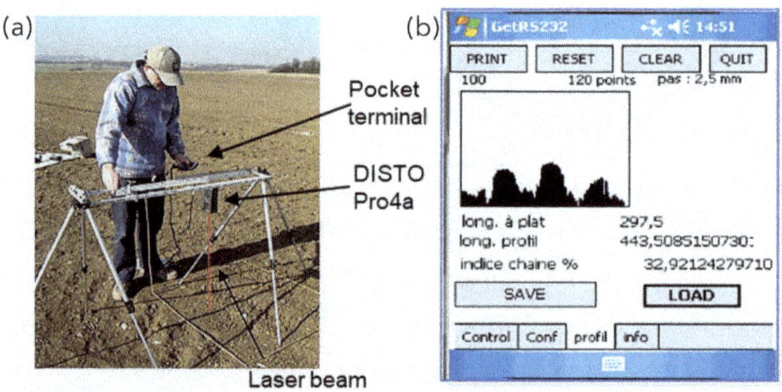

Figure 17 Profile meter. (a) Leica Disto Pro 4a laser distance meter and (b) a dedicated application on a windows pocket terminal.

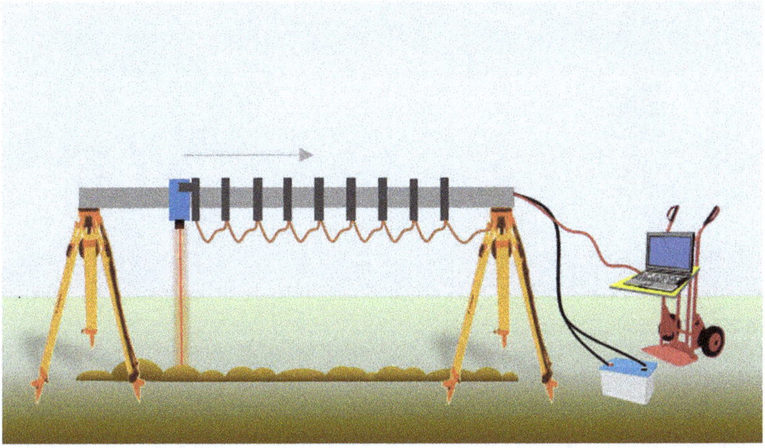

Figure 18 Profile meter motorized with a step-by-step electric motor. Source: Adapted from Fanigliulo et al. (2020).

Figure 19 Microsoft Kinect depth camera mounted on a tractor. (a) The infrared laser projector, (b) RGB camera and (c) infrared camera; the red dots symbolize IR dots (visible only in infrared).

3D soil models with a resolution of 1 mm were produced with the Kinect sensor, with tractor speeds higher than 2.5 m s^{-1}. Realsense D400 (Intel, 2018) and Asus Xtion 2 (Asus, 2022) are depth cameras equivalent to Kinect and have also been used by the authors to study soil roughness (Thomsen et al., 2015; Grundy et al., 2020).

TLS is a 3D laser scanner for terrestrial use, which uses the TOF measurement principle; for each point, it emits a laser pulse and measures the time taken for the beam to travel from the TLS to the target and back (Fig. 20). Knowing the

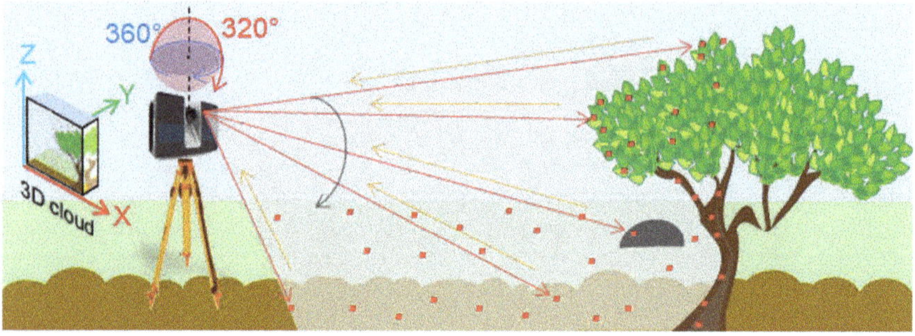

Figure 20 3D terrestrial laser scanner.

speed of light, time is directly transformed into distance, which gives the depth information. The TLS usually acquire several million measurements to create a 3D point cloud; photos are also often taken to make photorealistic renderings. The TLS is mounted on a tripod, and its motorized head rotates to scan the environment around it in 3D. The accuracy of this type of device varies from one to a few millimetres, and the range can reach a few hundred metres but its cost remains quite high (between 20 k€ and 50 k€).

Martinez-Agirre et al. (2020) compared TLS and SfM photogrammetry SSR measurements with 2D laser profiles. The TLS used was the FARO Focus 3D, and photogrammetry photos were taken from a height of 6–8 m above ground using a lifting platform. Both TLS and SfM provide very detailed 3D data. Six roughness parameters were analysed from both types of data; similar results were obtained for TLS and SfM, allowing a fine characterization of agricultural soil roughness.

2.3 Soil thin-section microscopy: soil property estimation by image analysis methods

2.3.1 Micromorphological approach and its input for soil science

Soil scientists used to analyse bulk and disturbed samples (e.g. granulometry, pH, soil organic matter and biomass contents) from the topsoil until depths of several metres to understand soil and its functioning (Baveye et al., 2018; Vogel et al., 2022; Fig. 21). According to Baveye et al. (2018), 'more appropriate macroscopic parameters are needed, which reflect better the spatial heterogeneity of soils at the microscale (i.e. the pore scale) that is commensurate with the habitat of many microorganisms'. There is thus a real need to understand and quantify how the soil architecture (i.e. soil structure as defined by Vogel et al. 2022) may change over time under the effect of the

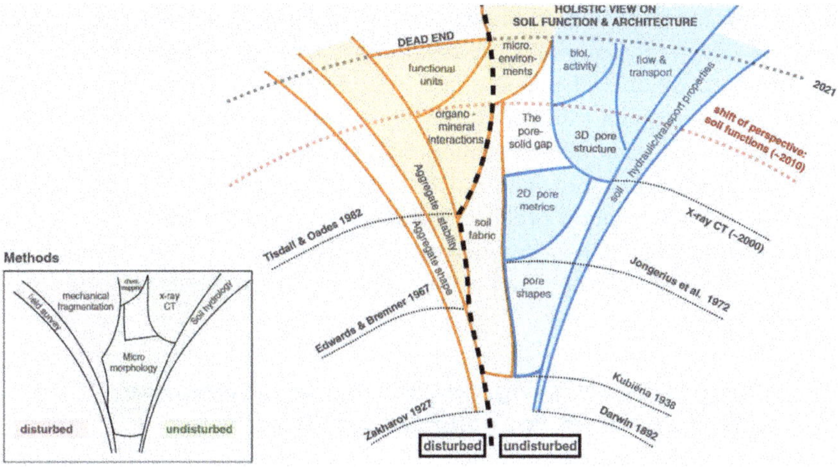

Figure 21 Historical development of exploring soil structure. The pore perspective is in blue and the solid perspective in ochre. Dark colours and light colours indicate quantitative and qualitative analysis, respectively. New branches of developments with respect to research foci are marked by milestone publications or new technical developments. Reprinted with permission from Vogel et al. (2022).

ever-changing environment due to global change and associated soil land use modifications. For instance, the total content of calcium carbonate can be assessed using traditional chemical methods (i.e. bulk sampling and disturbed state) or using non-destructive micromorphology able to distinguish between lithogenic or pedogenic calcite or aragonite present e.g. as coatings or nodules (Poch and Stoops, 2022).

In parallel to the use of the above-mentioned methods based on disturbed soil samples, so-called structural analysis methods based on the multi-scale analysis of structural relationships between different volumes of soil that are homogeneous in terms of the nature and arrangement of their constituents have been used to identify soil processes controlling the transformation and redistribution of matter in soils (Fritsch et al., 1992). Structural analysis approaches focusing on the identification and quantification of the abundance of pores and pedofeatures can help to quantify the intensity of soil processes and properties. Although some pores and pedofeatures are detectable in the field by observing soil pits and soil horizons with the naked eye, their characterization is much more precise at the microscopic scale. To observe how soil constituents are organized at a microscopic scale, i.e. between field and submicroscopic observations, Walter L. Kubiëna developed soil micromorphology based on impregnated soil samples (Kubiena, 1939, Fig. 21). According to Poch and Stoops (2022), soil micromorphology is 'the observation and interpretation of undisturbed, oriented samples of soil material by microscopic and ultramicroscopic techniques to identify their constituents (including voids) and to determine their mutual relations in space and time'. This technique allowed researchers to investigate the pore space and the composition of the solid phase (groundmass and pedofeatures) at the same time for a given sample with minimal disturbance (Vogel et al., 2022). For instance, this scale of observation makes it possible to distinguish with a minimum of ambiguity between clay coatings (Sauzet et al., 2016, Case Study 1) linked to the illuviation process and coatings linked to other pedological processes such as shrinkage/swelling processes (Kemp, 1998; Kühn et al., 2018) and assess how much their nature and abundance evolve regarding land use changes.

2.3.2 Thin-section sampling, fabrication and observation techniques (from Poch and Stoops, 2022)

As soil micromorphology concerns the study of undisturbed samples, special sampling techniques are necessary. In the case of materials with enough coherence, blocks can be carved out of the profile and carefully packed with newspaper or plastic wrap, while for less stable and loose soil materials, the so-called Kubiëna boxes were developed. These are metal boxes consisting of a frame that can be opened and with two covers (top and bottom). Most soil micromorphology research is done on soil thin sections, although some can be done on polished block faces, whole blocks or soil aggregates. To prepare a thin section, a hard material is needed. This can be obtained by impregnating the undisturbed soil sample, once removed from the box, under vacuum conditions with a cold-setting resin, mostly a diluted polystyrene. As the resin is incompatible with water, the latter has to be removed, either by air or oven drying or by acetone replacement. After hardening, a slice of the impregnated block is cut with a diamond saw cooled with a special oil (to avoid solution of constituents by water). The slice is then ground and mounted on a support glass of corresponding sizes (e.g. 48 mm × 28 mm for a petrographic size thin section, up to 180 mm × 120 mm for a so-called 'mammoth' thin section). By further sawing and grinding, the thickness of the slice is reduced to 20–30 μm (standard thickness of a thin section). At that thickness, most minerals are transparent and can be determined by optical techniques. For grinding, a slurry of fine corundum or diamond powder and oil is used. Thin sections are sometimes covered with a very thin cover-glass for protection, unless staining tests, cathodoluminescence or microprobe studies are planned. The most common observation technique is optical microscopy, using a polarizing microscope (petrographic microscope). For specific research, a range of additional techniques are available. The use of oblique incident light is needed to distinguish between opaque constituents and to determine their internal fabric. Incident ultraviolet (UV) or blue light can be used to detect specific minerals such as phosphates (e.g. bone splitters) and cellulose. When a fluorochrome (a photoreactive chemical) is added to the impregnating resin, observation in UV or blue light shows a clear image of the porosity. For more details and references, see Poch and Stoops (2022) and Verrecchia and Trombino (2021).

2.3.3 History of quantitative micromorphology

Soil thin sections are generally analysed qualitatively and described according to internationally accepted concepts as a lot of efforts have been put into a well-defined terminology (Bullock et al., 1985; FitzPatrick, 1993; Stoops, 2021; Stoops and Vepraskas, 2003). The terminology is available in 19 languages at https://www.isric.org/explore/ISRIC-collections/micromulti (accessed April 2022). Research on quantitative soil micromorphology started in the 1940s and has been based either on visual estimates with reference to appropriate empirical charts of Bullock et al. (1985) and Thompson et al. (1990) or on counting procedures. Point counting used to be the most common procedure to determine the frequency (modal analysis), but it is considered tedious and error prone (Aydemir et al., 2004; Marschallinger, 1997). Both quantitative techniques tend to be too operator dependent (Ulery et al., 2008; Zhang et al., 2014). Increasing computing power led to the first tools for automated image analysis, such as the Quantimet Image Analysis Computer (since 2014 a trademark of Leica Microsystems CMS GmbH), previously developed by geologists and first used for soil analysis by Jongerius et al. (1972) (as reviewed In Vogel et al., 2022) (Fig. 21). With such a tool, the study of the pore space in 2D images visible on thin sections and polished blocks could be raised to a quantitative level (Murphy et al., 1977 (as reviewed in Vogel et al., 2022)). Digital image processing has been then developed and used for quantitative micromorphology as it seems to be potentially superior over point counting in accuracy and the speed of data extraction (Grove and Jerram, 2011). Image analysis methods moreover afford distinct advantages over conventional microscopic ones in that they can provide a range of previously unobtainable quantitative information, including measurements of morphological aspects and spatial relationships of different soil components (Adderley et al., 2006; Bielders et al., 1996; Bruneau et al., 2004; Bryant and Davidson, 1996; Gargiulo et al., 2013; Ringrose-Voase, 1996; Taina and Heck, 2010).

2.3.4 Microstructure characterization and pore quantification in two-dimensional images

In most studies of image analysis-based porosity quantification, the authors first convert the colour image of thin sections to greyscale before thresholding. The images follow a thresholding operation that separates the area of interest to be analysed later from the rest of the image. Segmentation, on the other hand, consists of partitioning the image by grouping the pixels into regions, which correspond to the different objects in the image. There are in fact three main segmentation methods (Kunt et al., 1993): thresholding, region growing and the multiresolution approach. Sometimes thresholding is used as a means, the simplest one at that, to segment an image (Ringrose-Voase, 1991). This process can be supervised that implies a part of subjectivity in the process, but today, the pore recognition on thin sections may be done in a more unsupervised way using, for example, the Fiji software and automatic thresholding like the Otsu one (Schindelin et al., 2012; Sauzet et al., 2017, Case Study 1). A way for isolating and classifying pores is to use a learning set consisting of pore classes (e.g. channels, planar voids and vughs) to teach the system how to recognize statistically soil pores in images using measurements made by Quantimet for example, as proposed by Ringrose-Voase and Bullock (1984) and Yagüe et al. (2016).

2.3.5 Towards three-dimensional pore quantification

With the access to synchrotrons in various countries, the development and widespread commercialization of x-ray computed tomography (CT) scanners or the use of medical x-ray CT at hospitals, the restriction to 2D sections out of the 3D reality was not a problem anymore (Vogel et al., 2022) (Fig. 21). Samples for this method do not need to be dried before analyses (Poch and Stoops, 2022) or even impregnated. Much of the very significant progress achieved in that context since the early 2000s has been reviewed in detail by Baveye et al. (2018) and Taina et al. (2008). The pioneering work carried out 30 years ago by Joschko et al. (1991) on earthworm burrow systems, monitored via low-resolution x-ray CT (1mm or 2 mm i.e. only macropores could be visualized), has inspired various other authors to investigate the effects of bioturbation by earthworms on pore space geometry sometimes using x-ray CT (Capowiez et al., 2011; Muñoz-Ortega et al., 2015; Sauzet et al., 2021, Case Study 2) or micron-scale x-ray CT (μCT) (Balseiro-Romero et al., 2020) in soil columns or undisturbed soil cores answering diverse questions such as of soil biopore structure under wetting and drying cycles (Cheik et al., 2021). Various mathematical techniques were developed to extract from 3D CT images a wealth of quantitative information about the tortuosity, connectivity and topology of the pore space (Vogel et al., 2022). After the turn of the millennium, table-top x-ray CT scanners became commercially available and increasingly common in soil science laboratories, enabling researchers to observe the geometry of the pore space at suitably fine resolutions (Baveye et al., 2022).

Associated thresholding methods have been reviewed in detail by Baveye et al. (2018):

> The '3-dimensional' soil images that CT scanners provide are in fact stacks of 2-dimensional, greyscale images associated with virtual slices within the soil sample. Very early on in the use of these images, researchers came to the conclusion that these greyscale images would not be very useful to quantify the geometry of the soil pore space and that it was necessary to derive binary (black and white) images from the original greyscale ones, a process alternatively referred to as 'thresholding' or 'segmentation'.

Significant progress has occurred over the years in how this thresholding is approached. Initially, it was carried out slice by slice, either manually by simple visual inspection ('eye-balling') or with the assistance of one of a number of available 2D algorithms (e.g. Nunan et al., 2006). The first improvement consisted of thresholding the whole 3D image at once, using an algorithm to calculate a unique, global threshold value. Then, various researchers showed that in the presence of textural heterogeneities (e.g. stones) within the samples, it was preferable to instead use local thresholds, which can vary from location to location within a sample (Iassonov et al., 2009; Schlüter et al., 2019). Up to that point, all thresholding algorithms required operator input, to adjust one or more parameters. This introduced unavoidable subjectivity in the process, which in principle would make it improbable for different individuals to threshold a given soil sample the same way, or even for a single individual to threshold different soil samples (e.g. associated with different agricultural practices or with successive times) in a consistent manner (Baveye et al., 2020). Nevertheless, recognition a few years ago that the subjectivity in thresholding operations and in the manipulation of CT images could be substantial prompted the development of a number of automated thresholding algorithms requiring no operator input (Schlüter et al., 2019; Hapca et al., 2015; Houston et al., 2013) regardless of the level of 'supervision' (learning from training data) adopted. Similar to the 2D-image analysis on thin sections, some plugins and software were developed specifically in order to identify and quantify pore space and differentiate it from the soil aggregates such as the ImageJ plugin developed by Koestel et al. (2021).

2.3.6 Soil pedofeature identification, characterization and quantification on soil thin sections

While porosity has been studied extensively, and researchers have succeeded in characterizing and classifying pores according to different morphological attributes, less studies deal with the characterization and quantification of pedofeatures by image analysis (Protz and VandenBygaart, 1998; Terribile and FitzPatrick, 1992; Terribile and Fitzpatrick, 1995). Recent developments in computer-assisted image analysis have made it possible to progressively move towards comprehensive quantifications of not only abundance but also size, shape or topology characteristics of different pedofeatures (Bryant and Davidson, 1996; Cornu et al., 2012; Taina and Heck, 2010).

Among diverse pioneering works regarding image analysis on thin sections for pedofeature recognition and abundance quantification Taina and Heck (2010) used computerized analysis of soil thin-section imagery to quantitatively estimate various soil features allowing the interpretation of soil properties in connection with soil genesis and evolution. They assessed an object-oriented software (that was created especially for remote sensing) in the study of soil thin sections. The software was based on a multiresolution segmentation function that enables the separation, at different levels of observation, of objects representing micromorphological features (Taina and Heck, 2010). Objects that corresponded to voids, coarse and fine fractions with variable compositions and redoximorphic pedofeatures were distinguished and grouped into classes arranged in a hierarchical manner. Classification was based not only on object spectral values but also on object shape characteristics, as well as class-related features (Taina and Heck, 2010).

As already underlined by Verrecchia and Trombino (2021), 'the advancement of technology opens up new opportunities to soil micromorphology'. Since the 1970s, it was possible to observe thin sections at high resolution (ultramicroscopic techniques), with the scanning electron microscope in its backscattered electron mode (or with transmission electron microscopy). It was also possible to generate chemical images with electron microprobes. The integrated automated mineralogy and petrography solution QEMSCAN (quantitative evaluation of minerals by scanning electron microscopy) provided the quantitative analysis of minerals, rocks and man-made materials (Verrecchia and Trombino, 2021). Until the mid-1990s, besides the standard bulk analytical methods, the only method that was available to researchers to determine local (bio)chemical properties of soils was thus based on energy-dispersive x-ray spectroscopy, either in the so-called 'electron microprobe' or SEM–EDX equipment or on electron energy loss spectroscopy (Watteau and Villemin, 2018 (cited in Baveye et al., 2018)).

Baveye et al. (2018) and Védère et al. (2022) reviewed extensively the 2D measurements that are now feasible to characterize the (bio)chemical nature of soils:

- X-ray absorption near-edge structure and near-edge x-ray absorption fine structure spectroscopies can provide information on element speciation while covering most a few square micrometres.

- X-ray micro-fluorescence helps to obtain elemental maps that cover bigger surface areas without providing information about speciation.
- Dynamic nanoscale secondary ion mass spectroscopy (nanoSIMS) uses a high-energy beam of ions (either Cs^+ or O^-) to eject secondary ions from a sample surface, which are then analysed using a mass spectrometer at a very high spatial resolution typically of the order of 100 nm for soil samples.
- Static or time-of-flight secondary ion mass spectroscopy can target ions and small-molecular fragments and is used for slightly larger areas.
- Laser desorption/ionization, laser ablation inductively coupled plasma, matrix-assisted laser desorption/ionization and desorption electrospray ionization spectroscopies work at spatial scales slightly larger than that of individual cells.

For instance, Manté et al. (2020) studied clay coatings, pedofeature indicating clay illuviation in soils and developed a mathematical/statistical method to 'automatically identify and map the minerals present in synchrotron X-ray diffraction maps. This method combines a subtraction of the background signal, an identification of the diffraction peaks and the attribution of the obtained peaks to the different minerals phases. The robustness of the method was tested for the two first steps that were critical. Most of the minerals present in the considered samples could be identified, including clay minerals and Fe oxides; however, their relative proportions were difficult to estimate within the experimental conditions'.

Thin sections of resin-impregnated soil samples were also useful for observing algae, fungal hyphae and bacteria, using either light or phase-contrast microscopy (Baveye et al., 2018), while staining techniques can help identify them. Unlike fungal hyphae, bacteria and archaea are not easily identified due to the difficulty, with traditional light microscopes focusing precisely on a specific depth (Baveye et al., 2018). An easier solution consists of using a confocal laser microscope, which can produce sharp 2D images at selected shallow depths within a soil thin section. With special software, 3D images can be reconstructed from a set of *z*-dependent 2D images (Baveye et al., 2018). Furthermore, many microbial cells may be undetectable against a very bright background of fluorescing soil constituents, but this problem was solved thanks to the development of very reliable fluorescence *in situ* hybridization (FISH) techniques, which use fluorescent-labelled oligonucleotide probes. Eickhorst and Tippkötter (2008) [In (Baveye et al., 2018)] have shown that when FISH is combined with tyramide signal amplification, in what is referred to as catalysed reporter deposition (CARDFISH), one typically obtains higher signal intensity and reduced interference of the background fluorescence of the soil. The combination of CARDFISH, or one of its variants, with confocal laser microscopy, affords a very powerful tool to visualize and quantify the distribution of microorganisms in soils, with a satisfactory depth resolution (Baveye et al., 2018).

As underlined by Verrecchia and Trombino (2021), both conventional and novel techniques significantly improve research on soil matter interactions. Not only do these techniques enhance spatial resolution with great precision, but they also provide chemical, mineralogical and microbiological visual evidence that allows us to considerably improve the accuracy of micromorphological diagnostics.

2.3.7 Combining two-dimensional images and three-dimensional computed tomography images for pedofeature characterization and quantification

Nowadays, soil structure can be studied via non-destructive methods, like x-ray CT or resonance magnetic imaging. Not only do these non-destructive methods allow the geometry, topology and connectivity of soil pores to be characterized at a variety of scales, relevant to a wide range of research questions, but, in addition, this information can now be combined with spatial data on (bio)chemical and microbiological properties of soils (Baveye et al., 2022; Védère et al., 2022).

Using x-ray tomography only, Capowiez et al. (2015) estimated directly on the CT images the volume of bioturbation due to earthworms inside soil cores following the method described by Capowiez et al. (2011). Bioturbated zones in images can indeed be assessed due to their intermediate grey-level values using a specific algorithm. This bioturbation volume corresponds to refilled macropores or soil lateral compaction around them (Capowiez et al., 2015).

Using synchrotron x-rays Pot et al. (2015) generated CT images of repacked aggregates in which histograms presented a good separation of voxels associated with the air, liquid and solid phases.

It is feasible to analyse soil architecture or other soil properties and pedofeatures by taking 2D images of sequences of thin sections in resin-impregnated blocks of undisturbed soil and then using dedicated software to reconstruct from these images a full 3D picture of the geometry of soil pores (Baveye et al., 2018). The idea is to perform multiple cuts through soil samples, analyse in turn the (bio)chemical make-up of each exposed surface within the soil and then, using an interpolation technique, generate a 3D picture from the data associated with the various surfaces as it is routinely used for biological samples, such as human tissues (Baveye et al., 2018). This tedious, time-consuming approach has been adopted with success by Vogel et al. (2022) [In (Baveye et al., 2018)]. Combining information from different sources (2D measurements) with tomographic images is a more promising and efficient way. In the first article exploring this route, Hapca et al. (2011) combined chemical maps produced by SEM-EDX with 3D x-ray computed microtomographic images (Vogel et al., 2022). Hapca et al. (2015) and Hapca et al. (2011) proposed a method to register 2D quantitative chemical information obtained, e.g. using energy-dispersive x-ray spectroscopy, with 3D data about soil pores derived from x-ray μCT images (Fig. 22) (Vogel et al., 2022). Hapca et al. (2015) proposed a method based on a regression-tree method and ordinary kriging applied to residuals and used it to predict the 3D spatial distribution of carbon, silicon, iron and oxygen at the microscale (Baveye et al., 2018). The spatial correlation between the x-ray greyscale intensities and the chemical maps made it possible to use a regression-tree model as an initial step to predict the 3D chemical composition. For chemical elements, e.g. iron, that have high attenuation and are sparsely distributed in a soil sample, the

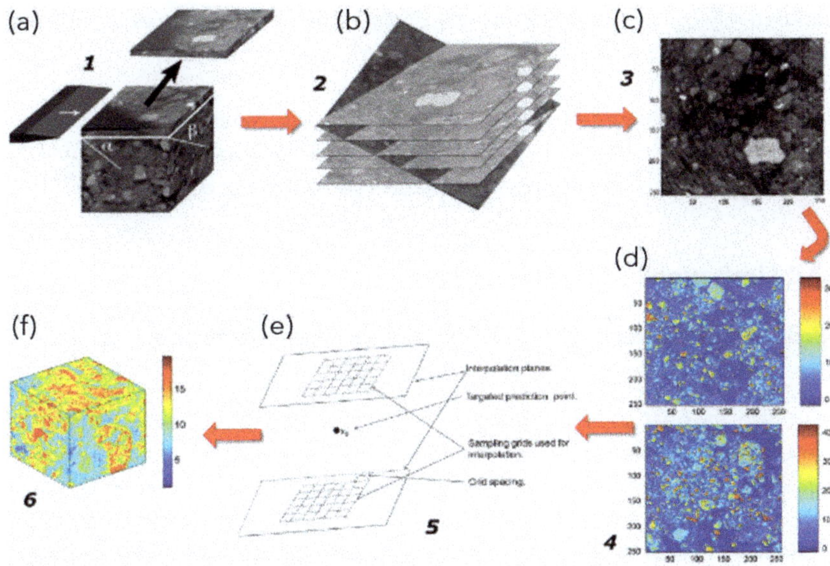

Figure 22 Schematic diagram of the successive steps in the 2D–3D interpolation method proposed by Hapca et al. (2011, 2015). (a) Illustration of a typical method to isolate a layer from a soil cube, with a microtome blade. The cut may be at angles *a* and *p* with the *x*-*y* plane in the *x* and *y* directions, respectively, resulting in layer surfaces that are not strictly parallel to each other, (b) rotation of the chemical analysis plane within the 3D CT image, (c) reconstituted CT image of the soil surface. The dotted lines correspond to the limits of the different masks applied to the successive layers during the zonation process, (d) spatial distribution, measured with SEM-EDX, of silicon in the top and bottom soil surfaces of an individual slice through the soil sample, (e) schematic representation of the interpolation layers and the corresponding sampling grids for the selection of the interpolation points and (f) 3D prediction of the silicon distribution in the soil sample. Reprinted with permission from Baveye et al. (2022).

regression-tree model provides a good prediction, explaining as much as 90% of the variability in some of the data for example (Baveye et al., 2018).

This research can be extended to a wide range of (bio)chemical parameters as well as microbiological ones (Védère et al., 2022). This correlative microscopy approach has since been adopted to address various research questions (Juyal et al., 2019; Schlüter et al., 2019) and will certainly support the development of the required holistic approach to soil architecture in the years to come (Vogel et al., 2022). Using a combination of x-ray μCT, fluorescence microscopy, scanning electron microscopy and nanoSIMS, Schlüter et al. (2019) were able to study the distribution of bacteria in soil and to show that they have a preference towards foraging near macropore surfaces and near fresh particulate organic matter (Baveye et al., 2022). Juyal et al. (2019, 2021) combined x-ray CT with biological thin sections to elucidate the impact of pore architecture on bacterial distribution in soil. This type of holistic perspective should prove particularly useful at a time when there is an urgent need to be able to describe quantitatively better than we currently do the functions (Baveye et al., 2022) to solve specific problems, especially in the field of organomineral interactions in soils.

Finally, soil micromorphology, based upon analytical and holistic approaches, will provide a foundation for future research which will be increasingly focused on nanoscale objects (Verrecchia and Trombino, 2021).

2.3.8 Limitations and perspectives

2.3.8.1 Spatial resolution and representative elementary area for pores

When using repeated 3D CT images to describe how earthworms modify the architecture of soil samples over time (Balseiro-Romero et al., 2020; Capowiez et al., 2015), soil samples need to be sufficiently large to accommodate several earthworms, which means that the resolution at which they are scanned has to be relatively coarse, of the order of 10 s or even hundreds of microns (Baveye et al., 2022). Pores whose size is smaller than the voxel size of the resulting images become invisible and, to date, the role that pores of various sizes fulfil in soil functioning is not quantitatively understood, and this needs to be urgently addressed according to Baveye et al. (2020, 2022).

To achieve a high resolution when scanning soil samples, it is necessary to restrict the size of samples (Baveye et al., 2018). Typically, with table-top scanners, samples cannot have a volume larger than 100 cm^3 to obtain micrometre resolution, and they need to be significantly smaller to achieve the highest resolution (of about 0.3 μm) that is advertised by manufacturers (Baveye et al., 2018). This small size of soil samples has occasionally raised questions in terms of the 'representativeness' of measurements stemming from the concept of representative elementary volume (REV). From that standpoint, soil samples scanned via x-ray CT or soil thin sections under an optic microscope, which are of a much smaller size than the REV, would not be sufficiently representative, and one should therefore try to work with larger samples, be it physically or virtually, by aggregating together mosaic-style images obtained on a number of small juxtaposed samples (e.g. Sauzet et al., 2017). As very often the REV of the objects under study is not determined, the best one can do is to carefully reference any observation that is made on a soil sample to the volume and shape of this sample (Baveye et al., 2018). 'Nevertheless, regardless of how one feels concerning the need to invoke the notion of REV, it is clear that observations made on cm^3-sized soil samples are not directly relevant to answering the questions raised by soil management, all of which relate to significantly larger spatial scales' (Baveye et al., 2018).

2.3.8.2 Representative elementary areas for pedofeatures

Due to technical difficulties during sampling and because manufacturing larger thin sections resulted in larger variations in thickness within single slides, biological activity and illuviation have been generally quantified on small areas (e.g. 10 mm × 10 mm) (Sauzet et al., 2017). Such sizes are in most cases smaller than the representative elementary areas, leading to variability in the results (Bruneau et al., 2004). In order to have accurate and representative measurements, high-resolution image mosaics covering a large contiguous area are required (Adderley et al., 2002). In light of the results of REV published by Sauzet et al. (2017), studies in the past have seldom worked on surfaces of sufficient sizes (larger than 1000 mm^2) with resolutions of some tens of squared micrometres (Bryk, 2016; Gargiulo et al., 2013; Jangorzo et al., 2013). It is thus likely that the analysed images were very often of insufficient size for most of the features of interest. This is particularly the case for previous quantitative studies of illuvial clay features [Murphy, 1983 In (Sauzet et al., 2017)], which were based on 2.5 cm × 2.5 cm areas (i.e. 625 mm^2).

2.3.8.3 Validation

In the work carried out to date, only a few authors have tried to validate their image analysis protocol (Sauzet et al., 2017). They have often compared their results with point counting results (Aydemir et al., 2004; Bruneau et al., 2004; Grove and Jerram, 2011; Piron et al., 2012; Terribile and Fitzpatrick, 1995). Alternatively, Protz and VandenBygaart (1998) has used direct visual assessment, while Terribile and Fitzpatrick (1992) used a comparison with a reference image (Sauzet et al., 2017).

2.3.8.4 Impregnation of soil samples

Micromorphology requires the very slow impregnation of soil samples with resin, less suited for routine measurements of a large number of soil samples (Baveye et al., 2022). If one uses 3D CT scans for 3D pore quantification, the samples do not need to be systematically impregnated (Sauzet et al., 2021; Case Study 2). The use of various spectroscopic types of equipment might be promising as they afford measurements of the concentration of various elements or their speciation but not at all of the physicochemical conditions in which specific (bio)chemical species are located. In particular, one would absolutely need information about the pH or redox potential locally in a porous medium, among other 'thermodynamic' variables, to have a full picture of what is going on. Unfortunately, once a block of soil has been impregnated with resin and has cured, none of these variables is accessible any more (Baveye et al., 2018).

2.3.8.5 Expensive equipment

The review article by Baveye et al. (2018) mentions a number of advances achieved by physicists in the last few years, such as laser wakefield accelerators not much larger than a shoebox yet producing synchrotron-quality, nearly monochromatic (single energy) tunable x-ray beams, that are paving the way for new, vastly better equipment in the not very distant future (Baveye et al., 2022). Progressively, these machines may thus become more affordable.

2.3.8.6 Affording four-dimensional measurements

Finally, 4D dynamical measurements are required to conduct research on the impact of these various processes on architecture dynamics, but their use has remained scarce to date, mostly due to technological limitations (Baveye et al., 2022).

3 Case studies

3.1 Photogrammetric approach applied to slaking mapping of an agricultural field

The 'Palmort' agricultural plot of 28 ha, located in the western region of Paris, France, was studied from March to May 2016. An area of 3.6 ha was delimited within the plot for this work. Forty-two ground control points (GCPs) were sampled within the 3.6 ha plot (Fig. 23). A centimetric Trimble Geo 7× differential global navigation satellite system (DGNSS) was used to determine precise GCP positions. On each GCP, the soil horizons were described, and the topsoil was sampled for standard physicochemical analysis. Soil apparent electrical conductivity measurements were carried out using a Geonics EM38 ground

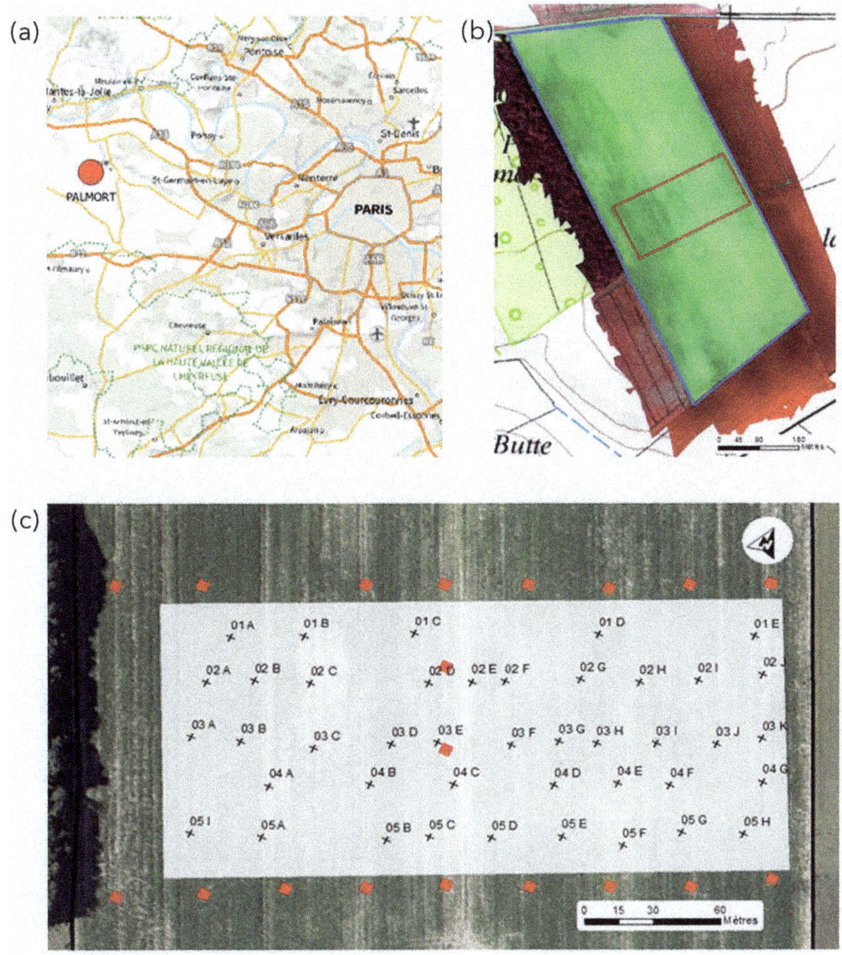

Figure 23 The 'Palmort' field case study. (a) Localization of the plot in the Paris region (France), (b) the study area inside the plot and (c) the 42 measurement points.

conductivity meter (Geonics, 2022) with the instrument in the vertical dipole mode, which corresponds to a measurement depth of 1.5 m. The measurements were made after soil tillage by the farmer just before sowing.

On each GCP, a series of photographs were taken on a plot of approximately 0.5 m^2 to characterize the soil roughness using a photogrammetric approach. Nine photos were taken from a height of about 1.5 m, eight from an oblique view around the plot and one from a top view (Fig. 24a). This series of measurements was used to spatialize the roughness at the scale of the 3.6 ha plot; for this, the STI index was calculated from the photogrammetric model, and then a geostatistical analysis was carried out to realize a kriging of this index on the study area.

A wooden square was placed next to each plot to serve as a 3D marker for the photographs. Four-coloured discs have been glued at known positions in (*x*,*y*,*z*) - three are in the same plane and one (the red one) is elevated by 33 mm to serve as a reference in *Z*. A Python script, named structure from motion for roughness (SFMr) (Fig. 25), has been developed to automate the processing of all the plots and uses VisualSFM for 3D point cloud creation and PMVS (Furukawa, 2022) for densification.

SFMr automatically detected the position of the four-coloured discs in the 3D point cloud (Fig. 25b), and then a linear regression was calculated between the position of the discs in the cloud reference system and the positions in centimetres relative to the square origin (Eq. 9); finally, the linear equation was applied to all the point cloud to reference them in centimetres with respect to the origin of the square.

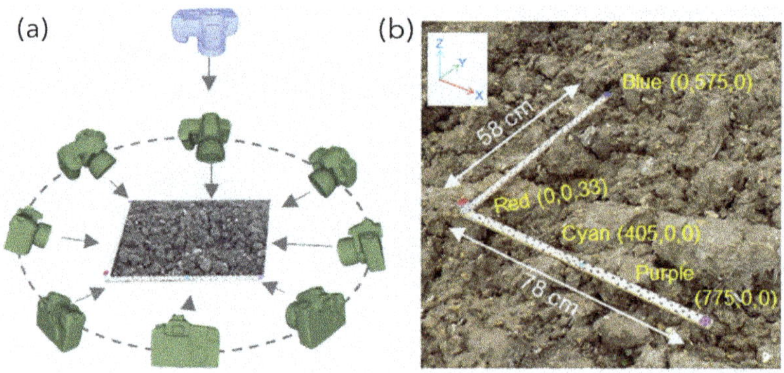

Figure 24 Photos acquisition. (a) Nine photos at a height of about 1.5 m and (b) wooden square as a three-dimensional marker.

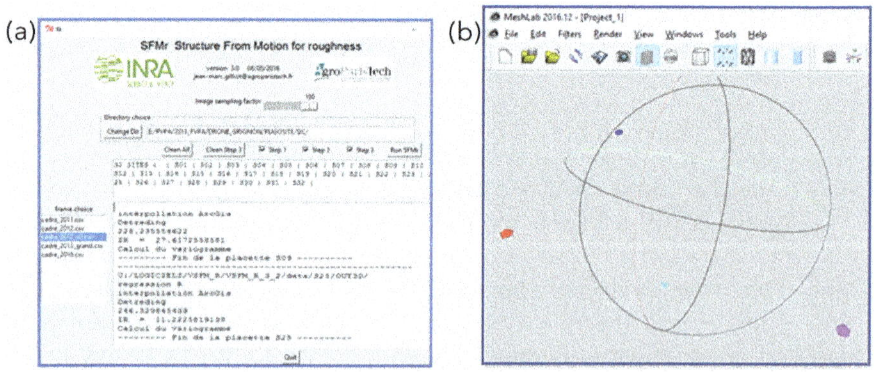

Figure 25 (a) SFMr software graphic user interface and (b) coloured disc automatic extraction.

$$\begin{cases} x = a_1x_c + a_2y_c + a_3z_c + a_0 \\ y = b_1x_c + b_2y_c + b_3z_c + b_0 \\ z = c_1x_c + c_2y_c + c_3z_c + c_0 \end{cases} \quad (9)$$

with *x*, *y*, and *z* coordinates in centimetres and x_c, y_c, and z_c coordinates in the original cloud. A detrending operation was also applied to coordinates to correct the local slope. Points were finally interpolated into a 2D DSM that represented the soil height. SFMr calculated roughness indices from this DSM and produced roughness graphs (Fig. 26).

A geostatistical analysis was conducted on the STI index from the 42 measurement points, and a model was fitted to the experimental variogram to interpolate the index by kriging on the whole plot. Figure 27 shows this interpolated STI map.

Figure 28 shows the comparison of the STI map with a map of the slaking sensitivity index (slak), which was calculated from laboratory analyses of the top horizon (Eq. 10).

$$slak = \begin{cases} \dfrac{(1.5 \times fine\_sil + 0.75 \times coarse\_silt)}{clay + 10 \times Organic\_Matter}, pH \leq 7 \\ \dfrac{(1.5 \times fine_{silt} + 0.75 \times coarse\_silt)}{clay + 10 \times Organic\_Matter} - 0.2 \times (pH - 7), pH > 7 \end{cases} \quad (10)$$

The higher the index, the more sensitive the soil is to slaking; soils with slak > 1.8 are considered to be very sensitive to slaking and those with slak < 1.4 are very little or not at all sensitive to slaking.

The plot was moderately slaked, but the most slaken areas corresponded well with the areas of lowest roughness.

3.2 Image analysis procedure to quantify biological porosity and illuvial clay in large soil thin sections and soil columns

3.2.1 Case study 1: A methodological innovation: development and validation of clay coatings abundance, thanks to quantification by image analysis on thin sections (Sauzet et al., 2017)

The clay-sized fraction is linked with many of the functions and services rendered by soils (Adhikari and Hartemink, 2016). Issues with the clay-sized fraction are particularly problematic in topsoil horizons. There, a decrease in clay-sized fraction is generally accompanied by degradation of the soil architecture and subsequent increases in the risks of soil erosion and compaction (Kay, 1997), all of which combine to induce negative impacts on crop yields (King et al., 2020). In non-eroded

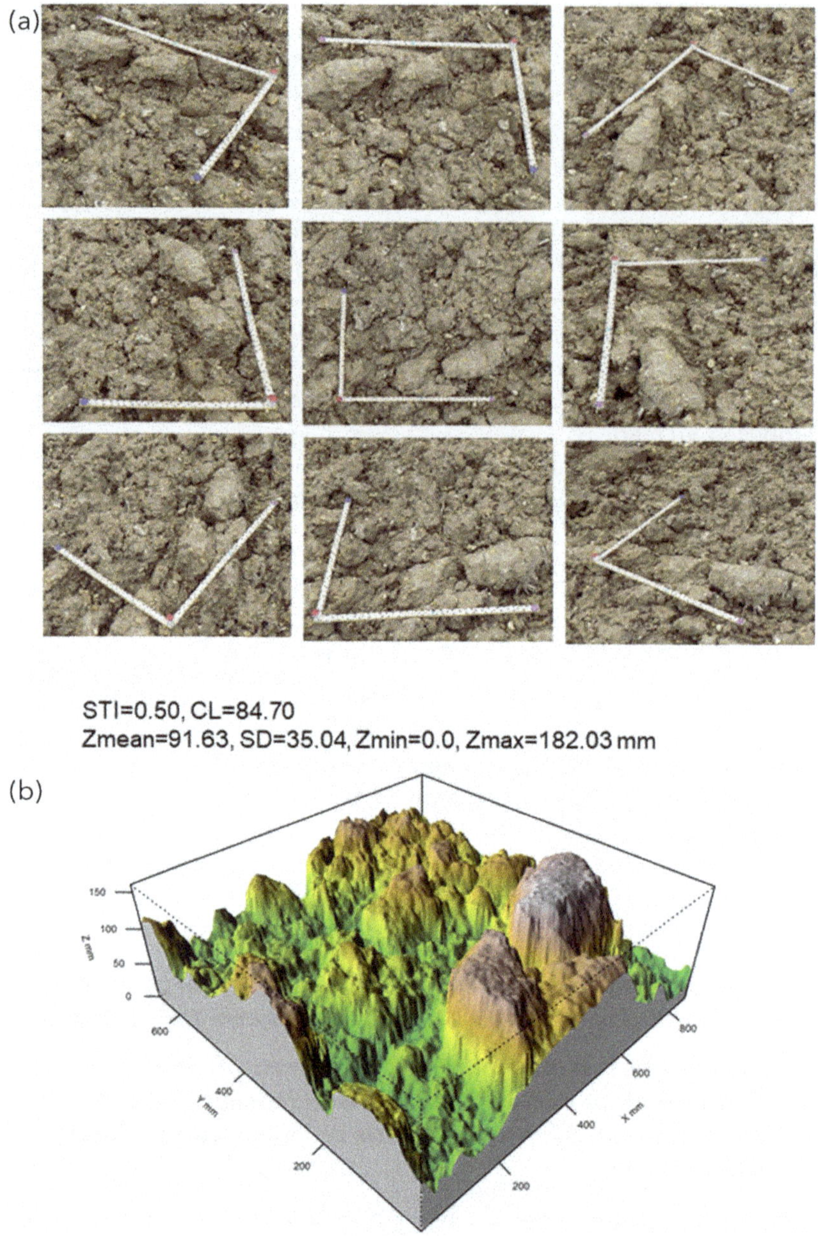

Figure 26 '1A' point model. (a) The nine input photos and (b) 3D view of the output model.

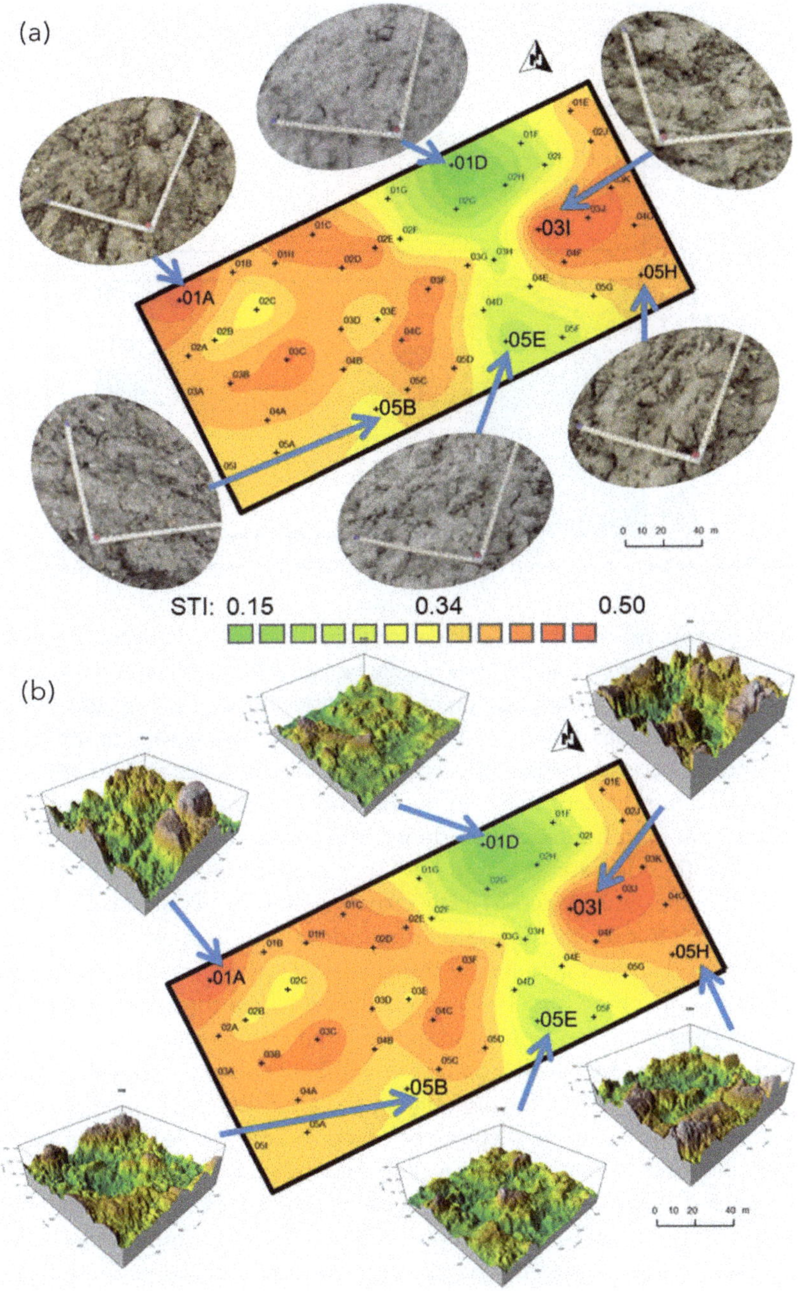

Figure 27 STI index map (kriging): (a) with six pictures in STI-contrasted places and (b) with 3D models at the same places.

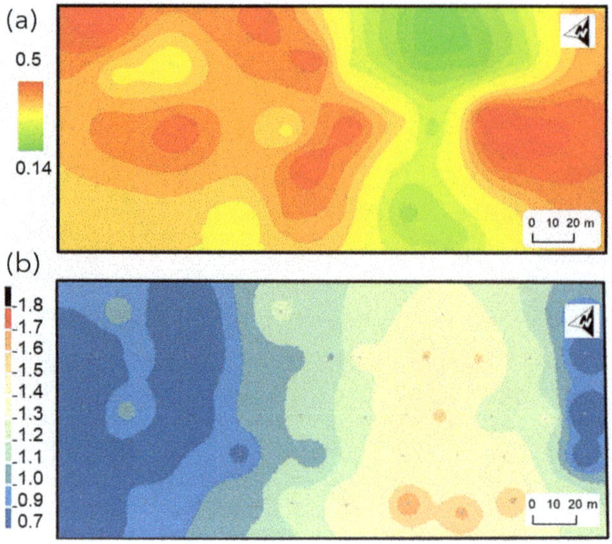

Figure 28 Comparison between (a) the STI I_R map and (b) the slaking index map.

temperate soils developed in a unique soil parent material, the distribution with a depth of the clay-sized fraction may mainly be controlled by illuviation. It consists of a substantial vertical translocation of fine particles up to 10 μm in size (Quénard et al., 2011) and results in the formation of an illuviated horizon in the soil profile due to the accumulation of the fine fraction called Bt-horizon (Anjos et al., 2016).

Observation, description and quantification of clay coatings on thin soil sections have long been used to identify the illuviation process and estimate its intensity (Miedema and Slager, 1972). However, the classical methods of quantifying the abundance of clay coatings by point counting have been criticized as tedious and operator dependent. At the same time, the rare attempts to identify and quantify the clay coatings by image analysis have most often concerned small areas of analysis; hence the great instability of the quantifications is obtained and most of them have never been the object of real validation. This is why Sauzet et al. (2017) aimed at developing and validating a new method for quantifying the abundance of macropores and clay coatings on soil thin sections sampled in soil profiles from a specific anthropo-chrono-sequence in the Paris basin (Sauzet et al., 2017). If clay coatings are easily recognizable because of their specific yellow to red colour, the use of a purely colorimetric approach proved impossible. Indeed, the choice of a large size of analysis surface results in a great variability of the thickness of the thin soil sections and consequently of the colour of the clay coatings (Fig. 29a and 29c). In addition, isolated concentrations can be observed within the matrix. Without being true clay coatings, they present nevertheless a comparable colour which can lead to confusion. Sauzet et al. (2017) then developed an image analysis method capable of taking into account the different levels of soil organization, thanks to (i) a locally adapted thresholding based on the *G/R* ratio mean within a window of size 1.6 mm × 1.6 mm allowing to take into account the structure of the samples in more or less thick zones (Fig. 29c) and (ii) the identification of a homogeneous texture within a window of about 40 μm × 40 μm typical of clay coatings (Fig. 29c). In addition to the clay coating recognition and quantification, macropores were quantified following a specific procedure, including an automatic triangle thresholding (Fig. 29b). Macropores related to earthworms were isolated based on shape and size criterion. The quality of the method was evaluated using a contingency matrix comparing the automatic extraction of clay coatings by image analysis to a manual and exhaustive digitization of all the clay coatings observed on a total surface of 300mm^2, i.e. nearly 1500 coatings. The 'producer' and 'user' accuracies obtained, which were always above 70% (Sauzet et al., 2017), indicate that this method (i) detects well the great majority of clay coating identified as such by the micromorphologist and (ii) makes few confusions between clay coatings and other pedological features. Once the abundance of clay coatings was quantified for each vertical thin section, the total mass of fraction <2 μm illuviated per unit area was calculated for each soil profile. Such quantification is useful and very promising to assess the consequences of land use change and management regarding soil processes such as illuviation, structuration and bioturbation and thus associated soil properties.

3.2.2 Case study 2: **Nicodrilus nocturnus** *and* **Allolobophora icterica** *effects on soil porosity: computed tomography insights (Sauzet et al., 2021)*

Machine vision techniques are often thought of using exclusively image-based sensors, but the application of CV techniques to CT images helps understand and quantify soil characteristics and processes. Earthworms are known to play a key role in soil ecological processes such as organic matter turnover, nutrient cycling and soil physical property engineering and soil fertility as a whole. However, their role in numerous soil physical properties is still poorly documented. Sauzet et al. (2021) aimed to test the effect of two earthworm species (*N. nocturnus* as anecic species and *A. icterica* as endogeic species and both as mixed population with 80% weight of *N. nocturnus* and 20% of *A. icterica*) on two different soils. A multi-scale approach based on shrinkage analysis and CT was used to assess the impact of earthworms on soil porosities and bulk volume. The study was performed for 23 weeks on microcosms, namely soil columns 30 cm height and 15 cm large, under controlled conditions (constant temperature and constant

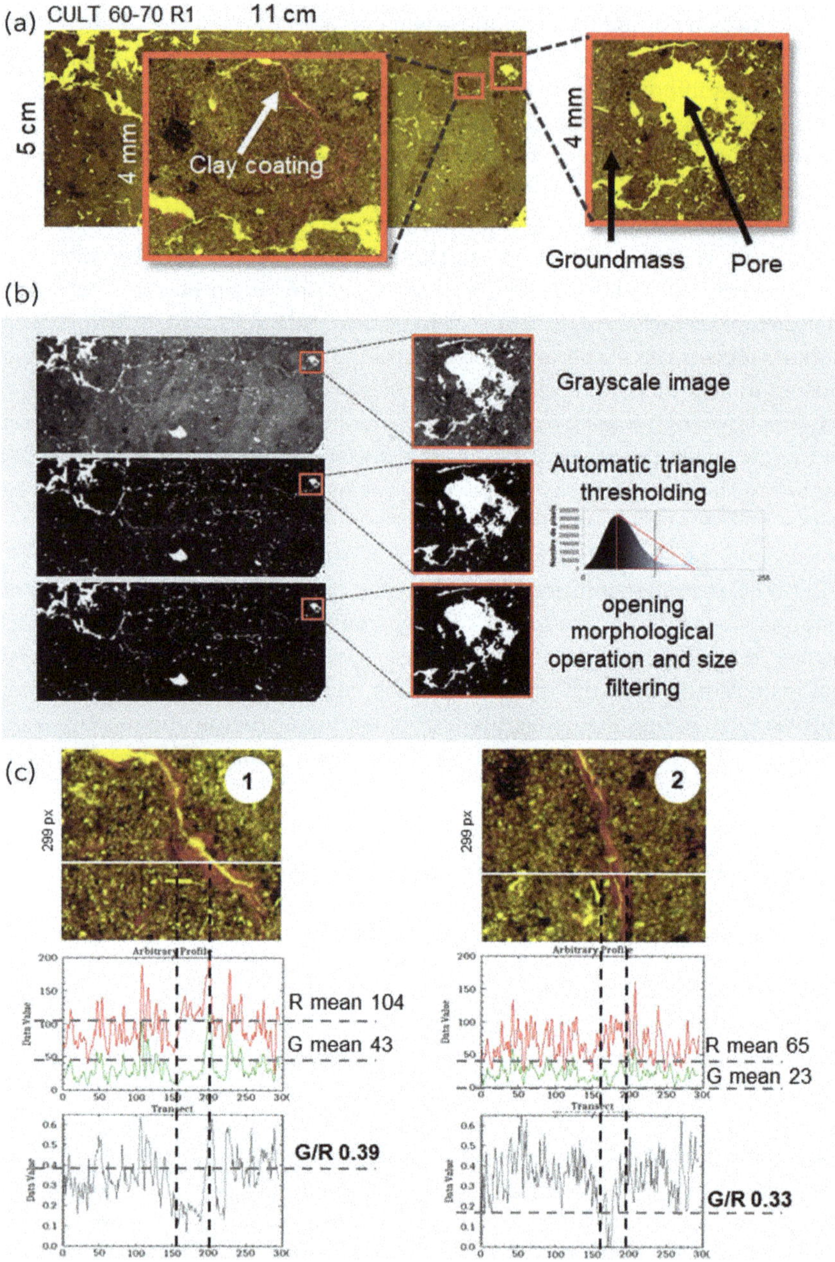

Figure 29 The identification and extraction of clay coatings. (a) Soil thin-section RGB image - Leica DM5500B (after mosaicing); resolution = 5.3 μm/pixel. (b) Procedure of macropore automatic recognition on soil thin sections. (c) Adaptative thresholding examples for clay coating recognition on soil thin sections.

soil matrix potential). The soils were a silt loam Luvisol, and a loamy Anthrosol repacked at field observed bulk density and at compacted bulk density. Therefore, the treatments were 2 soils × 2 levels of compaction × 3 worm species plus controls, three replicates each. Physical measurements were performed at the end of 5 months. The soil bulk volume (i.e. the inverse soil bulk density) in the whole mesocosms was measured by filling the top of the column with calibrated sand, and the soil bulk volume was then equalled to the column volume minus the sand volume added (Abedine el and Robinson, 1971; Favre et al., 1997). This allowed calculating the soil bulk density at mesocosm scale since the dry mass of soil introduced in the columns was known. The columns were then analysed by x-ray CT, on a LightSpeed Ultra 8b-unit (General Electric Medical Systems, Milwaukee, WI, USA). The x-ray source was perpendicular to the mesocosms and set to 150 mA at 120 kV, which allowed characterizing 1.25 mm thick layers with a 0.625 mm overlapping. The image resolution was 0.4 mm per pixel.

The images were processed with General Electric Advantage Windows Workstation 4.3 (General Electric Medical Systems) to get 3D images. Images were transformed into eight-bit images by calibrating the grey levels between -100 and 200 Hounsfield values (Fig. 30). The images were then binarized by choosing a threshold value of 225 that clearly separated the two peaks (soil matrix and air-filled porosity) in the grey-level histogram. Isolated pores with a volume of less than 300 voxels (1 voxel = 0.4 × 0.4 × 0.625 mm^3) were discarded. This lonly left pores that were large enough to be created by earthworms (Jégou et al., 2001). To characterize the earthworm burrow systems, we first computed the volume of macroporosity and the percentage of the burrow system in the upper and lower half of the core, respectively. Then, burrow diameters were assessed on the most circular pores in the 2D images (pores whose circularity was higher than 0.8), assuming that burrows are perfect cylinders. Finally, the

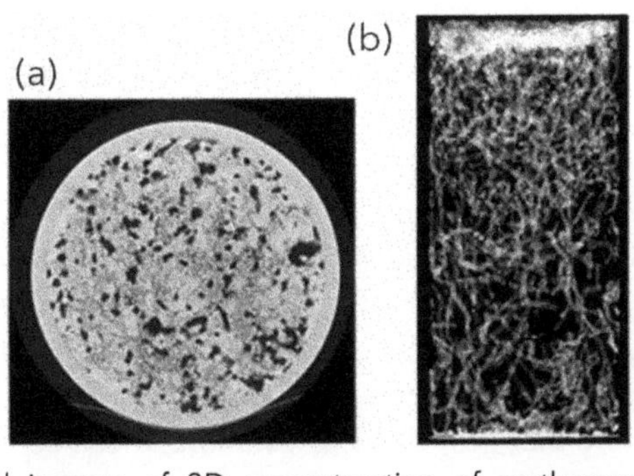

Figure 30 Selected images of 3D reconstruction of earthworm burrow system in repacked soil cores (30 cm in height and 15 cm in diameter) after 5 months as assessed by computed tomography imaging: (a) cross-section and (b) longitudinal section. Reprinted with permission from Sauzet et al. (2021).

skeleton of the burrow systems using the AnalyzeSkeleton plugin in ImageJ was computed. This enabled the assessment of burrow length and branching rate (number of branching pattern per metre of burrow). The specific volumes of the soil columns (cm^3 g^{-1}) (i.e. the inverse soil bulk density, g cm^{-3}) after subtraction of the burrow volumes were then calculated and referred to as the soil matrix specific volume. At mesocosm scale, no bulk soil volume change was recorded when using endogeics. However, when removing the burrow volumes, the soil matrix was significantly compacted by earthworms, particularly the endogeic, and the compacted soil matrix was not decompacted.

Finally, it was shown that the decompaction effect of earthworms was due to the opening of burrows at bulk soil scale, but the matrix volume was either compacted (in the case of loosened soil) with a decrease of the larger structural pores (>150 μm equivalent radius) or unchanged (in the case of compacted soil), and the compaction of the matrix is larger with endogeic species. The mixed-species treatment induced intermediate effects.

4 Conclusion and future trends

To manage crop production on a plot, the farmer must have a soil diagnosis to know the agronomic potential and to adapt his practices. Regularly carrying out chemical analyses of soil plots are essential for the proper management of soil amendments as well as mineral and organic fertilizers. However, as seen in the chapter, the soil structure, both on the surface and in depth, also determines successful plant development, in particular a good water supply, or because of the loss of nutrients due to erosion; soil workability (referring to ease of tillage) is also an important consideration for the farmer. Soil potential studies can be redone regularly after several years to track the evolution of the soil, in particular in the case of the implementation of new practices such as the adoption of NTi practices instead of CTi within the framework of SSM, which have impacts on soil structure. Generally, the study of the agronomic potential of soils at the plot scale will be carried out in a limited number of points supposed to be representative of the diversity of soils; the description of a soil profile and detailed chemical and physical analyses will be realized there. With the development of precision agriculture, more and more farmers are opting for a spatialized (cartographic) management of practices (site-specific farming). It is challenging to assess soil information with high spatial density for reasons of cost and time. Sensors offer a solution to spatially densify the soil information and better capture its intra-plot variability. According to the principles of measurement most often used as well as the soil parameters that they allow to estimate, these sensors are: (i) optical and radiometric sensors (organic matter, texture and roughness), which have been detailed in this chapter, to which we can add ground-penetrating radar and microwave sensors; (ii) electrical resistivity/conductivity sensors (salinity, texture, thickness and water content); (iii) acoustic sensors (compaction and clay content); (iv) pneumatic sensors (structure and water content); (v) mechanical sensors (compaction and mechanical resistance); (vi) electrochemical sensors (pH, K, NO_3 and Na). For field mapping purposes, preferable sensors are those that give an almost instantaneous response and that (coupled with a GNSS position receptor) can be installed on a tractor to map the plot during return trips; they are often referred to as 'on-the-go soil sensors'. The sensor can also be installed on a robot or a UAV. The measurement that is provided by a sensor is most often an indirect estimate of the soil variable that is sought; for example the clay content of a soil will be estimated by measuring its impact on the soil colour and brightness; this estimation is less precise than the laboratory analysis, but it can be done a large number of times on the whole plot. The precision that is lost at the analytical level is gained at the spatial level. The response of these sensors is also often dependent on several factors, which are not easy to decouple; an electrical resistivity sensor for example is sensitive to the thickness, texture and humidity of a horizon and to the temperature at the time of measurement. These elements make it very often necessary to calibrate a sensor to be able to interpret its response according to the studied soil parameter. This is usually done with a few soil samples on which the reference laboratory analysis is done. Then, the calibration consists in establishing by regression the equation between the sensor and the soil parameter, which can then be applied to the entire plot in prediction mode. The development of soil sensors that require little or no calibration would be a great advance for their practical use in agriculture. The current trend is to develop multi-sensor systems. Wijewardane et al. (2020) developed a Vis-NIR-integrated multi-sensing penetrometer for vertical soil sensing, which measured reflectance spectra and penetration resistance. Knadel et al. (2015) used a Veris mobile sensor platform, developed by Veris Technologies (Kansas, USA), to collect simultaneously VIS-NIR spectra, electrical conductivity and temperature measurements for soil organic carbon and particle size prediction.

Finally, Fig. 31 proposes a simplified five-step process to produce a map of a soil parameter, such as clay content, starting from an on-the-go soil sensor.

5 Where to look for further information

- Many references cited in the chapter provide detailed information on the chapter subject.
- An abridged, student-oriented edition of Hillel's earlier published Environmental Soil Physics, this is a more succinct elucidation of the physical principles and processes governing the behaviour of soil and the vital role it plays in both natural and managed ecosystems. Hillel, D. (2003). Introduction to Environmental Soil Physics. Elsevier Science.

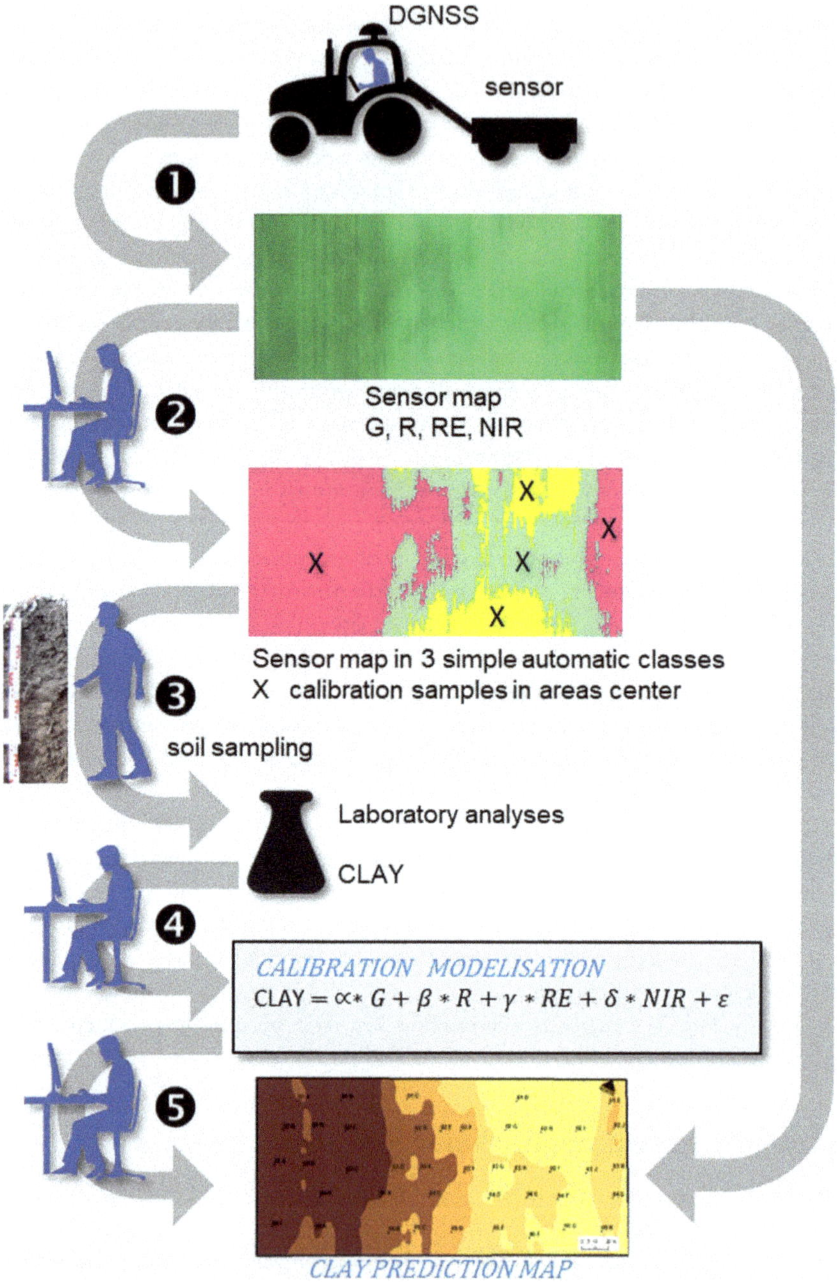

Figure 31 Simplified five-step process to produce a soil parameter map from an on-the-go soil sensor.

- Detailed information about sustainable soil management (SSM) can be found in this FAO report: Revised World Soil Charter https://www.fao.org/documents/card/en/c/e60df30b-0269-4247-a15f-db564161fee0/.

6 Acknowledgements

The authors are grateful to Joël Michelin for being in charge of the pedological study and Dalila Hadjar for her help in the various field surveys as part of the 'Palmort' case study.

7 References

Abedine el, A. Z. and Robinson, G. H. (1971). Study on cracking in some vertisols of Sudan. *Geoderma* 5(3), 229–241. https://doi.org/10.1016/0016-7061(71)90012-7.

Adamchuk, V. I., Hummel, J. W., Morgan, M. T. and Upadhyaya, S. K. (2004). On-the-go soil sensors for precision agriculture. *Computers and Electronics in Agriculture* 44(1), 71–91. https://doi.org/10.1016/j.compag.2004.03.002.

Adderley, W. P., Simpson, I. A. and Davidson, D. A. (2002). Colour description and quantification in mosaic images of soil thin sections. *Geoderma* 108(3–4), 181–195. https://doi.org/10.1016/S0016-7061(02)00123-4.

Adderley, W. P., Simpson, I. A. and Davidson, D. A. (2006). Historic landscape management: A validation of quantitative soil thin-section analyses. *Journal of Archaeological Science* 33(3), 320–334. https://doi.org/10.1016/j.jas.2005.07.016.

Adhikari, K. and Hartemink, A. E. (2016). Linking soils to ecosystem services—A global review. *Geoderma* 262, 101–111. https://doi.org/10.1016/j.geoderma.2015.08.009.

Agisoft (2022). *Agisoft Metashape*. Available at: https://www.agisoft.com/.

Aitkenhead, M., Coull, M., Gwatkin, R. and Donnelly, D. (2016). Automated soil physical parameter assessment using smartphone and digital camera imagery. *Journal of Imaging* 2(4), Article 4. https://doi.org/10.3390/jimaging2040035.

Aliah Baharom, S. N., Shibusawa, S., Kodaira, M. and Kanda, R. (2015). Multiple-depth mapping of soil properties using a visible and near infrared real-time soil sensor for a paddy field. *Engineering in Agriculture, Environment and Food* 8(1), 13–17. https://doi.org/10.1016/j.eaef.2015.01.002.

AliceVision. (2022). Vision|Meshroom–3D reconstruction software. Available at: https://alicevision.org/#meshroom.

Anjos, L., Gaistardo, C. C., Deckers, J., Dondeyne, S., Eberhardt, E., Gerasimova, M., Harms, B., Jones, A., Krasilnikov, P., Reinsch, T., Vargas, R. and Zhang, G.-L. (2016). World reference base for soil resources 2014 international soil classification system for naming soils and creating legends for soil maps. Joint Research Center Publications Repository. Available at: https://publications.jrc.ec.europa.eu/repository/handle/JRC91947.

Asus. (2022). *ASUS Xtion* (Vol. 2). Available at: http://xtionprolive.com/asus-xtion2.

Aydemir, S., Keskin, S. and Drees, L. R. (2004). Quantification of soil features using digital image processing (DIP) techniques. *Geoderma* 119(1–2), 1–8. https://doi.org/10.1016/S0016-7061(03)00218-0.

Azadnia, R., Jahanbakhshi, A., Rashidi, S., Khajehzadeh, M. and Bazyar, P. (2022). Developing an automated monitoring system for fast and accurate prediction of soil texture using an image-based deep learning network and machine vision system. *Measurement* 190, 110669. https://doi.org/10.1016/j.measurement.2021.110669.

Baize, D., Duval, O. and Richard, G. (2013). *Soils and Their Structures: Observations at Different Scales (Les Sols et Leurs Structures: Observations à Différentes Échelles)* (1er édition). Editions QUAE GIE.

Ball, B. C., Batey, T. and Munkholm, L. J. (2007). Field assessment of soil structural quality – a development of the Peerlkamp test. *Soil Use and Management* 23(4), 329–337. https://doi.org/10.1111/j.1475-2743.2007.00102.x.

Balseiro-Romero, M., Mazurier, A., Monoshyn, D., Baveye, P. C. and Clause, J. (2020). Using X-ray microtomography to characterize the burrowing behaviour of earthworms in heterogeneously polluted soils. *Pedobiologia* 83, 150671. https://doi.org/10.1016/j.pedobi.2020.150671.

Barman, U. and Choudhury, R. D. (2020). Soil texture classification using multi class support vector machine. *Information Processing in Agriculture* 7(2), 318–332. https://doi.org/10.1016/j.inpa.2019.08.001.

Baveye, P. C., Balseiro-Romero, M., Bottinelli, N., Briones, M., Capowiez, Y., Garnier, P., Kravchenko, A., Otten, W., Pot, V., Schlüter, S. and Vogel, H.-J. (2022). Lessons from a landmark 1991 article on soil structure: Distinct precedence of non-destructive assessment and benefits of fresh perspectives in soil research. *Soil Research*. https://doi.org/10.1071/SR21268.

Baveye, P. C., Otten, W., Kravchenko, A., Balseiro-Romero, M., Beckers, É., Chalhoub, M., Darnault, C., Eickhorst, T., Garnier, P., Hapca, S., Kiranyaz, S., Monga, O., Mueller, C. W., Nunan, N., Pot, V., Schlüter, S., Schmidt, H. and Vogel, H. J. (2018). Emergent properties of microbial activity in heterogeneous soil microenvironments: Different research approaches are slowly converging, yet major challenges remain. *Frontiers in Microbiology* 9, 1929. Available at: https://www.frontiersin.org/article/10.3389/fmicb.2018.01929.

Baveye, P. C., Schnee, L. S., Boivin, P., Laba, M. and Radulovich, R. (2020). Soil organic matter research and climate change: Merely re-storing carbon versus restoring soil functions. *Frontiers in Environmental Science* 8. Available at: https://www.frontiersin.org/article/10.3389/fenvs.2020.579904.

Bertuzzi, P., Caussignac, J. M., Stengel, P., Morel, G., Lorendeau, J. Y. and Pelloux, G. (1990). An automated, noncontact laser profile meter for measuring soil roughness in situ. *Soil Science* 149(3), 169–178.

Bielders, C. L., Baveye, P., Wilding, L. P., Drees, L. R. and Valentin, C. (1996). Tillage-induced spatial distribution of surface crusts on a sandy Paleustult from Togo. *Soil Science Society of America Journal* 60(3), 843–855. https://doi.org/10.2136/SSSAJ1996.03615995006000030024X.

Boiffin, J. (1984). *Structural degradation of the soil surface by the action of rainfall* [Ph.D. Dissertation]. Institut National Agronomique Paris-Grignon.

Borselli, L. (1999). Segmentation of soil roughness profiles. *Earth Surface Processes and Landforms* 24(1), 71–90. https://doi.org/10.1002/(SICI)1096-9837(199901)24:1<71::AID-ESP953>3.0.CO;2-T.

Bretar, F., Arab-Sedze, M., Champion, J., Pierrot-Deseilligny, M., Heggy, E. and Jacquemoud, S. (2013). An advanced photogrammetric method to measure surface roughness: Application to volcanic terrains in the Piton de la Fournaise, Reunion Island. *Remote Sensing of Environment* 135, 1–11. https://doi.org/10.1016/j.rse.2013.03.026.

Bruneau, P. M. C., Davidson, D. A. and Grieve, I. C. (2004). An evaluation of image analysis for measuring changes in void space and excremental features on soil thin sections in an upland grassland soil. *Geoderma* 120(3–4), 165–175. https://doi.org/10.1016/J.GEODERMA.2003.08.012.

Bryant, R. G. and Davidson, D. A. (1996). The use of image analysis in the micromorphological study of old cultivated soils: An evaluation based on soils from the Island of Papa Stour, Shetland. *Journal of Archaeological Science* 23(6), 811–822. https://doi.org/10.1006/jasc.1996.0076.

Bryk, M. (2016). Macrostructure of diagnostic B horizons relative to underlying BC and C horizons in podzols, luvisol, cambisol, and arenosol evaluated by image analysis. *Geoderma* 263, 86–103. https://doi.org/10.1016/j.geoderma.2015.09.014.

Bullock, P., Fedoroff, N. and Jongerius, A. (1985). *Handbook for Soil Thin Section Description*. Available at: https://library.wur.nl/WebQuery/wurpubs/487492.

Bünemann, E. K., Bongiorno, G., Bai, Z., Creamer, R. E., De Deyn, G., de Goede, R., Fleskens, L., Geissen, V., Kuyper, T. W., Mäder, P., Pulleman, M., Sukkel, W., van Groenigen, J. W. and Brussaard, L. (2018). Soil quality – A critical review. *Soil Biology and Biochemistry* 120, 105–125. https://doi.org/10.1016/j.soilbio.2018.01.030.

Capowiez, Y., Bottinelli, N., Sammartino, S., Michel, E. and Jouquet, P. (2015). Morphological and functional characterisation of the burrow systems of six earthworm species (Lumbricidae). *Biology and Fertility of Soils* 51(7), 869–877. https://doi.org/10.1007/s00374-015-1036-x.

Capowiez, Y., Sammartino, S. and Michel, E. (2011). Using X-ray tomography to quantify earthworm bioturbation non-destructively in repacked soil cores. *Geoderma* 162(1–2), 124–131. https://doi.org/10.1016/j.geoderma.2011.01.011.

Cheik, S., Jouquet, P., Maeght, J.-L., Capowiez, Y., Tran, T. M. and Bottinelli, N. (2021). X-ray tomography analysis of soil biopores structure under wetting and drying cycles. *European Journal of Soil Science* 72(5), 2128–2132. https://doi.org/10.1111/ejss.13119.

Chung, S.-O., Cho, K.-H., Kong, J.-W., Sudduth, K. A. and Jung, K.-Y. (2010). Soil texture classification algorithm using RGB characteristics of soil images. *IFAC Proceedings Volumes* 43(26), 34–38. https://doi.org/10.3182/20101206-3-JP-3009.00005.

Coblinski, J. A., Giasson, É., Demattê, J. A. M., Dotto, A. C., Costa, J. J. F. and Vašát, R. (2020). Prediction of soil texture classes through different wavelength regions of reflectance spectroscopy at various soil depths. *CATENA* 189, 104485. https://doi.org/10.1016/j.catena.2020.104485.

Colmap. (2022). COLMAP - Structure-from-motion and multi-view stereo. Available at: https://demuc.de/colmap/.

Cornu, S., Montagne, D., Daroussin, J. and Cousin, I. (2012). Image-analytically derived conceptual model of Albeluvisol morphological degradation induced by artificial drainage in France. *Geoderma* 189–190, 296–303. https://doi.org/10.1016/j.geoderma.2012.05.024.

Dalla Rosa, J., Cooper, M., Darboux, F. and Medeiros, J. C. (2012). Soil roughness evolution in different tillage systems under simulated rainfall using a semivariogram-based index. *Soil and Tillage Research* 124(0), 226–232. https://doi.org/10.1016/j.still.2012.06.001.

Darboux, F. (1999). Modélisation numérique et expérimentale du ruissellement. Effet de la rugosité sur les distances de transfert. *Université de Rennes* 1. Available at: http://prodinra.inra.fr/record/63638.

Darboux, F., Davy, Ph., Gascuel-Odoux, C. and Huang, C. (2002). Evolution of soil surface roughness and flowpath connectivity in overland flow experiments. *CATENA* 46(2–3), 125–139. https://doi.org/10.1016/S0341-8162(01)00162-X.

de Oliveira Morais, P. A., de Souza, D. M., Madari, B. E. and de Oliveira, A. E. (2020). A computer-assisted soil texture analysis using digitally scanned images. *Computers and Electronics in Agriculture* 174, 105435. https://doi.org/10.1016/j.compag.2020.105435.

Dhond, U. R. and Aggarwal, J. K. (1989). Structure from stereo-a review. *IEEE Transactions on Systems, Man, and Cybernetics* 19(6), 1489–1510. https://doi.org/10.1109/21.44067.

Dji. (2022). Phantom 4 RTK - DJI. *DJI Official*. Available at: https://www.dji.com/phantom-4-rtk.

Domsch, H. and Giebel, A. (2004). Estimation of soil textural features from soil electrical conductivity recorded using the EM38. *Precision Agriculture* 5(4), 389–409. https://doi.org/10.1023/B:PRAG.0000040807.18932.80.

Eickhorst, T. and Tippkötter, R. (2008). Improved detection of soil microorganisms using fluorescence in situ hybridization (FISH) and catalysed reporter deposition (CARD-FISH). *Soil Biology and Biochemistry* 40(7), 1883–1891.

Emmet-Booth, J. P., Forristal, P. D., Fenton, O., Ball, B. C. and Holden, N. M. (2016). A review of visual soil evaluation techniques for soil structure. *Soil Use and Management* 32(4), 623–634. https://doi.org/10.1111/sum.12300.

Fajardo, M., Jones, E. and Wittig, R. (2020). Assessing soil aggregate stability with mobile phones (No. EGU2020-3240). *Copernicus Meetings*. Available at: https://doi.org/10.5194/egusphere-egu2020-3240.

Fajardo, M., McBratney, A. B., Field, D. J. and Minasny, B. (2016). Soil slaking assessment using image recognition. *Soil and Tillage Research* 163, 119–129. https://doi.org/10.1016/j.still.2016.05.018.

Fanigliulo, R., Antonucci, F., Figorilli, S., Pochi, D., Pallottino, F., Fornaciari, L., Grilli, R. and Costa, C. (2020). Light drone-based application to assess soil tillage quality parameters. *Sensors* 20(3), Article 3. https://doi.org/10.3390/s20030728.

FAO. (2015a). *Revised World Soil Charter*. Food and Agriculture Organization. Available at: https://www.fao.org/documents/card/en/c/e60df30b-0269-4247-a15f-db564161fee0/.

FAO. (2015b). Status of the World's soil resources. Main Report. *Policy Support and Governance| Food and Agriculture Organization of the United Nations*. Available at: https://www.fao.org/policy-support/tools-and-publications/resources-details/en/c/435200/.

FAO. (2017). *Voluntary Guidelines for Sustainable Soil Management*. Food and Agriculture Organization. Available at: https://www.fao.org/documents/card/en/c/5544358d-f11f-4e9f-90ef-a37c3bf52db7/.

Favre, F., Boivin, P. and Wopereis, M. C. S. (1997). Water movement and soil swelling in a dry, cracked vertisol. *Geoderma* 78(1–2), 113–123. https://doi.org/10.1016/S0016-7061(97)00030-X.

Ffmpeg. (2022). *FFmpeg*. Available at: https://ffmpeg.org/.

Fisher, P., Aumann, C., Chia, K., O'Halloran, N. and Chandra, S. (2017). Adequacy of laser diffraction for soil particle size analysis. *PLoS ONE* 12(5), e0176510. https://doi.org/10.1371/journal.pone.0176510.

FitzPatrick, E. A. (1993). *Soil Microscopy and Micromorphology*. John Wiley and Sons.

Fritsch, E., Peterschmitt, E. and Herbillon, A.-J. (1992). A structural approach to the regolith: Identification of structures, analysis of structural relationships and interpretations/Une approche structurale appliquée aux couvertures pédologiques: Identification des structures, analyse des relations structurales et interprétations. *Sciences Géologiques. Bulletin* 45(2), 77-97. https://doi.org/10.3406/sgeol.1992.1885.

Furukawa, Y. (2022). PMVS2. Available at: https://www.di.ens.fr/pmvs/.

Furukawa, Y. and Ponce, J. (2008). Accurate camera calibration from multi-view stereo and bundle adjustment. In: *IEEE Conference on Computer Vision and Pattern Recognition* (Vol. 2008), pp. 1-8. https://doi.org/10.1109/CVPR.2008.4587681.

García Moreno, R., Saa Requejo, A., Tarquis Alonso, A. M., Barrington, S. and Díaz, M. C. (2008). Shadow analysis: A method for measuring soil surface roughness. *Geoderma* 146(1-2), 201-208. https://doi.org/10.1016/j.geoderma.2008.05.026.

Gargiulo, L., Mele, G. and Terribile, F. (2013). Image analysis and soil micromorphology applied to study physical mechanisms of soil pore development: An experiment using iron oxides and calcium carbonate. *Geoderma* 197-198, 151-160. https://doi.org/10.1016/j.geoderma.2013.01.008.

Geonics. (2022). Geonics limited–Electromagnetic geophysical instrumentation. Available at: http://www.geonics.com/.

Gilliot, J. M., Michelin, J. and Vaudour, E. (2012). An automatic processing chain for estimating the roughness of agricultural soils by 3D photogrammetry from photographs taken without constraints for radiometric soil monitoring (in French). In: *Colloque des 11e Journées d'Étude des Sols*, pp. 244-245.

Gilliot, J. M., Vaudour, E. and Michelin, J. (2017). Soil surface roughness measurement: A new fully automatic photogrammetric approach applied to agricultural bare fields. *Computers and Electronics in Agriculture* 134, 63-78. https://doi.org/10.1016/j.compag.2017.01.010.

Grove, C. and Jerram, D. A. (2011). jPOR: An ImageJ macro to quantify total optical porosity from blue-stained thin sections. *Computers and Geosciences* 37(11), 1850-1859. https://doi.org/10.1016/j.cageo.2011.03.002.

Grundy, L., Ghimire, C. and Snow, V. (2020). Characterisation of soil micro-topography using a depth camera. *MethodsX* 7, 101144. https://doi.org/10.1016/j.mex.2020.101144.

Guillobez, S. and Arnaud, M. (1998). Regionalized soil roughness indices. *Soil and Tillage Research* 45(3-4), 419-432. https://doi.org/10.1016/S0933-3630(97)00031-7.

Hapca, S., Baveye, P. C., Wilson, C., Lark, R. M. and Otten, W. (2015). Three-dimensional mapping of soil chemical characteristics at micrometric scale by combining 2D SEM-EDX data and 3D X-ray CT images. *PLoS ONE* 10(9), e0137205. https://doi.org/10.1371/journal.pone.0137205.

Hapca, S. M., Wang, Z. X., Otten, W., Wilson, C. and Baveye, P. C. (2011). Automated statistical method to align 2D chemical maps with 3D X-ray computed micro-tomographic images of soils. *Geoderma* 164(3-4), 146-154. https://doi.org/10.1016/j.geoderma.2011.05.018.

Hillel, D. (2003). Introduction to environmental soil physics. Elsevier Science. Available at: https://univ-scholarvox-com.ezproxy.universite-paris-saclay.fr/catalog/book/docid/45001580?searchterm=soil.

Houston, A. N., Otten, W., Baveye, P. C. and Hapca, S. (2013). Adaptive window-indicator kriging: a thresholding method for computed tomography images of porous media. *Computers and Geosciences* 54, 239–248.

Hudson, N. W. (1993). Field measurement of soil erosion and runoff. *Soils Bulletin* 68, 160.

Iassonov, P., Gebrenegus, T. and Tuller, M. (2009). Segmentation of computed tomography images of porous media. *Water Resources Research* 45(9), 1–12.

Intel. (2018). Introducing the Intel® RealSenseTM D400 product family. Intel® RealSense™ Depth and Tracking Cameras. Available at: https://www.intelrealsense.com/introducing-intel-realsense-d400-product-family/.

Jangorzo, N. S., Watteau, F. and Schwartz, C. (2013). Evolution of the pore structure of constructed Technosols during early pedogenesis quantified by image analysis. *Geoderma* 207–208, 180–192. https://doi.org/10.1016/j.geoderma.2013.05.016.

Jégou, D., Capowiez, Y. and Cluzeau, D. (2001). Interactions between earthworm species in artificial soil cores assessed through the 3D reconstruction of the burrow systems. *Geoderma* 102(1–2), 123–137. https://doi.org/10.1016/S0016-7061(00)00107-5.

Jester, W. and Klik, A. (2005). Soil surface roughness measurement—Methods, applicability, and surface representation. *CATENA* 64(2–3), 174–192. https://doi.org/10.1016/j.catena.2005.08.005.

Jongerius, A., Schoonderbeek, D., Jager, A. and Kowalinski, St. (1972). Electro-optical soil porosity investigation using Quantimet-B equipment. *Geoderma* 7, 177–198.

Joschko, M., Graff, O., Müller, P. C., Kotzke, K., Lindner, P., Pretschner, D. P. and Larink, O. (1991). A non-destructive method for the morphological assessment of earthworm burrow systems in three dimensions by X-ray computed tomography. *Biology and Fertility of Soils* 11(2), 88–92. https://doi.org/10.1007/BF00336369.

Juyal, A., Otten, W., Baveye, P. C. and Eickhorst, T. (2021). Influence of soil structure on the spread of Pseudomonas fluorescens in soil at microscale. *European Journal of Soil Science* 72(1), 141–153. https://doi.org/10.1111/ejss.12975.

Juyal, A., Otten, W., Falconer, R., Hapca, S., Schmidt, H., Baveye, P. C. and Eickhorst, T. (2019). Combination of techniques to quantify the distribution of bacteria in their soil microhabitats at different spatial scales. *Geoderma* 334, 165–174. https://doi.org/10.1016/j.geoderma.2018.07.031.

Kamphorst, E. C., Jetten, V., Guérif, J., Pitka¨nen, J., Iversen, B. V., Douglas, J. T. and Paz, A. (2000). Predicting depressional storage from soil surface roughness. *Soil Science Society of America Journal* 64(5), 1749–1758. https://doi.org/10.2136/sssaj2000.6451749x.

Kay, B. D. (1997). Soil structure and organic carbon: A review. In: *Soil Processes and the Carbon Cycle*. CRC Press.

Kemp, R. A. (1998). Role of micromorphology in paleopedological research. *Quaternary International* 51–52, 133–141. https://doi.org/10.1016/S1040-6182(97)00040-2.

King, A. E., Ali, G. A., Gillespie, A. W. and Wagner-Riddle, C. (2020). Soil organic matter as catalyst of crop resource capture. *Frontiers in Environmental Science* 8. Available at: https://www.frontiersin.org/articles/10.3389/fenvs.2020.00050.

Knadel, M., Thomsen, A., Schelde, K. and Greve, M. H. (2015). Soil organic carbon and particle sizes mapping using vis-NIR, EC and temperature mobile sensor platform.

Computers and Electronics in Agriculture 114, 134–144. https://doi.org/10.1016/j.compag.2015.03.013.

Koestel, J., Fukumasu, J., Garland, G., Larsbo, M. and Nimblad Svensson, D. (2021). Approaches to delineate aggregates in intact soil using X-ray imaging. *Geoderma* 402, 115360. https://doi.org/10.1016/j.geoderma.2021.115360.

Konen, M. E., Burras, C. L. and Sandor, J. A. (2003). Organic carbon, texture, and quantitative color measurement relationships for cultivated soils in North Central Iowa. *Soil Science Society of America Journal* 67(6), 1823–1830. https://doi.org/10.2136/sssaj2003.1823.

Kraus, K. and Waldhäusl, P. (1998). *Manuel de Photogrammétrie: Principes et Procédés Fondamentaux (Photogrammetry Manual: Basic Principles and Processes)*. Hermès, Paris, France.

Krishna Murti, G. S. R. and Satyanarayana, K. V. S. (1971). Influence of chemical characteristics in the development of soil colour. *Geoderma* 5(3), 243–248. https://doi.org/10.1016/0016-7061(71)90013-9.

Kubiena, W. L. (1939). Micropedology. *Soil Science* 47(2), 163.

Kühn, P., Aguilar, J., Miedema, R. and Bronnikova, M. (2018). Chapter 14: Textural pedofeatures and related horizons. In: Stoops, G., Marcelino, V. and Mees, F. (Eds.), *Interpretation of Micromorphological Features of Soils and Regoliths* (2nd edn., pp. 377–423). Elsevier. https://doi.org/10.1016/B978-0-444-63522-8.00014-0.

Kunt, M., Granlund, G. and Kocher, M. (1993). *Traitement numérique Des Images*. Presses polytechniques et universitaires romandes; CNET-ENST.

Lazaros, N., Sirakoulis, G. C. and Gasteratos, A. (2008). Review of stereo vision algorithms: From software to hardware. *International Journal of Optomechatronics* 2(4), 435–462. https://doi.org/10.1080/15599610802438680.

Lowe, D. G. (2004). Distinctive image features from scale-invariant keypoints. *International Journal of Computer Vision* 60(2), 91–110. https://doi.org/10.1023/B:VISI.0000029664.99615.94.

Malobane, M. E., Nciizah, A. D., Mudau, F. N. and Wakindiki, I. I. C. (2019). Discrimination of soil aggregates using micro-focus X-ray computed tomography in a five-year-old no-till natural fallow and conventional tillage in South Africa. *Heliyon* 5(5), e01819. https://doi.org/10.1016/j.heliyon.2019.e01819.

Manté, C., Borschneck, D., Mocuta, C., Bogaert, R. van den, Montagne, D., Cammas, C. and Cornu, S. (2020). Combining wavelets with statistical inference to map the mineralogical composition of pedological features from synchrotron X-ray diffraction data. *SN Applied Sciences* 2(7), 1–20. https://doi.org/10.1007/s42452-020-2971-1.

Marinello, F., Pezzuolo, A., Gasparini, F., Arvidsson, J. and Sartori, L. (2015). Application of the Kinect sensor for dynamic soil surface characterization. *Precision Agriculture* 16(6), 601–612.

Marschallinger, R. (1997). Automatic mineral classification in the macroscopic scale. *Computers and Geosciences* 23(1), 119–126. https://doi.org/10.1016/S0098-3004(96)00074-X.

Martinez-Agirre, A., Álvarez-Mozos, J., Milenković, M., Pfeifer, N., Giménez, R., Valle, J. M. and Rodríguez, Á. (2020). Evaluation of terrestrial laser scanner and structure from motion photogrammetry techniques for quantifying soil surface roughness parameters over agricultural soils. *Earth Surface Processes and Landforms* 45(3), 605–621. https://doi.org/10.1002/esp.4758.

Marzahn, P., Seidel, M. and Ludwig, R. (2012). Decomposing dual scale soil surface roughness for microwave remote sensing applications. *Remote Sensing* 4(7), 2016-2032. https://doi.org/10.3390/rs4072016.

Miedema, R. and Slager, S. (1972). Micromorphological quantification of clay illuviation. *Journal of Soil Science* 23(3), 309–314. https://doi.org/10.1111/j.1365-2389.1972.tb01662.x.

Muñoz-Ortega, F. J., San José Martínez, F. and Caniego Monreal, F. J. (2015). Volume, surface, connectivity and size distribution of soil pore space in CT images: Comparison of samples at different depths from nearby natural and tillage areas. *Pure and Applied Geophysics* 172(1), 167–179. https://doi.org/10.1007/s00024-014-0897-5.

Murphy, C. P. (1983). Point counting pores and illuvial clay in thin section. *Geoderma* 31(2), 133–150.

Murphy, C. P., Bullock, P. and Turner, R. H. (1977). The measurement and characterization of voids in soil thin sections by image analysis. *European Journal of Soil Science* 28(3), 509–518.

Naimi, S., Ayoubi, S., Di Raimo, L. A. D. L. and Dematte, J. A. M. (2022). Quantification of some intrinsic soil properties using proximal sensing in arid lands: Application of Vis-NIR, MIR, and pXRF spectroscopy. *Geoderma Regional* 28, e00484. https://doi.org/10.1016/j.geodrs.2022.e00484.

nix. (2022). *Nix Color Sensors*. Nix Sensor Ltd. Available at: https://www.nixsensor.com/.

Norouzi, S., Sadeghi, M., Liaghat, A., Tuller, M., Jones, S. B. and Ebrahimian, H. (2021). Information depth of NIR/SWIR soil reflectance spectroscopy. *Remote Sensing of Environment* 256, 112315. https://doi.org/10.1016/j.rse.2021.112315.

Nunan, N., Ritz, K., Rivers, M., Feeney, D. S. and Young, I. M. (2006). Investigating microbial micro-habitat structure using x-ray computed tomography. *Geoderma* 133(34): 398–407.

Pham, V., Weindorf, D. C. and Dang, T. (2021). Soil profile analysis using interactive visualizations, machine learning, and deep learning. *Computers and Electronics in Agriculture* 191, 106539. https://doi.org/10.1016/j.compag.2021.106539.

Piron, D., Peres, G., Hallaire, V. and Cluzeau, D. (2012). Morphological description of soil structure patterns produced by earthworm bioturbation at the profile scale. *European Journal of Soil Biology* 50, 83–90. https://doi.org/10.1016/j.ejsobi.2011.12.006.

Poch, R. M. and Stoops, G. (2022). Soil micromorphology. In: *Reference Module in Earth Systems and Environmental Sciences*. Elsevier. https://doi.org/10.1016/B978-0-12-822974-3.00034-3.

Pot, V., Peth, S., Monga, O., Vogel, L. E., Genty, A., Garnier, P., Vieublé-Gonod, L., Ogurreck, M., Beckmann, F. and Baveye, P. C. (2015). Three-dimensional distribution of water and air in soil pores: Comparison of two-phase two-relaxation-times lattice-Boltzmann and morphological model outputs with synchrotron X-ray computed tomography data. *Advances in Water Resources* 84, 87–102. https://doi.org/10.1016/j.advwatres.2015.08.006.

Protz, R. and VandenBygaart, A. J. (1998). Towards systematic image analysis in the study of soil micromorphology. *Sciences of Soils* 3(1), 34–44. https://doi.org/10.1007/s10112-998-0004-0.

Quénard, Samouëlian, A., Laroche, B. and Cornu, S. (2011). Lessivage as a major process of soil formation. *Geoderma* 167-8, 135–147.

Ringrose-Voase, A. J. (1991). Micromorphology of soil structure—Description, quantification, application. *Soil Research* 29(6), 777–813. https://doi.org/10.1071/SR9910777.

Ringrose-Voase, A. J. (1996). Measurement of soil macropore geometry by image analysis of sections through impregnated soil. *Plant and Soil* 183(1), 27–47.

Ringrose-Voase, A. J. and Bullock, P. (1984). The automatic recognition and measurement of soil pore types by image analysis and computer programs. *Journal of Soil Science* 35(4), 673–684. https://doi.org/10.1111/j.1365-2389.1984.tb00624.x.

Robinson, G. W. (1922). A new method for the mechanical analysis of soils and other dispersions. *The Journal of Agricultural Science* 12(3), 306–321. https://doi.org/10.1017/S0021859600005360.

Römkens, M. J. M., Singarayar, S. and Gantzer, C. J. (1986). An automated non-contact surface profile meter. *Soil and Tillage Research* 6(3), 193–202. https://doi.org/10.1016/0167-1987(86)90454-X.

Römkens, M. J. M. and Wang, J. Y. (1986). Effect of tillage on surface roughness. *Transactions of the ASAE* 29(2), 429–433.

Saleh, A. (1993). Soil roughness measurement: Chain method. *Journal of Soil and Water Conservation* 48(6), 527–529.

Sauzet, O., Cammas, C., Barbillon, P., Etienne, M.-P. and Montagne, D. (2016). Illuviation intensity and land use change: Quantification via micromorphological analysis. *Geoderma* 266, 46–57. https://doi.org/10.1016/j.geoderma.2015.11.035.

Sauzet, O., Cammas, C., Gilliot, J. M., Bajard, M. and Montagne, D. (2017). Development of a novel image analysis procedure to quantify biological porosity and illuvial clay in large soil thin sections. *Geoderma* 292, 135–148. https://doi.org/10.1016/j.geoderma.2017.01.004.

Sauzet, O., Kohler-Milleret, R., Füllemann, F., Capowiez, Y. and Boivin, P. (2021). Nicodrilus nocturnus and Allolobophora icterica drill compacted soils but do not decrease their bulk density – A laboratory experiment using two contrasted soils at two different compaction levels. *Geoderma* 402, 115164. https://doi.org/10.1016/j.geoderma.2021.115164.

Schindelin, J., Arganda-Carreras, I., Frise, E., Kaynig, V., Longair, M., Pietzsch, T., Preibisch, S., Rueden, C., Saalfeld, S., Schmid, B., Tinevez, J. Y., White, D. J., Hartenstein, V., Eliceiri, K., Tomancak, P. and Cardona, A. (2012). Fiji: An open-source platform for biological-image analysis. *Nature Methods* 9(7), 676–682. https://doi.org/10.1038/nmeth.2019.

Schlüter, S., Eickhorst, T. and Mueller, C. W. (2019). Correlative imaging reveals holistic view of soil microenvironments. *Environmental Science and Technology* 53(2), 829–837. https://doi.org/10.1021/acs.est.8b05245.

Schlüter, S., Sammartino, S. and Koestel, J. (2020). Exploring the relationship between soil structure and soil functions via pore-scale imaging. *Geoderma* 370, 114370. https://doi.org/10.1016/j.geoderma.2020.114370.

Smith, M. L., Smith, L. N. and Hansen, M. F. (2021). The quiet revolution in machine vision—A state-of-the-art survey paper, including historical review, perspectives, and future directions. *Computers in Industry* 130, 103472. https://doi.org/10.1016/j.compind.2021.103472.

Snapir, B., Hobbs, S. and Waine, T. W. (2014). Roughness measurements over an agricultural soil surface with structure from motion. *ISPRS Journal of Photogrammetry and Remote Sensing* 96, 210–223. https://doi.org/10.1016/j.isprsjprs.2014.07.010.

Snavely, N. (2010). Bundler 0.4. Available at: http://phototour.cs.washington.edu/bundler/.

Stoops, G. (2021). *Guidelines for Analysis and Description of Soil and Regolith Thin Sections*. John Wiley & Sons.

Stoops, G. and Vepraskas, M. J. (2003). *Guidelines for Analysis and Description of Soil and Regolith Thin Sections*. Soil Science Society of America.

Sudarsan, B., Ji, W., Adamchuk, V. and Biswas, A. (2018). Characterizing soil particle sizes using wavelet analysis of microscope images. *Computers and Electronics in Agriculture* 148, 217–225. https://doi.org/10.1016/j.compag.2018.03.019.

Taconet, O. and Ciarletti, V. (2007). Estimating soil roughness indices on a ridge-and-furrow surface using stereo photogrammetry. *Soil and Tillage Research* 93(1), 64–76. https://doi.org/10.1016/j.still.2006.03.018.

Taina, I. A. and Heck, R. J. (2010). Utilization of object-oriented software in the image analysis of soil thin sections. *Soil Science Society of America Journal* 74(5), 1670–1681. https://doi.org/10.2136/sssaj2008.0359.

Taina, I. A., Heck, R. J. and Elliot, T. R. (2008). Application of X-ray computed tomography to soil science: A literature review. *Canadian Journal of Soil Science* 88(1), 1–19. https://doi.org/10.4141/CJSS06027.

Terribile, F. and FitzPatrick, E. A. (1992). The application of multilayer digital image processing techniques to the description of soil thin sections. *Geoderma* 55(1–2), 159–174. https://doi.org/10.1016/0016-7061(92)90011-U.

Terribile, F. and FitzPatrick, E. A. (1995). The application of some image-analysis techniques to recognition of soil micromorphological features. *European Journal of Soil Science* 46(1), 29–45. https://doi.org/10.1111/j.1365-2389.1995.tb01810.x.

Thien, S. J. (1979). A flow diagram for teaching texture-by-feel analysis. *Journal of Agronomic Education* 8(1), 54–55. https://doi.org/10.2134/jae.1979.0054.

Thompson, M. L., Fedoroff, N. and Fournier, B. (1990). Morphological features related to agriculture and faunal activity in three loess-derived soils in France. *Geoderma* 46(4), 329–349. https://doi.org/10.1016/0016-7061(90)90023-3.

Thomsen, L. M., Baartman, J. E. M., Barneveld, R. J., Starkloff, T. and Stolte, J. (2015). Soil surface roughness: Comparing old and new measuring methods and application in a soil erosion model. *Soil* 1(1), 399–410. https://doi.org/10.5194/soil-1-399-2015.

Ulery, A. L., Drees, L. R. and America, S. S. S. (2008). *Methods of Soil Analysis: Mineralogical Methods. Part 5*. CSSA, ASA-SSSA.

Ullman, S. (1979). The interpretation of structure from motion. *Proceedings of the Royal Society of London. Series B, Biological Sciences* 203(1153), 405–426. https://doi.org/10.1098/rspb.1979.0006.

van Egmond, F. M., van der Veeke, S., Knotters, M., Koomans, R. I., Walvoort, D. J. J. and Limburg, J. (2018). *Mapping Soil Texture with a Gamma-Ray Spectrometer: Comparison Between UAV and Proximal Measurements and Traditional Sampling: Validation Study*. Statutory Research Tasks Unit for Nature & the Environment. https://doi.org/10.18174/466037.

Vázquez, E. V., Vieira, S. R., De Maria, I. C. and González, A. P. (2009). Geostatistical analysis of microrelief of an oxisol as a function of tillage and cumulative rainfall. *Scientia Agricola* 66(2), 225–232. https://doi.org/10.1590/S0103-90162009000200012.

Vázquez-Arellano, M., Griepentrog, H. W., Reiser, D. and Paraforos, D. S. (2016). 3-D imaging systems for agricultural applications–A review. *Sensors* 16(5), Article 5. https://doi.org/10.3390/s16050618.

Védère, C., Vieublé-Gonod, L., Nunan, N. and Chenu, C. (2022). Opportunities and limits in imaging microorganisms and their activities in soil microhabitats. *Soil Biology and Biochemistry* 174, 108807. https://doi.org/10.1016/j.soilbio.2022.108807.

Verhoeven, G., Doneus, M., Briese, Ch. and Vermeulen, F. (2012). Mapping by matching: A computer vision-based approach to fast and accurate georeferencing of archaeological aerial photographs. *Journal of Archaeological Science* 39(7), 2060–2070. https://doi.org/10.1016/j.jas.2012.02.022.

Vermang, J., Norton, L. D., Huang, C., Cornelis, W., Vermang, J., Norton, L. D., Huang, C., Cornelis, W. M., da Silva, A. M. and Gabriels, D. (2015). Characterization of soil surface roughness effects on runoff and soil erosion rates under simulated rainfall. *Soil Science Society of America Journal* 79(3), 903–916. https://doi.org/10.2136/sssaj2014.08.0329.

Verrecchia, E. P. and Trombino, L. (2021). The future of soil micromorphology. In: Verrecchia, E. P. and Trombino, L. (Eds.), *A Visual Atlas for Soil Micromorphologists* (pp. 151–155). Springer International Publishing. https://doi.org/10.1007/978-3-030-67806-7_6.

Vidal Vázquez, E., Vivas Miranda, J. G. and Paz González, A. (2005). Characterizing anisotropy and heterogeneity of soil surface microtopography using fractal models. *Ecological Modelling* 182(3–4), 337–353. https://doi.org/10.1016/j.ecolmodel.2004.04.012.

Viscarra Rossel, R. and Adamchuk, V. (2013). Proximal soil sensing. In: *Precision Agriculture for Sustainability and Environmental Protection* (pp. 99–118). https://doi.org/10.13140/2.1.2799.1847.

Viscarra Rossel, R. A., Adamchuk, V. I., Sudduth, K. A., McKenzie, N. J. and Lobsey, C. (2011). Proximal soil sensing: An effective approach for soil measurements in space and time. *Advances in Agronomy* 113, 243–291. https://doi.org/10.1016/B978-0-12-386473-4.00005-1.

Viscarra Rossel, R. A. and McBratney, A. B. (1998). Laboratory evaluation of a proximal sensing technique for simultaneous measurement of soil clay and water content. *Geoderma* 85(1), 19-39. https://doi.org/10.1016/S0016-7061(98)00023-8.

Viscarra Rossel, R. A., Walvoort, D. J. J., McBratney, A. B., Janik, L. J. and Skjemstad, J. O. (2006). Visible, near infrared, mid infrared or combined diffuse reflectance spectroscopy for simultaneous assessment of various soil properties. *Geoderma* 131(1–2), 59–75. https://doi.org/10.1016/j.geoderma.2005.03.007.

Vitton, S. and Sadler, L. Y. (1997). Particle-size analysis of soils using laser light scattering and X-ray absorption technology. *Geotechnical Testing Journal* 20(1). https://doi.org/10.1520/GTJ11421J.

Vogel, H.-J., Balseiro-Romero, M., Kravchenko, A., Otten, W., Pot, V., Schlüter, S., Weller, U. and Baveye, P. C. (2022). A holistic perspective on soil architecture is needed as a key to soil functions. *European Journal of Soil Science* 73(1), e13152. https://doi.org/10.1111/ejss.13152.

Watteau, F. and Villemin, G. (2018). Soil microstructures examined transmission electron microscopy. *Frontiers in Environmental Science*. DOI: 10.3389/fenvs.2018.00106.

Wijewardane, N. K., Hetrick, S., Ackerson, J., Morgan, C. L. S. and Ge, Y. (2020). VisNIR integrated multi-sensing penetrometer for in situ high-resolution vertical soil sensing. *Soil and Tillage Research* 199, 104604. https://doi.org/10.1016/j.still.2020.104604.

Witkin, A. P. (1983). Scale-space filtering 2, 1019–1022.

Wu. (2022). *VisualSFM: A Visual Structure From Motion System. Changchang*. Available at: http://ccwu.me/vsfm/index.html.

Yagüe, M. R., Domingo-Olivé, F., Bosch-Serra, À. D., Poch, R. M. and Boixadera, J. (2016). Dairy cattle manure effects on soil quality: Porosity, earthworms, aggregates and soil organic carbon fractions. *Land Degradation and Development* 27(7), 1753–1762. https://doi.org/10.1002/ldr.2477.

Zhang, X., Liu, B., Wang, J., Zhang, Z., Shi, K. and Wu, S. (2014). Adobe Photoshop quantification (PSQ) rather than point-counting: A rapid and precise method for quantifying rock textural data and porosities. *Computers and Geosciences* 69, 62–71. https://doi.org/10.1016/j.cageo.2014.04.003.

Zhang, Z. (2012). Microsoft Kinect sensor and its effect. *IEEE Multimedia* 19(2), 4–10. https://doi.org/10.1109/MMUL.2012.24.

Zhu, Z. and Stein, M. (2002). Parameter estimation for fractional brownian surfaces. *Statistica Sinica* 12, 863–883.

Zribi, M., Ciarletti, V., Taconet, O., Paillé, J. and Boissard, P. (2000). Characterisation of the soil structure and microwave backscattering based on numerical three-dimensional surface representation: Analysis with a fractional brownian model. *Remote Sensing of Environment* 72(2), 159–169. https://doi.org/10.1016/S0034-4257(99)00097-8.

Zribi, M. and Dechambre, M. (2003). A new empirical model to retrieve soil moisture and roughness from C-band radar data. *Remote Sensing of Environment* 84(1), 42–52. https://doi.org/10.1016/S0034-4257(02)00069-X.

Chapter 9

Using machine learning to identify and diagnose crop disease

Megan Long, John Innes Centre, UK

1 Introduction

As the population of the world continues to grow, one of the biggest challenges facing food security is crop disease (Strange and Scott, 2005). Disease can have devastating effects on the yield of a crop, in some cases causing major losses where food quality is concerned. Not only is this a problem on a global scale, but it also has effects on a local scale for individual farmers. In poorer areas of the world, farming is the main or only source of income for many families

For any farmer, being able to detect and identify diseases in their plants is hugely important for the mitigation of potential losses. The problem with this, however, is that identifying crop diseases often requires specialist knowledge which is not always readily available to all farmers and can be expensive. Even with specialist knowledge, there are multiple factors that make it even more challenging. Many diseases appear with similar symptoms, meaning they are easily confused with one another. Furthermore, it is not uncommon for multiple diseases to be present at any one time, making the task of distinguishing them even more difficult.

http://dx.doi.org/10.19103/AS.2022.0107.22

Over recent years, machine learning techniques have been developed to assist with the identification and classification of diseases on a number of different crop types. Machine learning uses algorithms to learn a specific task from data without the solution or the process being explicitly defined by a human. It learns to recognise patterns in the data and make predictions based on this knowledge.

Multiple different methods have been used over the years, but in this chapter, we will focus on deep learning. This has become one of the most widely used machine learning methods over recent years, tackling problems in healthcare, self-driving cars and natural language processing as well as crop disease detection and a vast array of other use cases. In this chapter, we introduce deep learning for image analysis and discuss the key successes and pitfalls of using these methods for the identification of plant diseases.

2 A quick introduction to deep learning

A common machine learning technique is artificial neural networks (ANN), which are inspired by the biological neural network of the brain. They follow a logic structure that is meant to mimic the way the human brain thinks and draws conclusions. Over recent years, deep learning has taken centre stage in the machine learning world. Deep learning takes the ANNs further, making them deeper by adding layers. A typical machine learning ANN contains an input layer, an output layer and perhaps one hidden layer in between. In comparison, deep learning ANNs or deep learning networks contain many hidden layers.

The deep learning networks that are used for the identification of crop diseases are often a type of convolutional neural network (CNN) (LeCun et al., 1999) trained to perform their task by image analysis and classification. As before, each network has an input layer where the data (in this case images) is fed into the network and an output layer, which is where any predictions are given. Between these are a number of hidden layers which perform feature extraction. CNNs are used because of their strong ability to extract useful features from the images.

Feature extraction is where the network learns features about the images throughout the training process, allowing it to make predictions about the data. Earlier hidden layers learn low-level features, for example, lines and edges. As the network progresses through the hidden layers, the features extracted become more complex, which helps it to make a prediction about the image, for example, a classification of which disease is present in an input image. Figure 1 shows a simplified representation of the deep learning workflow for image analysis.

Training a network for image analysis requires a large dataset of images to work with, ideally tens of thousands of images depending on the problem. The image dataset is split into smaller datasets, usually train, validation and test sets; however, some studies use only a train set and test set. The split of the dataset between these sets varies over different works, but the bulk of the images is always contained within the train set, with a smaller amount in the validation and test sets.

When training a model, the train set of images is fed into the network in small batches (e.g. 64 images at a time). Once all the images have been through once, this is an epoch. The network will be trained for a certain number of epochs, as defined by the programmer. This number is often picked by taking an educated guess based on previous research and experiments in similar fields. Many studies will try multiple experiments with different numbers of epochs before finding the number that yields the best results for their data.

Between every epoch, the current network parameters are evaluated against the validation set to ensure that the training process is not overfitting to the data in the train set. If it were overfitting, then it would be learning features that are specific to the images in the train set and would not be able to generalise to new data of the same type, as in the test set. For example, if there were photos of a certain disease taken with light-coloured soil in the background, and these all ended up in the train set of images, the validation set with photos of that disease taken with dark soil in the background would highlight this. The network parameters would then readjust to ensure that it is not using that soil information from the train set for classifying that disease, but rather the disease information on the plant instead. Ideally, the train, validation and test tests would all contain both colours of soil.

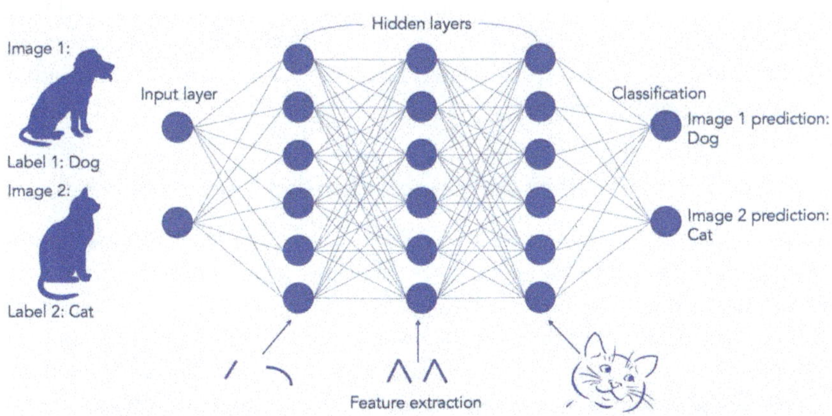

Figure 1 An artistic representation of a deep learning workflow for image classification. There is an input layer that takes in labelled images and an output layer that gives classification predictions. In between, there are multiple hidden layers that perform feature extraction. Earlier in the network basic features are learned, such as lines or edges, and further through the network they get more and more complicated until they learn features which allow the network to make its predictions.

Throughout training, the network is constantly adjusting its internal parameters after each batch and epoch to allow it to better make predictions about the images. As it learns, the accuracy of the predictions increases until it reaches a peak at the end of training. At this point, the network can be evaluated on the test set of images. This is a set of images of the same kind as contained in the train and validation sets, but that the network has never seen before. This shows how the trained network performs on brand new images to give a final accuracy rating. The whole process from start to finish can take a long time, from hours to days, to even months! This depends on multiple factors, such as the computing power available, the size of the network and the size of the dataset.

A lot of studies begin their experiments by using transfer learning with their datasets. This is a method that takes the knowledge learned by a previously trained network and applies it to the new problem. The main advantage of this is that it is relatively quick compared to training a deep learning network from scratch. Some examples of networks often used for transfer learning are AlexNet (Krizhevsky et al., 2012), GoogLeNet (Szegedy et al., 2014), VGGNet (Simonyan and Zisserman, 2015), ResNet (He et al., 2016), Inception V4 (Szegedy et al., 2016a) and MobileNet (Howard et al., 2017).

Transfer learning uses a network which has been fully trained on a large dataset (often the ImageNet dataset described in section 3). The pre-trained network is divided into two parts; the convolutional base, which is the part that performs feature extraction on the images, and the fully connected classifier, which forms predictions about the images. Depending on the method used, all or parts of the network are repurposed for the new dataset. In some cases, only the network structure is used and is retrained for the new problem without using the pre-trained knowledge.

3 Preparation of data for deep learning experiments

One of the biggest challenges for the successful application of machine learning techniques for the identification of plant diseases (or any image classification task for that matter) is the availability of data. The majority of these methods require large datasets of labelled or annotated images, which can be time-consuming to collect and process. For example, with plant-disease detection, it is necessary to have a large number of images for each disease for each plant species that is being modelled.

One of the most famous, and largest, datasets used for image analysis with deep learning is the ImageNet dataset (Deng et al., 2009). This dataset was created for use with object recognition software. The full dataset contains more than 14 million images with over 20 000 categories; however, a smaller subset of this has been used in the ImageNet Large Scale Visual Recognition Challenge (ILSVRC) (Russakovsky et al., 2015). This challenge ran annually between 2010 and 2017, encouraging participants to develop and improve computer vision techniques for image classification and object recognition. Multiple winning networks created for this competition over the years are now used as the starting point for hundreds of deep learning problems, including the problem of crop disease detection.

Collecting a dataset of images for use with any deep learning problem is not quite as easy as simply gathering as many images as possible by any means. It is important to ensure that the dataset contains appropriate information for the required use case. The rest of this section will discuss the factors to consider when preparing a dataset for plant-disease recognition and classification. These factors include a range of conditions, controlled versus uncontrolled capture conditions, image quality issues, the number of images required and labelling and annotation requirements.

The most widely known, and one of the only openly available datasets used for the recognition of plant diseases, is the Plant Village dataset (Hughes and Salathe, 2016). This is a collection of almost 88 000 images taken in controlled conditions with 38 categories, each corresponding to a plant-disease pair. Each image contains a single diseased or healthy leaf taken from the plant and placed on a neutral background and photographed under different lighting conditions. While this dataset was ground-breaking in the field of plant-disease detection when it was first created, the use of controlled conditions in the photos means that it is not comprehensive enough to be useful for an automated system in field conditions.

The Plant Village dataset was useful for demonstrating the potential of deep learning methods for the classification of plant diseases; however, in order to create a model that will be useful in realistic growth conditions, it is now important to collect datasets which accurately represent those conditions. PlantDoc (Singh et al., 2020) is a dataset created to cover many of the diseases present in the Plant Village dataset, but the images instead cover real field conditions. Here, images were downloaded from the internet and checked by members of the team before being added to the new dataset. This resulted in almost 3000 images, spanning 27 of the categories from Plant Village (any classes with fewer than 50 samples were removed for this dataset). This is a step in the right direction, but there is a distinct possibility of misclassified samples within the dataset due to them being taken from internet searches. Also, with it still being a relatively small dataset spanning a lot of classes, there is still a high chance that not all conditions are being covered.

For studies that are looking at building a model for a certain crop, it is unlikely that there are already datasets openly ready and available for use. This means that, for each case, there will be a large collection operation required prior to any numerical experiments. The result is usually a relatively small dataset with few categories (in some cases only two: diseased or healthy). While these can be useful for the problem at hand, there is still some way to go with generating a larger dataset to be used in a wider variety of cases.

The collection of a dataset which sufficiently covers each category can be a time-consuming task, often requiring the specialist knowledge of an expert pathologist and multiple volunteers to take the pictures. Furthermore, it is not simply a case of capturing a large number of images for each category, but also including a representative range of conditions. If the model is to be used for identifying diseases in the field, then the range of typical conditions that could be encountered in the field need to be represented. This includes:

- the variation in crop varieties/species - for example, different leaf colours or sizes;
- state of the crop - seedling, adult, flowering, mature with seed;
- stage and severity of the disease - early to late infection, mild to severe symptoms;
- weather and lighting conditions - full sun, sun and cloud, overcast, rain, etc.;
- background information - this needs to be consistent throughout the dataset. Having one class with different background information to the rest (e.g. glasshouse instead of field) will cause issues in training;
- image qualities - focus, depth of field, range of angles.

The main point to remember when creating a dataset for deep learning is that the conditions present need to be consistent between classes. Any class containing conditions which are not present in the others, for example, one class having sky in the background whereas no other does, will cause the network to learn the wrong information about that class and classify it based on the presence of sky, rather than the disease information.

Another factor to consider if working with real condition images is the diversity of background information, which might contribute negatively to the training process by distracting from the features that are of interest. If the images are collected in a field, for example, this may not be too much of an issue as the field conditions are likely to be relatively uniform. However, if the images are of a plant species which grows in various wild locations, then a vast array of background information can be expected. Where possible, the full range of diverse background conditions should be represented in images across all classes.

The number of images is also important. The number of images to aim for per category will depend on the complexity of the problem at hand. A simpler problem, for example, a binary classification problem of healthy or diseased, will require fewer images than a classification problem with multiple diseases

Published by Burleigh Dodds Science Publishing Limited, 2023.

with similar symptoms. However, the general rule of thumb with deep learning datasets is the more data, the better (ideally hundreds, if not thousands, of images per category in our opinion). The more images the network has to learn from, the better its performance is likely to be. It is also best if the data is relatively well balanced between each category, so the network does not learn a bias towards one class due to it having significantly more training samples than the others.

One technique to increase the number of images in a dataset where it is not possible to collect more is data augmentation. Augmenting the data involves performing multiple transformations on each image to add new samples to the dataset. For example, an image may be mirrored, flipped horizontally or vertically, rotated, or shifted to create tens of new images from a single sample. The main drawback of this is that there is no actual new data created, just variations of existing data. This means that the original dataset still needs to contain enough variation so that the network can learn enough to form predictions. There are other methods for working with smaller datasets; however, where possible it is always better to collect more data.

After collecting all available data, it will then need to be labelled and collated into a full dataset. For best results, a pathologist will need to label each image with the correct category, either as the images are taken, or by going through all data and assigning categories later. This of course can be incredibly time consuming and can result in misclassifications within the dataset if a pathologist is not available. In cases where different visualisation techniques are being used with the dataset (see Section 5), it may also be necessary to annotate the data with further information (e.g. a bounding box around a disease lesion). This often has to be done manually on each image and is a huge undertaking. Once all labelling and annotation is complete, the data can be sorted into the train, validation and test sets and start being used with a deep learning model.

4 Crop disease classification

A common use of deep learning methods for crop disease detection is to classify images of diseased plants into pre-defined classes. This can be a hugely difficult task for multiple reasons. For most, if not all, diseases there is a wide variation in visible symptoms throughout the life cycle of the disease, with some symptoms being more common than others. Furthermore, there can be a lot of similarities in visible symptoms of multiple diseases. Consequently, less common symptoms can easily be misclassified as another disease.

CNNs (LeCun et al., 1999) are a type of deep learning network which have become popular for image classification of plant diseases (Boulent et al., 2019). Many studies utilise pre-defined CNN structures for their work. A few examples that occur over and over again throughout the literature are AlexNet (Krizhevsky et al., 2012), GoogLeNet (Szegedy et al., 2014) and Inception (Szegedy et al., 2016b); however, there are plenty of others which provide a starting point for almost all of the studies we will discuss. These pre-defined networks are all CNNs with different numbers of layers and additional features to aid with feature extraction.

A good place to start here is with the Plant Village dataset. Multiple studies use the whole or part of this dataset with their work. Mohanty et al. (2016) aimed to show the viability of deep learning networks for the classification of a range of different diseases. They performed the first deep learning experiments using the Plant Village dataset with two different pre-trained networks: AlexNet and GoogLeNet. Through training these two networks both from scratch and using transfer learning, with a range of image processing techniques and train-test splits (the split of data between training and testing), they returned near-perfect accuracy with their best-performing method. Using a pre-trained GoogLeNet, full-colour images and an 80-20 train-test split the accuracy reached 99.34%.

Much like Mohanty et al., Brahimi et al. (2017) also used transfer learning with the two pre-trained networks AlexNet and GoogLeNet. In this study, however, rather than using the entire Plant Village dataset, a subset of images containing only diseased tomato leaves was used. Both studies utilised the networks by transfer learning and by training from scratch in an attempt to compare the results from both methods. In the same way, as in the work of Mohanty et al., the best results gained in Brahimi et al.'s work were of extremely high accuracy, reaching 99.18% accuracy in classifying tomato diseases. Again, this result came from the use of GoogLeNet with pre-training, although they do not specify the train-test split.

Another study that made use of a subset of the Plant Village dataset is that of Amara et al. (2017). They used only the banana leaf images in their work with the LeNet (Lecun et al., 1998) architecture. Although using a previously defined network, they did not use a pre-trained version, rather the architecture was trained from scratch with the banana leaf images. They used a range of train-test splits with both coloured and grayscale images. It was shown that the networks that used coloured images always outperform those without, thus showing the importance of colour information for the problem. Using a train-test split of 80-20, the network achieved an accuracy of 98.61%, another extremely promising result.

Too et al. (2019) took the whole of the Plant Village dataset and evaluated the performance of multiple pre-trained networks in classifying the diseases. They used transfer learning with some fine-tuning of VGG16, Inception V4, Resnet with 50, 101 and 152 layers and DenseNets (Huang et al., 2018). DenseNets was the best performer having gained an almost perfect accuracy of 99.75%.

This almost perfect accuracy is a common occurrence in studies which use only images from the Plant Village dataset. Although comprehensive in that it covers a wide range of diseases and plant species, the images within are not representative of those which would be found in real growth situations. They contain images of leaves taken from the plant and placed on a plain background, thus eliminating any background information, which obviously would not be the case in the field. The high accuracies gained in these studies are impressive; however, it is unknown how any of the models would perform when confronted with real field data.

Ferentinos (2018) demonstrated the issues of the Plant Village dataset for field use in their work. They made use of multiple pre-trained networks within his study; AlexNet, AlexNetOWTBn (Krizhevsky, 2014), GoogLeNet, Overfeat (Sermanet et al., 2013) and VGGNet. The dataset used contains images taken from Plant Village as 'lab condition' images, and was supplemented with more images taken in the field. This resulted in a dataset of 87 848 images sorted into 58 classes, some that contained just lab conditions, others that contained just field conditions and some with both. The most successful architecture in this study was the VGG network, which gained an accuracy of 99.53% on unseen images. Due to the presence of both lab condition and field condition images within the dataset used, Ferentinos (2018) experimented with training on laboratory condition images and testing on field condition images and vice versa. The accuracy of classification in these experiments was significantly lower than with the mixture of images for training. Training on field images and testing on laboratory images resulted in an accuracy of 65.69%, whereas the other way around resulted in an accuracy of only 33.27%. These figures emphasise the importance of including all relevant conditions within a training set for use in practice.

Although the Plant Village dataset is used regularly throughout the literature, there are plenty of studies which make use of data acquired elsewhere. Sladojevic et al. (2016) created a large dataset of images (over 30000 in 15 classes) by taking pictures from internet searches. The dataset included a class for just healthy leaves and also a class with just background images. The reason for this was to train their network to differentiate leaves from their surroundings. The network used for this study was the pre-trained CaffeNet (Jia et al., 2014) model. Using this method, they gained a classification accuracy on their dataset of 96.3%. They concluded that the accuracy for individual categories was slightly lower in the classes which contained fewer images. Another thing to note about this study is how the images were collected. As they were taken straight from the internet, it is possible that some of the images have been wrongly classified which would have affected the accuracy of the network.

A study by Lu et al. (2017) used a relatively small dataset of rice disease images (500 images) to train CNNs inspired by LeNet and AlexNet architectures. Although they did not use the actual networks for either training from scratch or transfer learning, they did create a very similar network to those already defined. The accuracy gained for this network was 95%; while still a very encouraging result, this is slightly lower than many of the results discussed before. A reason for this could be to do with the size of the dataset used; with only 500 images spanning 10 categories, it could be hard for the network to learn all the characteristics present in each of the categories.

Alongside their own network modelled on a combination of AlexNet and GoogLeNet, Liu et al. (2018) utilised four pre-trained networks on their apple leaf disease identification problem; AlexNet, GoogLeNet, VGGNet and ResNet. They compared their network results to those obtained through transfer learning with the pre-trained networks and found that their model outperforms the known networks. The final accuracy recorded for their network was 97.62%, a percentage point higher than the next best performer VGGNet. Many studies make use of pre-defined networks; however, Liu et al. show that in some cases, defining a new network will gain a better performance. Often new networks will be inspired by one or several of the widely known networks (like in Lu et al., 2017), but this might be the best way to get all the best components for tackling the problem.

The train-test split is important for ensuring a network has enough data to learn from, while also having enough to for evaluating its performance. It is also important to include validation where possible. Often the validation is incorporated into the train part of the split when described in the literature.

Oppenheim et al. (2019) experimented with different train-test splits to find the best combination for their work detecting potato tuber disease. Their dataset contained 2465 images with 4 diseased and 1 uninfected category. They found that, unsurprisingly, more training data increased accuracy. The model that performed best on the test data used a 90-10 train-test split and gained an accuracy of 95.8%. Many studies elect to stick to an 80-20 split in the training and test data, in this case the higher amount of training images may improve training, but the lower amount of test images may not have contained enough images to fully show the performance of the network considering the size of the original dataset. A 90-10 split may be more suited to a larger dataset where the test set would contain more images.

The studies discussed in this section have shown the great potential for deep learning to be used for crop disease detection. The Plant Village dataset was a breakthrough in the field, which has seen multiple networks classify its images with incredibly high accuracy. Furthermore, other works have used more complex images while still gaining promising results. There is a lot of room for expanding these techniques for use with more diseases and more crop types.

5 Different visualisation techniques

Not all studies use only the image and the class label in their work. There are plenty of studies which use different visualisation techniques to try and improve the abilities of deep learning for plant-disease recognition tasks. As there are many different techniques being used throughout the field, we will discuss only a few in this section.

The first technique we will mention is the use of segmentation of images. This is where images are taken as usual and are cropped into smaller pieces. Ramcharan et al. (2017) collected a dataset of cassava leaf images with disease or pest damage in real field conditions. They ended up with two usable datasets: the first containing 2756 original cassava leaf images and a second with 15 000 segmented images containing smaller leaf sections (leaflets). Transfer learning with Inception V3 was used with three different classifiers on both datasets, with the best performer giving an accuracy of 93% on the leaflet dataset. It is not surprising that the larger dataset yielded better results as the networks were able to train on more data, plus the images were a lot less complex than those of the full cassava leaves.

In a slightly different approach, Ma et al. (2018) used image segmentation techniques to cut out the visible symptoms in images of cucumber diseases. These techniques were able to distinguish between lesion and background information and so crop away everything but the lesion. This resulted in a collection of images of only the visible symptoms placed on plain black backgrounds. They collected their dataset using images from the Plant Village dataset and from forestry images (https://www.forestryimages.org/) as well as supplementing them with their own images collected in real field conditions. After segmenting, they augmented the data to give 14 208 images for training a deep CNN. They gained an accuracy of 93.4% using their model, which outperformed AlexNet when they were compared. This is one of the lowest accuracy scores in all the studies using Plant Village images. It seems that the segmentation methods used here do not add anything to the process other than a higher preparation time for the dataset.

From the two approaches taken by Ramcharan et al. (2017) and Ma et al. (2018), it is clear that in some cases segmentation of images can be a useful tool. For complex images with lots of background information, segmenting the images to create smaller, less complicated samples will likely yield better classification results. However, for images taken in controlled conditions, segmenting may be adding an unnecessary, time-consuming step to the process. It is best to look at each problem on a case-by-case basis and decide based on the data whether the additional time injection for segmentation will be worthwhile.

Published by Burleigh Dodds Science Publishing Limited, 2023.

Zhang et al. (2019a) also used a different method for classifying diseased plant images. They took the three channels of RBG images and fed them each into a different CNN for each channel. The outputs of the three CNNs were then sent into a fully connected classifier together to get a single classification. The idea was to utilise all the colour information in each image for better classification power. Each of the CNNs had the same architecture; however, different learned weights were obtained due to the different colour information in each of the channels; red, green and blue. Different CNN architectures (their own model, DNN, LeNet-5 (Lecun et al., 1998) and GoogLeNet) were tested on the Plant Village tomato disease images, while a cucumber disease image dataset containing 500 images were used to test different train-test splits. The best results gained in this study used a 70-30 train-test split with their own network architecture and gave a classification accuracy of 94.27%. Although this is still a high disease recognition rate, it is not as high as some of the other studies show suggesting that better results can be gained when all three channels are fed into the same network.

Some studies like to include lesion location information in their work. For example, DeChant et al. (2017) created a dataset containing 1834 images of healthy and northern leaf blight-infected leaves of maize plants. In the infected images, each lesion was annotated with a line which was used as extra information for training the networks. They used a three-step method for classifying their images, first taking small portions of the images and training to detect the presence of lesions in these images. The second step used the networks trained in step one to create heat maps of the probability of parts of the image containing a lesion. The final step used these heatmaps to train a network to classify an image into either containing lesions (infected) or not (healthy). This method managed to get an accuracy in the full image classification of 96.7%. This result is extremely encouraging considering the use of images containing much background information. It does however only evaluate the method on two classes, healthy or diseased. It would be interesting to see how the accuracy would be affected by using this technique with more classes.

The work by DeChant et al. (2017) also made use of heatmaps. These can be an especially useful tool for ensuring that a network is functioning correctly. A heatmap shows the parts of the image that the network predicts to be most likely to contain disease information. These parts appear 'hotter', so on a blue-red scale, red sections are more likely to contain a lesion, whereas blue is the background material. From these heatmaps it is easy to pick out issues in the data; for example, if there is a red patch on a piece of background material in an image, it indicates that the network is probably not using the correct information to drive its classifications but is using background information instead.

There are various other visualisation techniques used for classifying crop diseases, each with their own merits and pitfalls. It would take a long time to go over them all! We have added a paper to the 'Where to look for more information' section which contains plenty of references to studies using different visualisation methods in their work for anyone who wishes to find out more.

6 Hyperspectral imaging for early disease detection

A relatively new addition to the field of deep learning is the use of hyperspectral imagery. Hyperspectral images capture information across the electromagnetic spectrum, not just the visual spectrum. In many cases there can be 'fingerprints' left by certain objects or substances (such as a disease) in different spectrums, which are not visible in regular conditions. It is for this reason that hyperspectral imagery has started to be deployed for the early detection of crop diseases. The idea is that, before there are visible symptoms on the plant, there may be unseen effects across other spectrums that would allow for early diagnosis and treatment before the disease got more severe.

Studies by Wang et al. (2019) and Jin et al. (2018) took hyperspectral images of sweet peppers with tomato spotted wilt virus and wheat with fusarium head blight, respectively. They both performed their experiments by taking the pixels of the images to train their networks. The pixels were labelled with background, diseased or healthy. This method required a small number of full hyperspectral images due to the number of pixels contained within each image. Both studies gain promising results with their small datasets, with Wang et al. gaining an accuracy of 96.25% on images of plants taken prior to disease symptoms being visible.

Nagasubramanian et al. (2018) worked with hyperspectral images of soybean crops with and without charcoal rot. They collected a dataset of 111 images, which were split into smaller data patches to create a larger dataset of 1090 training images and 539 test images. There appears to be a large bias towards the healthy class in their dataset; however, this did not appear to cause any problems with training. They used a CNN as their model, which gained an accuracy of 95.73% when presented with the test images.

In a similar experiment to Nagasubramanian, but with *Aphis gossypii* Glover infection of cotton leaves, Yan et al. (2021) used CNNs with hyperspectral images. They performed multiple experiments with RBG images and full hyperspectral images, comparing the performance of the CNNs against other machine learning methods. In all cases, they found that the CNNs gained the highest accuracies. They also found that the hyperspectral images gave better results than just RGB images.

An interesting study by Zhang et al. (2019b) utilised an unmanned aerial vehicle for collecting their hyperspectral images. Five images were taken of full wheat plots, with and without yellow rust, from a height of 30 m. These images were segmented into 15 000 smaller hyperspectral image blocks which were used for training and evaluating their network, which combined features of Inception and ResNet. Each block was labelled either as containing rust, healthy plants or other (e.g. soil or road). After the model had been trained on 10000 of the data blocks, it achieved a classification accuracy of 85% on the remaining 5000 test data blocks. The next step in their process involved mapping the data blocks back onto the original images. Here the sections that were predicted as rust areas were highlighted in red to show an infection map over the entire plot. This is extremely useful for showing the infection levels in a full plot and methods like this could possibly be utilised for scoring the amount of disease present as well as classification.

Although there is not a great number of studies already utilising this technology for plant-disease detection, it shows great promise and there is plenty of room for growth in the future. As always, the availability of data is a bottleneck for advancement. Hyperspectral imagining requires a lot more preparation than simply collecting photos with a regular camera. For example, environmental factors such as light and heat levels can have an effect on the different wavelengths, thus affecting the collected images. Furthermore, hyperspectral imaging cameras can be a costly investment, meaning they are not readily available for everyone to be able use in their work. A positive, however, is that for some methods, these studies tend not to require as large a dataset as those using regular visible spectrum images. A smaller number of images can be segmented or taken at pixel level to give a large dataset for training and evaluation.

7 Case study: identification and classification of diseases on wheat

In this section we will discuss some recent advances. In this work we used deep learning methods for the identification and classification of wheat diseases in field conditions. Wheat is a staple crop, vital for feeding many people across the world. Therefore, it is important to be able to control any diseases and mitigate any yield losses.

Alongside a healthy category, we included four of the most commercially important wheat diseases for the UK and other countries in our dataset: Yellow rust, otherwise known as stripe rust, caused by the basidiomycete fungus *Piccunia striiformis* f.sp. *tritici (Pgt)* (Liu and Hambleton, 2010). Septoria, otherwise known as Septoria leaf blotch or just Septoria, caused by the ascomycete fungus *Zymoseptoria tritici* (formerly *Mycosphaerella graminicola)*

(Hardwick et al., 2001). Brown rust, caused by *Puccinia triticina* (Goyeau et al., 2006; Bolton et al., 2008) and powdery mildew, caused by *Blumeria graminis* (Dubin and Duveiller, 2011).

The most familiar symptoms of yellow rust are the easily recognisable yellow/orange pustules which form in stripe patterns on the leaves of wheat. Later in its life cycle, however, when the yellow/orange pustules fall off, necrotic lesions with black telia remain on the leaf. It is at this stage that yellow rust can be easily mistaken for mature Septoria which appears as necrotic lesions which follow the veins of the leaf containing many small, black pycnidia. To make matters even more complicated, brown rust appears as orange/brown pustules on wheat leaves, which can be similar in appearance to the early-stage yellow rust pustules. Figure 2 shows examples of these diseases on wheat leaves. Although visibly different from the other three diseases, the white powdery pustules of powdery mildew still cause problems in many parts of the world, making it an important foliar wheat disease and worthy of inclusion in a wheat disease detection model.

Our first important step to creating our model was the collection of a viable dataset of images to use for training. For our model to be useful in the field, it needed to be trained using images which cover the range of conditions which would be encountered in the field, for example, different weather and light conditions, different varieties and colours of wheat, growth stages of the plants and life cycle stage of the diseases. In this experiment, we were looking at only a

Figure 2 Examples of wheat leaf diseases. (a) Brown rust appears with orange/brown pustules along the leaf, which can be confused with the yellow/orange pustules of (b) yellow rust. (c) Yellow rust where the orange pustules have fallen off leaves necrotic lesions that follow the veins of the leaf. This is easily mistaken with (d) mature Septoria, which also appears as necrotic lesions that follow the veins of the leaf.

Published by Burleigh Dodds Science Publishing Limited, 2023.

Figure 3 Example wheat disease images collected for the dataset. The images contain complex background information including soil, other plants and shadow. From left to right: top - brown rust, healthy, mildew; bottom - yellow rust, Septoria.

single disease at a time, therefore the sites for photographing diseases needed to be carefully selected by a pathologist to contain only one disease at a time.

Having collected a significant number of images for each category, the photos taken went through manual quality control. Any images which were blurry, contained no important information or where the important information was obstructed were removed. Images which were thought to contain multiple diseases or where it was not clear which disease was present were also removed. The resulting dataset contained between 2000 and 5000 images per category, providing a vast array of conditions and complex background information, as would be expected in the field. Figure 3 shows example images from the dataset.

We first used transfer learning with four pre-trained networks: VGG16 (Simonyan and Zisserman, 2015), Inception V3 (Szegedy et al., 2016b), Mobilenet (Howard et al., 2017), Xception (Chollet, 2017). Each had been pre-trained using the ImageNet dataset. These experiments allowed us to determine whether a deep learning model would be able to learn to classify these diseases with such complex input data. The results for each pre-trained model were between 85% and 92% classification accuracy.

Published by Burleigh Dodds Science Publishing Limited, 2023.

These results gave us the confidence that deep learning methods would work for this problem. We then developed a bespoke model for this problem using a CNN architecture. The model took several days to train on all of the available training data, before being evaluated on the test dataset. When challenged with the new images from the test dataset, our trained model performed with a classification accuracy of over 97%.

We decided to compare the performance of our model against that of a selection of trained pathologists. Five participants, with differing backgrounds and specialisations, were included in the experiment. A smaller subset of the dataset was used taken from the test set, including a number of images that were incorrectly classified by the network. Each participant was shown the images one by one on a computer screen, they were asked to assign a tag corresponding to their classification of the image. Their classifications were collected and compared with those from the network and the results showed that the network outperformed each member of the group, gaining the highest classification accuracy.

Deep learning networks with the appropriate architecture have the power to deal with real field images containing complex background information. Our resultant wheat disease classification network will be useful for identifying a disease in the field, and so making it easier to take appropriate action. For the purpose of breeding for disease resistance, it would be beneficial for a model to be able to quantify the amount of disease as well as identify and classify different symptoms.

8 Conclusion and future trends

The control of crop diseases is becoming more important than ever as the population of the world continues to increase. For farmers and agronomists, the first step to controlling a disease is identification, which, without access to a trained pathologist, can be difficult in itself. Deep learning methods open the door to automated crop disease detection, which will be game-changing for the treatment and prevention of diseases across the globe.

Recent research had made great strides in classifying images of diseased plants taken in controlled conditions, gaining some extremely high accuracy results. There has also been plenty of progress in using real field images to train networks. It is clear that deep learning techniques are more than capable of handling the task of disease detection and classification. The main obstacle holding back this area of research is the availability of data. Collecting a useful dataset with sufficient samples, covering enough conditions can be a challenging and time-consuming task.

There are a few directions that this research could take in the future. The first is looking at crops which are infected with multiple diseases. At present, very few, if any, studies use images that contain multiple diseases, and it is easy to see why as the presence of multiple diseases complicates the problem substantially. However, it is a very common occurrence in the field to have more than one disease present. Therefore, it is important that deep learning networks are able to handle this if they are to be deployed in the field to aid farmers and agronomists.

Another direction that could be taken in the future is the use of deep learning for quantifying the amount of disease present as well as classifying it. This will be an extremely valuable tool for breeders looking to breed varieties of staple crops with resistance to important diseases. Currently, pathologists are required to spend a lot of time scoring the diseases manually on thousands of plots, so automating this process would be beneficial for them in freeing up time for other important tasks. There are currently a few studies that have started working on this problem, but there is plenty of room for growth and improvement.

This work was supported by the BBSRC Norwich Research Park Biosciences Doctoral Training Partnership as a CASE Award, [grant number BB/S507428/1], in collaboration with Limagrain UK, KWS UK Ltd and RAGT Seeds Ltd.

9 Where to look for more information

- A clear introduction to deep learning and how to create your own deep learning networks – 'Deep Learning with Python' by Francois Chollet.
- About dataset size and variation – Barbedo (2018) 'Impact of dataset size and variety on the effectiveness of deep learning and transfer learning for plant disease classification' in *Computers and Electronics in Agriculture*.
- More references to different visualisation techniques – Saleem et al. (2019) 'Plant Disease Detection and Classification by Deep Learning' in *Plants*.

10 References

Amara, J., Bouaziz, B. and Algergawy, A. (2017). A deep learning-based approach for banana leaf diseases classification. Datenbanksysteme für Business, Technologie und Web – Workshopband, pp. 79–88.

Bolton, M. D., Kolmer, J. A. and Garvin, D. F. (2008). Wheat leaf rust caused by Puccinia triticina, *Molecular Plant Pathology* 9(5), 563–575. doi:10.1111/j.1364-3703.2008.00487.x.

Boulent, J., Foucher, S., Théau, J. and St-Charles, P. L. (2019). Convolutional neural networks for the automatic identification of plant diseases, *Frontiers in Plant Science* 10, 941. doi:10.3389/fpls.2019.00941.

Brahimi, M., Boukhalfa, K. and Moussaoui, A. (2017). Deep learning for tomato diseases: classification and symptoms visualization, *Applied Artificial Intelligence* 31(4), 299–315. doi:10.1080/08839514.2017.1315516.

Chollet, F. (2017). Xception: deep learning with depthwise separable convolutions, *arXiv:1610.02357 [cs]* [Preprint]. Available at: http://arxiv.org/abs/1610.02357 (Accessed: 3 March 2021).

DeChant, C., Wiesner-Hanks, T., Chen, S., Stewart, E. L., Yosinski, J., Gore, M. A., Nelson, R. J. and Lipson, H. (2017). Automated identification of northern leaf blight-infected maize plants from field imagery using deep learning, *Phytopathology* 107(11), 1426–1432. doi:10.1094/PHYTO-11-16-0417-R.

Deng, J., Dong, W., Socher, R., Li, L., Li, K. and Fei-Fei, L. (2009). ImageNet: a large-scale hierarchical image database. *2009 IEEE Conference on Computer Vision and Pattern Recognition. 2009 IEEE Conference on Computer Vision and Pattern Recognition*, pp. 248–255. doi:10.1109/CVPR.2009.5206848.

Dubin, H. J. and Duveiller, E. (2011). Fungal, bacterial and nematode diseases of wheat: breeding for resistance and other control measures. In: *The World Wheat Book, A History of Wheat Breeding* Angus, W., Bonjean, A. and van Ginkel, M. (Eds), pp. 1131–1191.

Ferentinos, K. P. (2018). Deep learning models for plant disease detection and diagnosis, *Computers and Electronics in Agriculture* 145, 311–318. doi:10.1016/j.compag.2018.01.009.

Goyeau, H., Park, R., Schaeffer, B. and Lannou, C. (2006). Distribution of pathotypes with regard to host cultivars in French wheat leaf rust populations, *Phytopathology* 96(3), 264–273. doi:10.1094/PHYTO-96-0264.

Hardwick, N. V., Jones, D. R. and Slough, J. E. (2001). Factors affecting diseases of winter wheat in England and Wales, 1989–98, *Plant Pathology* 50(4), 453–462. doi:10.1046/j.1365-3059.2001.00596.x.

He, K., Zhang, X., Ren, S. and Sun, J. (2016). Deep residual learning for image recognition. *2016 IEEE Conference on Computer Vision and Pattern Recognition (CVPR). 2016 IEEE Conference on Computer Vision and Pattern Recognition (CVPR)*, pp. 770–778. doi:10.1109/CVPR.2016.90.

Howard, A. G., Zhu, M., Chen, B., Kalenichenko, D., Wang, W., Weyand, T., Andreetto, M. and Adam, H. (2017). MobileNets: efficient convolutional neural networks for mobile vision applications, *arXiv:1704.04861 [cs]* [Preprint]. Available at: http://arxiv.org/abs/1704.04861 (Accessed: 3 March 2021).

Huang, G., Liu, Z., van der Maaten, L. and Weinnberger, K. Q. (2018). Densely connected convolutional networks, *arXiv:1608.06993 [cs]* [Preprint]. Available at: http://arxiv.org/abs/1608.06993 (Accessed: 28 September 2021).

Hughes, D. P. and Salathe, M. (2016). An open access repository of images on plant health to enable the development of mobile disease diagnostics, *arXiv:1511.08060 [cs]* [Preprint]. Available at: http://arxiv.org/abs/1511.08060 (Accessed: 5 March 2021).

Jia, Y., Shelhamer, E., Donahue, J., Karayev, S., Long, J., Girshick, R., Guadarrama, S. and Darrell, T. (2014). Caffe: convolutional architecture for fast feature embedding, *arXiv:1408.5093 [cs]* [Preprint]. Available at: http://arxiv.org/abs/1408.5093 (Accessed: 27 September 2021).

Jin, X., Jie, L., Wang, S., Qi, H. and Li, S. (2018). Classifying wheat hyperspectral pixels of healthy heads and fusarium head blight disease using a deep neural network in the wild field, *Remote Sensing* 10(3), 395. doi:10.3390/rs10030395.

Krizhevsky, A. (2014). One weird trick for parallelizing convolutional neural networks, *arXiv:1404.5997 [cs]* [Preprint]. Available at: http://arxiv.org/abs/1404.5997 (Accessed: 27 September 2021).

Krizhevsky, A., Sutskever, I. and Hinton, G. E. (2012). ImageNet classification with deep convolutional neural networks, *Proceedings of the 25th International Conference on Neural Information Processing Systems - Volume 1*. Red Hook, NY, USA: Curran Associates Inc. (NIPS'12), pp. 1097–1105.

Lecun, Y., Bottou, L., Bengio, Y. and Haffner, P. (1998). Gradient-based learning applied to document recognition, *Proceedings of the IEEE* 86(11), 2278–2324. doi:10.1109/5.726791.

LeCun, Y., Haffner, P., Bottou, L. and Bengio, Y. (1999). *Shape, Contour and Grouping in Computer Vision. International Workshop on Shape, Contour and Grouping in Computer Vision*, Springer Verlag, pp. 319–345. doi:10.1007/3-540-46805-6_19.

Liu, B., Zhang, Y., He, D. and Li, Y. (2018). Identification of apple leaf diseases based on deep convolutional neural networks, *Symmetry* 10(1), 11. doi:10.3390/sym10010011.

Liu, M. and Hambleton, S. (2010). Taxonomic study of stripe rust, Puccinia striiformis sensu lato, based on molecular and morphological evidence, *Fungal Biology* 114(10), 881-899. doi:10.1016/j.funbio.2010.08.005.

Lu, Y., Yi, S., Zeng, N., Liu, Y. and Zhang, Y. (2017). Identification of rice diseases using deep convolutional neural networks, *Neurocomputing* 267, 378–384. doi:10.1016/j.neucom.2017.06.023.

Ma, J., Du, K., Zheng, F., Zhang, L., Gong, Z. and Sun, Z. (2018). A recognition method for cucumber diseases using leaf symptom images based on deep convolutional neural network, *Computers and Electronics in Agriculture* 154, 18–24. doi:10.1016/j.compag.2018.08.048.

Mohanty, S. P., Hughes, D. P. and Salathé, M. (2016). Using deep learning for image-based plant disease detection, *Frontiers in Plant Science* 7, 1419. doi:10.3389/fpls.2016.01419.

Nagasubramanian, K., Jones, S., Singh, A. K., Singh, A., Ganapathysubramanian, B. and Sarkar, S. (2018). *Explaining Hyperspectral Imaging Based Plant Disease Identification: 3D CNN and Saliency Maps*, arXiv:1804.08831 [cs.CV]. Available at: https://arxiv.org/abs/1804.08831 (Accessed: 7 September 2022).

Oppenheim, D., Shani, G., Erlich, O. and Tsror, L. (2019). Using deep learning for image-based potato tuber disease detection, *Phytopathology* 109(6), 1083–1087. doi:10.1094/PHYTO-08-18-0288-R.

Ramcharan, A., Baranowski, K., McCloskey, P., Ahmed, B., Legg, J. and Hughes, D. P. (2017). Deep learning for image-based cassava disease detection, *Frontiers in Plant Science* 8, 1852. doi:10.3389/fpls.2017.01852.

Russakovsky, O., Deng, J., Su, H., Krause, J., Satheesh, S., Ma, S., Huang, Z., Karpathy, A., Khosla, A., Bernstein, M., Berg, A. C. and Fei-Fei, L. (2015). ImageNet large scale visual recognition challenge, arXiv:1409.0575 [cs] [Preprint]. Available at: http://arxiv.org/abs/1409.0575 (Accessed: 21 September 2021).

Sermanet, P., Eigen, D., Zhang, X., Mathieu, M., Fergus, R. and LeCun, Y. (2013). OverFeat: integrated recognition, localization and detection using convolutional networks, *International Conference on Learning Representations (ICLR) (Banff)* [Preprint].

Simonyan, K. and Zisserman, A. (2015). Very deep convolutional networks for large-scale image recognition, *arXiv:1409.1556 [cs]* [Preprint]. Available at: http://arxiv.org/abs/1409.1556 (Accessed: 3 March 2021).

Singh, D., Jain, N., Jain, P., Kayal, P., Kumawat, S. and Batra, N. (2020). PlantDoc: a dataset for visual plant disease detection. 249–253. doi:10.1145/3371158.3371196.

Sladojevic, S., Arsenovic, M., Anderla, A., Culibrk, D. and Stefanovic, D. (2016). Deep neural networks based recognition of plant diseases by leaf image classification, *Computational Intelligence and Neuroscience* 2016, 3289801. doi:10.1155/2016/3289801.

Strange, R. N. and Scott, P. R. (2005). Plant disease: a threat to global food security, *Annual Review of Phytopathology* 43(1), 83–116. doi:10.1146/annurev.phyto.43.113004.133839.

Szegedy, C., Liu, W., Jia, Y., Sermanet, P., Reed, S., Anguelov, D., Erhan, D., Vanhoucke, V. and Rabinovich, A. (2014). Going deeper with convolutions, *arXiv:1409.4842 [cs]* [Preprint]. Available at: http://arxiv.org/abs/1409.4842 (Accessed: 28 May 2021).

Szegedy, C., Ioffe, S., Vanhoucke, V. and Alemi, A. (2016a). Inception-v4, inception-resnet and the impact of residual connections on learning, *arXiv:1602.07261 [cs]* [Preprint]. Available at: http://arxiv.org/abs/1602.07261 (Accessed: 27 September 2021).

Szegedy, C., Vanhoucke, V., Ioffe, S., Shlens, J. and Wojna, Z. (2016b). Rethinking the inception architecture for computer vision, doi:10.1109/CVPR.2016.308.

Too, E. C., Yujian, L., Njuki, S. and Yingchun, L. (2019). A comparative study of fine-tuning deep learning models for plant disease identification, *Computers and Electronics in Agriculture*, 161, pp. 272–279. doi:10.1016/j.compag.2018.03.032.

Wang, D., Vinson, R., Holmes, M., Seibel, G., Bechar, A., Nof, S. and Tao, Y. (2019). Early detection of tomato spotted wilt virus by hyperspectral imaging and outlier removal auxiliary classifier generative adversarial nets (OR-AC-GAN), *Scientific Reports* 9(1), 4377. doi:10.1038/s41598-019-40066-y.

Yan, T., Xu, W., Lin, J., Duan, L., Gao, P., Zhang, C. and Lv, X. (2021). Combining multi-dimensional convolutional neural network (CNN) With visualization method for detection of aphis gossypii Glover infection in cotton leaves using hyperspectral imaging, *Frontiers in Plant Science* 12, 604510. doi:10.3389/fpls.2021.604510.

Zhang, S., Huang, W. and Zhang, C. (2019a). Three-channel convolutional neural networks for vegetable leaf disease recognition, *Cognitive Systems Research* 53, 31–41. doi:10.1016/j.cogsys.2018.04.006.

Zhang, X., Han, L., Dong, Y., Shi, Y., Huang, W., Han, L., González-Moreno, P., Ma, H., Ye, H. and Sobeih, T. (2019b). A deep learning-based approach for automated yellow rust disease detection from high-resolution hyperspectral UAV images, *Remote Sensing* 11(13), 1554. doi:10.3390/rs11131554.

Chapter 10

Advances in proximal sensor fusion and multi-sensor platforms for improved crop management

David W. Franzen and Anne M. Denton, North Dakota State University, USA

1 Introduction

Other chapters in this book will describe numerous examples of the use of proximal and remote sensors as a tool to more easily evaluate crop growth and diagnose the need for application of amendments necessary to achieve improved plant health and ultimately more sustainable and profitable crop yield. Some of these sensors are also used to predict crop yield for its own value to crop marketing strategies and farm financial planning (Doraiswamy et al., 2007; Sabini et al., 2017; Donohue et al., 2018) or to use the information in formulas constructed to relate to in-season application of nitrogen (N) or other nutrients (Franzen et al., 2016; Sharma and Franzen, 2016; Ransom et al., 2019). In all of these evaluations, the relationship between sensor reading and crop health indicator or crop yield predictor is not perfect. It is therefore logical to consider whether the addition of other layers of sensor input from the area of interest would add strength to the relationship between sensor outputs and therefore increase predictability. The concept of using multiple layers of sensor information is not new and is supported by studies of methods to construct management zones for use in site-specific nutrient sampling and management (Khosla et al., 2002; Franzen et al., 2011). The use of a single method, such as

http://dx.doi.org/10.19103/AS.2022.0107.18

the use of topography, satellite imagery, aerial imagery, electrical conductivity sensor data, or multi-year crop yield map data, to produce zone boundary delineation resulted in significant correlation with intensively gridded soil sampling values in these studies; however, adding the zone boundary delineation information from two or more of these methods resulted in greater correlation and predictability.

Although there have been many studies and research publications on the use of sensors to direct the application of crop amendments, these have not resulted in acceptance of the technology in mainstream farmer use. According to Lowenberg-DeBoer and Erickson (2019), the three obstacles to adoption are cost, improved reliability of variable-rate technology tools, and demonstrated economic value.

Sensors to aid in crop management are either proximal (Viscarra Rossel and Adamchuk, 2013) or remote sensing tools. Sensors provide either a direct measurement, such as pH measurement at spaced points on-the-go with a proximal sensor, or an indirect measurement, such as relative crop growth using normalized differential vegetative indexes. A single sensor has unique capabilities and strengths of the data it generates, but it also is subject to limitations. To increase the reliability of variable-rate technology, the combined use of multiple sensors has been investigated to increase the predictability of sensors with yield and associated crop traits. The base sensor in most multi-sensor experiments tends to be proximal or remote electromagnetic radiation sensors and thermal sensors. Sensor data that has been added to the base sensor data to increase relationships of the base sensor output with yield, input need, or crop characteristic has been crop height, weather data, directed soil sampling data, electrical/magnetic conductivity proximal sensors, and combine/harvester yield sensors.

The development of high-speed personal computers has enabled the development of computer-modeling techniques such as neural network analysis and machine-learning algorithms to combine different sources of data that are somewhat related to the characteristic of interest, and to make improved predictions of the item with multiple sensor input (Salvador et al., 2020). An example of these techniques for non-yield prediction is relating satellite imagery and non-imagery sensor data to predict residual soil nitrate in northwest Minnesota, USA, after sugar beet (Gautam et al., 2011). Here, satellite imagery in the RGB range of vegetative indexes of the recently harvested sugar beet root crop is combined with data from an electrical conductivity sensor, active-optical sensor readings within the previously growing sugar beet crop, sugar beet canopy height measurements using a meter-stick, root yield, and surface soil elevation from differential GPS readings relative to other points in the study. These data were compared individually to residual soil nitrate values to a 60 cm depth after harvest, and together, using a radial basis function neural network (RBFNN) with 100 iterations. In addition, a back propagation neural network (BPNN) was examined for its utility, along with a modular neural network. The resulting prediction with the RBFNN had an r^2 of about 0.74 and a root mean square error (RMSE) of about 13%. The BPNN model had an r^2 of about 84% and an RMSE of about 10.6%, making the BPNN procedure most predictive. The modular neural network was not as predictive as the RBFNN or the BPNN. The prediction of soil nitrate after sugar beet using either of the neural network techniques was superior to any remote or ground sensor data input.

In this chapter, examples of the use of multi-sensor approaches to crop production management and prediction will be explored. The ultimate goal for the use of these sensors is to increase crop yield with an economy of inputs for the financial advantage of the grower and the conservation enhancements from their use that lead toward a more environmentally friendly and sustainably productive agriculture.

2 Use of plant height and proximal/remote sensing

Differences in crop height intuitively indicate that a taller crop is healthier and requires fewer amendments compared to a shorter crop. Several studies have evaluated the use of corn height alone with the prediction of corn yield (Boomsma et al., 2009; Yin et al., 2011a,b). Crop height differences themselves may not be a direct result of a single variable. In a series of Indiana experiments over tillage treatments, there were sometimes differences in yield and corn height between no-till compared to deeper tillage treatments (Boomsma et al., 2009); however, the differences in height were not consistently related to yield. The differences within row of height due to delayed emergence appeared to have a greater relationship to yield than the mean height over a length of row. In Yin et al. (2011a,b), differences in corn height were determined over a range of N rates; thus, the differences in corn height were related to yield, with the implication that corn height might be used to direct in-season N fertilization if relative height to a sufficient N area was considered. These experiments were conducted using meter sticks or similar analog measurements that would not be practical for most commercial agricultural management. The use of an acoustic sensor, which has industrial uses such as measuring liquid volumes in storage tanks and for assembly-line purposes to help maintain quality, has been investigated in some studies (Shrestha et al., 2002; Sui et al., 2013; Sharma and Franzen, 2016; Yuan et al., 2018). Other methods of sensor-based plant height measurements include use of 3D cameras (Hämmerle and Höfle, 2016) and LiDAR and UAS-based platforms (Verela et al., 2017), combined with previously determined digital elevation models. These additional methods have not been combined with other sensors to provide relationships to crop yield to date. Acoustic height measurements were combined with active-optical sensor data in corn (Sharma and Franzen, 2014) and in sunflower (Schultz et al., 2018). In corn, the relationships of RedNDVI and RedEdgeNDVI with yield were increased with height data, while in sunflower, the relationship of active-optical sensor data with yield was low in oil-seed sunflower and the sunflower height relationship with yield was high. In confection sunflower, which generally has a broader leaf surface, the RNDVI and RENDVI relationship with yield was stronger and sunflower height increased the yield and sensor relationship. The combination of active-optical sensor data and sunflower height did not result in a stronger relationship of the height data with sunflower seed yield in oil-seed sunflower. In Oklahoma (Martin et al., 2012), a formula using the combination of corn plant height, distance between plants, and active-optical sensor NDVI resulted in improved corn yield prediction.

In sugar beet, prediction of root yield and sugar yield is important to aid in sugar processing planning for sugar cooperatives in the Red River Valley of Minnesota and North Dakota. Production is monitored to make certain there is enough sugar to meet demand while limiting production so as not to glut the market, which results in low prices and unprofitable production. The use of satellite imagery has been useful by the industry to produce regional yield estimates (Beeri et al., 2004).Also, the use of satellite imagery alone has been useful since the late 1990s for sugar beet farmers to anticipate the nitrogen release from sugar beet leafy-top residues for the subsequent crop (Daberkow et al., 2003; Franzen, 2004). The use of active-optical sensors was examined for use in estimating root yield and sugar yield at multiple harvest dates, since root yield increases from early pre-harvest (mid-late August) until the soil is frozen, typically anytime from late October to early November (Bu et al., 2016). Although red NDVI and red-edge NDVI were useful yield predictors at most locations and harvest timings, the addition of canopy height increased the prediction of root yield and sucrose yield at most harvests, particularly when the measurements were made early in the season from V6 to V8.

3 Sensors and weather data

Most regional and national crop yield forecasts combine remote sensing from satellites and weather station data (AMIS, 2016). Some models also include additional input from crop growth models and background soil information (Paudel et al., 2021). Crop growth models use weather information as an important part of their formulas. Weather information comes from sensors for rainfall, solar radiation, temperature, and other crop growth influencing measurements. Weather data is generally from sensors placed at some distance to specific land. Therefore, the weather data is site-specific only to the extent of the spatial distance between sensor locations. Satellite imagery or other remote sensing tools are spatial in nature, so combining imagery within a model produces a spatial crop growth model (Kasampalis et al., 2018; Zhao et al., 2020). The scale of the sensor tools defaults ultimately to the smallest scale input, with the assumption that the weather data is relatively accurate at that scale even though it was not measured at that smaller scale. For temperature, relative humidity, and solar radiation, that assumption is probably correct; however, for rainfall, particularly in regions where seasonal rainfall is largely dominated by thunderstorms, that assumption is not correct (Patrick and Stephenson, 1990; Hatfield et al., 1999). It is important, therefore, that if weather data is included in a multi-sensor approach the weather station should ideally be within or adjacent to the field where a spatial decision on inputs is intended to be made.

In a study of US national maize yield, the base model for yield was weather-based only (Peng et al., 2018), with the weather-based model based on temperature, precipitation, and vapor pressure deficit. Adding satellite imagery-derived enhanced vegetative index increased the predictability of the weather-only-based model.

Adding rainfall data as average precipitation from date of planting to date of harvest most consistently improved the relationship of corn yield to active-optical sensor values alone or sensor values together with corn height, and potato yield with active-optical sensor reading (Sharma et al., 2018). The interesting aspect of this study, conducted in North Dakota (corn) and Maine (potato), was the most often the weather station with rainfall data considered in the study was up to 50 km away from the experimental area. So although the seasonal rainfall at the experiment site was more or less than that of the nearest weather station whose rainfall data was considered, the data was nonetheless helpful to the analysis.

A three-US-state study (Thompson et al., 2015) was conducted to explore differences between an active-optical sensor approach to in-season N application compared to N rate predictions using the MAIZE-N model (Setiyono et al., 2011). The MAIZE-N model incorporates weather data together with soil data, particularly an estimate of soil organic matter mineralization, along with crop price and N costs into an in-season N rate prediction. Although the active-optical sensor used (RapidSCAN CS-45 Handheld Crop Sensor - Holland Scientific, Lincoln, NE, USA) correctly predicted N rate at 7 of 11 sites, the model approach predicted 9 of 11 sites correctly. The authors suggested that the use of a combination of the two approaches may make the model approach more responsive to in-season variations in N availability to corn.

In an eight-US state study on exploring commonalities in corn N rate recommendations, active-optical sensor readings, soil moisture measurements, satellite imagery, on-site weather station data, and soil sampling preseason and in-season were evaluated using machine-learning methodology (Qin et al.,

2018). An important feature of the most successful machine-learning-derived models in this effort was including soil hydrological status. The soil hydrological status was measured using the available soil moisture holding capacity, dictated by soil texture with depth, and the ratio of in-season rainfall to the available water holding capacity. Sites that did not perform well in the model probably did not adequately measure soil moisture changes through the growing season. One site that did not perform as well with the model was the Durbin site in North Dakota, which has about 50% clay content (Smectite dominant, illite about 20% of total clay content). The North Dakota climate differs from other states in the study because summer transpiration often exceeds rainfall during the same period. That results in capillary water movement from groundwater at depths below the depth of soil moisture measurements. The deeper groundwater may also hold nitrate that would not be considered in the model. A greater water supply than that predicted by rainfall and soil moisture measurements would also result in greater yield than that predicted. For most sites, however, the use of active-optical sensors, on-site weather data, and soil moisture measurements were adequate to predict economically optimum N rates.

In the same eight-US state study, the economic optimum nitrogen rate (EONR) algorithms used in each state were compared among the states (Ransom et al., 2020). In this analysis, there was no state algorithm that was common to all states. The study concluded that additional factors needed to be considered to produce a more regionally effective algorithm. In subsequent analysis, these algorithms were evaluated with the addition of active-optical sensor readings, satellite image values, weather data, and soil characteristics (Ransom et al., 2021). The active-optical sensor and satellite image-based algorithms were improved in performance when weather data (from at-site weather stations) and soil characteristics were included. Weather parameters that were most useful were evenness of rainfall and abundant and well-distributed rainfall.

Winter wheat yield prediction at the in-field scale has been conducted in the southern US Great Plains since about 2000 using red normalized difference vegetation index (NDVI) standardized using growing degree days from planting. The standardization of NDVI using growing degree days is referred to as in-season estimate of yield (INSEY). Including weather data, consisting of total rainfall from September to December and temperature from September to December, improved the r^2 of the relationship between measured and predicted yield from 0.620 to 0.768 at one location and 0.476 to 0.698 at a second location (Aula et al., 2021).

Using a multi-spectral camera mounted on a UAV to collect red-green-blue (RGB) imagery from fields in Finland, a sequential series of flights, timed according to a certain interval of cumulative temperatures, were conducted to produce a temporal map of RGB during crop growth (Nevavuori et al., 2020). Yield monitor sensor data from field harvests were used to train the spatial neural network-derived model for the relationship between sensor readings and crop yield. In this study, the conclusion was that data from the first month of crop growth was sufficient to build a yield model from a series of UAV flight data obtained using cumulative temperature from weather data. Time-series remote sensor readings have also been important in rice prediction models. Setiyono et al. (2014) used satellite imagery based on synthetic aperture radar (SAR), which is radar-based, and unaffected by cloud cover, which is often a problem in major rice-growing regions of Asia. The rice growth SAR model is developed using temporally separated multi-image acquisition. Leaf area index (LAI) is a common product of an SAR image. Another rice yield estimation model was developed using soil data to estimate soil nitrogen dynamics and weather data from a regional source, corrected to more local weather stations when available. Although each of the models, the SAR-based model and the weather/soil-based model, was effective in predicting rice yield, the combination of models was more effective than either by itself.

Using Sentinel-2 satellite imagery, 109 wheat fields in NE Australia were scanned during the growing season for a relationship to wheat yields. Although the red-edge chlorophyll index and optimized soil-adjusted vegetation index (OSAVI), using satellite imagery only explained over 70% of yield variation, addition of weather data, specifically the crop stress index (SI) developed for this region in Australia, increased the prediction to over 90% of yield variation. The SI consisted of the actual evapotranspiration divided by the potential evapotranspiration. Both parameters are dependent on rainfall, wind speed, temperature, and other weather-related variables from weather stations in the area.

Crop yield modeling was examined in Nebraska, USA, over 134 irrigated and 94 rainfed maize fields (Sibley et al., 2014). Use of the satellite-derived Moderate Resolution Image Spectroradiometer (MODIS) sensor, a crop model (Hybrid-Maize) with either MODIS or Landsat imagery, and the Hybrid-Maize model applied to MODIS and Landsat data was compared. The MODIS methods were consistently poor at yield prediction, whereas the best was the Hybrid-Maize model applied to Landsat imagery. The Hybrid-Maize model includes maize growth related to soil and weather parameters, while the Landsat imagery, R-band NDVI, was the best predictor (R^2 0.54–0.63).

In machine-learning exercises relating satellite imagery NDVI and weather data to potato yield in autumn-winter and spring-summer crops in Mexico, the most successful approach for the autumn-winter crop was using a random forest (RF) approach, while for the spring-summer crop a support vector machine linear (svml) approach was best. For the RF approach, the most important variables supporting the model were the NDVI, total cloud cover, consideration of previous crop yield, solar radiation, and evaporation. For the svml approach, LAI measured each month, NDVI, total cloud cover, precipitation, and evaporation were most important in yield prediction. A combination of satellite imagery and weather data was important for national potato production prediction.

Predicting sunflower seed yield is a greater challenge. As seen in the failure of active-optical sensor red NDVI and red-edge NDVI to relate to sunflower yield in North Dakota (Franzen et al., 2019), similar problems have also been observed in Spain (Trepos et al., 2020). The sunflower yield crop model used (SUNFLO) to predict yield based on time, soil, and climate; management practices including crop nitrogen content, plant transpiration, leaf expansion, senescence and biomass accumulation as a result of nitrogen deficits; and weather data including temperature, radiation, and water availability. The crop forecast models without LAI from satellite imagery resulted in poor yield estimates with large errors, generally overestimating yield. Including LAI tended to reduce errors, with a smoothed LAI over the study area reducing errors the most. The resulting model was an improvement, but it still resulted in general yield overestimation. The explanation for yield overestimation was the lack of consideration of weeds, diseases, and other pests in the fields. Disease and insect pests can particularly attack formally healthy sunflowers at or near heading, late in the season (Berglund, 2007).

Soybean yield is also best predicted using a combination of satellite imagery NDVI and climatic data. Soybean yield is greatly influenced by late-season environmental conditions, particularly soil moisture stress (Licht et al., 2013). If the season is favorable in soil moisture for late season growth and pod-fill, then NDVI can be a good predictor of yield. However, in a study of soybean fields in Russia, the satellite NDVI saturated in values mid-season when the rows closed over the soil, and climatic data was important to add to the model to predict soil moisture condition and ultimately yield (Stepanov et al., 2020). A similar approach was also taken in a study of Brazil soybean yield prediction (Schwalbert et al., 2020). Satellite imagery and weather data were significant, but relatively poor predictors of yield 70 days before harvest; however, the prediction was greatly improved at 40 days before harvest with updated weather data.

Canola is mostly grown as a spring crop in Canada, North Dakota, and some northern areas in Montana and Minnesota. Yield prediction is important for the canola oil industry in this region. Knowing that canola grows as a rosette for the first 30 days of the season, then bolts to form heads and flowers, with the length of flowering important for final yield, largely dictated by temperature and soil moisture status, a study across the Canadian Prairies was conducted using as a primary tool estimates of soil moisture obtained from the SM Ocean Salinity Mission (SMOS) satellite (White et al., 2020). Additional tools were climatic variables and NDVI derived from the Advanced Very High Resolution Radiometer (AVHRR) platform, currently used as an input for canola yield models. The NDVI-based model alone predicted 41.2% of yield variation, while adding satellite-derived soil moisture increased prediction to 74.2%.

4 Multi-sensor approaches

Most studies that investigate yield prediction or traits associated with yield prediction using multiple tools utilize one set of electromagnetic sensor-based vegetative indexes and the output from an unrelated tool, such as weather data or crop height. Studies are now being conducted that utilize machine learning to combine multiple vegetative indexes in their analyses. Osco et al. (2020) combined red NDVI, red-edge NDVI, green NDVI, and soil-adjusted vegetative index (SAVI) in several machine-learning analyses for the prediction of leaf N concentration and maize plant height at V12. This study was conducted on an N rate X maize hybrid experiment in maize in NE Mato Grosso do Sul, Brazil. The dimensions of the study were approximately 60 m × 30 m. The algorithms were conducted on spectral bands alone, on the vegetative indices derived from the spectral bands, and the models compared were RF, REPTree, K-Nearest Neighbor with K = 1, 5, and 10, singular boundary method-radial basis function (SBM-RBF), support vector machine polynomial (SVMP), linear regression, and radial basis function regression. Of all the methods, the RF method using the vegetative indexes performed best, with a RMSE of 1.9 g kg^{-1} for leaf N concentration and 0.17 m for plant height.

A multiple vegetative index approach was also applied to an N rate study in maize in Mississippi, USA. Although yield prediction using a single vegetative index was optimized using the OSAVI or the Simplified Canopy Chlorophyll Content Index (SCCCI), the combination of Green Atmospherically Resistant Index (GARI), red-edge NDVI, and green NDVI was the best yield predictor at V6-7 ($r^2 = 0.70$); the SCCCI and SAVI were best at V10-11 ($r^2 = 0.90$); and SCCI, Green Leaf Index (GLI), and Visible Atmospherically Resistant Index ($VARI_{green}$) were the best predictors at tasseling ($r^2 = 0.93$).

Leaf nitrogen estimation in maize was studied using a 'fusion' of bands and 18 vegetative indices from a hybrid by nitrogen rate experiment in Beijing, China (Xu et al., 2021). The analysis was based on development of cover-adjusted spectral indices (CASIs), where $CASI = VI/(1+FV_{cover})$, where VI is the vegetative index selected, and FV_{cover} indicates the fraction of vegetative cover. The vegetative indices were extracted from multispectral imagery with high spatial resolution. The FV_{cover} was also calculated from the RBG imagery, as area of vegetation divided by total area. A random frog algorithm was used to identify the five most optimal characteristics among the VIs. Then a partial least squares method was utilized to investigate relationships between leaf nitrogen concentration and an optimal set of CASIs/Vis at three growth stages of corn (V12; R1; R3). The CASIs at the R1 growth stage were most related to leaf nitrogen content with an r^2 of 0.59, an RMSE of about 22%, and an NRMSE (normalized root mean square error) of about 8.4%. The removal of soil information from the analysis resulted in greater relationships between leaf nitrogen concentration and CASIs/VIs.

Use of all spectral bands in a wheat variety experiment involving 1170 lines in Spain was superior in yield prediction compared to individual vegetative indexes (Montesinos-Lopez et al., 2017). The bands used were 250 discrete narrow bands that ranged from 392 nm to 851 nm. Mid-season imagery was more predictive than early or late-season flights. Also, the use of all bands was most predictive over all environments. Prediction of yield was best under early heat and under irrigation, and poorest under the drought environment.

A combination of vegetative indices was explored individually for ground proximal sensors (chlorophyll meters) and multispectral imagery from an airplane flying at 330 m elevation and a drone operated at 80 m elevation was explored in maize (Gabriel et al., 2017). The ground proximal sensors individually or combined were more related to experiment nitrogen availability than the aerial sensors. In the suite of possible aerial-based VIs, the TCARI (transformed chlorophyll absorption in reflectance index) = $3[(R_{700}\text{-}R_{670}) - 0.2(R_{700}-R_{550})/(R_{700}/R_{700})]$and OSAVI = $(1+0.16) \times ((R_{800}-R_{670})/(R_{800}+R_{670} + 0.16)$ were most related, particularly when evaluated as TCARI/OSAVI. In the case of aerial image relationship, the best relationship with maize nitrogen concentration was achieved when soil interference was eliminated. The ground proximal sensors were applied to the leaves only, so soil had no effect; however, obtaining information from the leaf-clipping derived proximal chlorophyll sensors was laborious and slow compared to either aerial data source (Barzin et al., 2020).

Thermal sensors have historically been studied for their use in identifying differences in transpiration/evaporation from soil or crop canopy surfaces, drought conditions (Jones et al., 2021), and pest/disease infestation (Pineda et al., 2021). Lately, thermal sensors have been linked directly or in combination with remote sensing and proximal light sensors. A recently developed proximal tool, Crop Circle Phenom (Holland Scientific, Inc., Lincoln, NE, USA) was studied for its use in improving corn N status prediction. The instrument includes sensors with red, red edge, and near-infrared wavelengths, a thermal sensor for determining the difference between crop canopy temperature and air temperature ΔT, and a sensor for determining the fractional photosynthetically active radiation (fPAR). The best prediction of crop nitrogen status was made using a machine-learning tool, eXtreme Gradient Boost (XGBoost) that included the vegetative indexes calculated by the sensor tool, the ΔT, the fPAR, and adding in drainage, tillage, and pre-plant nitrogen rate (Cummings et al., 2021).

A suite of proximal soil sensing tools was utilized in a study on a small pasture in Brazil to characterize its soils. The tools consisted of an apparent magnetic susceptibility sensor, an apparent electrical conductivity sensor, a volumetric water sensor, a portable gamma-ray radiation sensor, a cone-penetrometer, and an x-ray fluorescence sensor. Readings were compared to laboratory analysis of soil clay content, water content, cation exchange capacity (CEC), organic carbon, and sum of bases. The use of the x-ray fluorescence sensor was most highly correlated with organic carbon, clay content, and bulk density, while other tools were more highly correlated with sum of bases, CEC, and water content. The use of a combination of tools was superior in correlation to the use of only one tool (Vasques et al., 2020).

Imagery from a UAV fit with sensors able to capture thermal, infrared, and RGB spectral bands, as well as plant height using Structure from Motion and LiDAR to predict soybean yield at R4/R5 in a large soybean variety trial. Machine-learning tools RF and XGBoost were used to obtain highly predictive algorithms ($r^2 > 0.9$) from the UAS-generated sensor data (Herrero-Huerta et al., 2020).

5 Statistical tools for fusing multi-sensor data

A common theme of all studies that include multi-sensor data in prediction of any crop attribute is the use or comparison of statistical tools to combine the data sources. Some options for combining data are:

- egression in some form, linear, quadratic, or some transformed model;
- neural networks;
- machine-learning tools; and
- deep learning tools.

Among statistical approaches for sensor fusion problems in agriculture, most fall into the group of supervised learning problems in that there is a ground truth available that can be measured for at least some of the data. This is unlike clustering or pattern detection problems where no labeled data is available. To be yet more specific, sensor fusion is often used in regression problems, where a variable is to be predicted that is continuous, for example, yield or crop health indicators. Classification, or the prediction of categorical target variables, can also be of interest for problems like identifying diseased plants (Moshou et al., 2011). A more general discussion that includes a broader set of machine-learning goals can be found in Liakos et al. (2018).

The most basic approaches to regression problems assume that a linear combination of basis functions can be used for prediction. Such approaches are called linear regression, even while the basis functions themselves do not have to be linear functions of the independent variables. Regression using quadratic or higher order basis functions can still be linear in the parameters that are optimized and would then be considered as linear regression in the statistics literature. Yin et al. (2011a) compare linear, quadratic, square root, logarithmic, and exponential basis functions for assessing the relationship of corn yield with plant height.

Linear regression problems are more straightforward to solve than nonlinear ones, and there is less ambiguity about the assumptions that were made. Except when physical reasons suggest selecting a specific model, it is normally recommended to use linear models where those are sufficient. Most available implementations of linear regression also return information on the importance of different independent variables as part of the result. Like other regression approaches, linear regression is vulnerable to overfitting, especially when many independent variables or basis functions are included. In such cases, regularization is recommended, which serves the purpose of reducing the number of non-zero fit parameters, such that only important independent variables and basis functions are considered in the final model. Least absolute shrinkage and selection operator (LASSO) is an example of a regularization and variable selection algorithm that is used for example in Aula et al. (2021). Using regularization is especially important when the independent variables are highly correlated.

Instead of using regression models directly, combined features can be precomputed. The simplest examples are indices like the NDVI or the LAI. A common statistical way of deriving combined features is principal component analysis (PCA), which returns orthogonal combinations of features ordered starting with highest variance. Basso and Liu (2019) give examples of the use of vegetation indices and PCA for the purpose of crop yield forecasting. They also discuss examples of using the output of physical models as input into statistical models such as used by Ratjen and Kage (2015).

Many sensor fusion problems are too complex to yield good results using only linear methods either directly or on derived features that were explicitly specified. Nonlinear regression algorithms can be useful for these cases, including artificial neural networks (ANN), support vector regression (SVR), and tree-based methods. SVR is a mathematically elegant approach to nonlinear regression that uses the concept of kernels for encoding assumptions about the learning problem. Unfortunately, in SVR the model size depends on the size of the training data, and in many learning contexts, it is desirable to prescribe the size of the model.

Tree-based techniques are relatively fast and are good for reducing noise due to less relevant features. Regression trees are constructed by identifying the most important feature at every level. Individual regression trees typically do not yield the highest prediction quality, but they can be useful for creating explainable models, since the number of features that contribute to a prediction is usually small, and its impact on the prediction straightforward, see, for example, Hamann et al. (2011). To achieve high accuracy, ensembles

Published by Burleigh Dodds Science Publishing Limited, 2023.

of trees should be used, such as RF. RF models are constructed by creating a large set of regression trees, each from a subsample of training data points, and then computing the predicted value as the average of the predictions of each individual tree. Osco et al. (2020) conclude that an RF model is a suitable technique for predicting leaf nitrogen and plant height in maize.

Among the most popular statistical prediction models are neural networks of various designs. The choice of model has a strong impact on what solution can be achieved, making it important to understand their characteristics. One of the simplest assumptions is that predictions should be based on proximity to known examples, which is used in RBFNN and SVR with Radial Basis Function kernels. While an assumption of overall proximity is straightforward, it suffers from the shortcoming that distances are calculated over all independent variables. For sensor fusion, the relevance of different input sources toward the predicted quantity may differ vastly, and it has been shown that RBFNN may be less suitable to such problems than, for example, backpropagation neural networks (BNNs) (Gautam et al., 2011).

Among the oldest and most versatile neural network designs are BNNs, which are also sometimes referred to as feed-forward neural networks or multilayer perceptrons. The most common design has a layer of input nodes that are connected to a layer of hidden nodes, which in turn is connected to a third layer that represents the outputs. The hidden layer allows these neural networks to represent nonlinearities in a way that does not have to be prescribed explicitly. The BNN was one of the top-performing networks in Gautam et al. (2011).

A breakthrough for the recognition of objects in images and for understanding text input was the development of deep learning (LeCun et al., 2015). Deep neural networks have substantially more than one hidden layer and can directly learn features within data that do not have to be structured, such as images or text, thereby resolving or reducing the need for encoding features explicitly. The success of deep neural networks relied on some general algorithmic improvements, in particular, the prevention of overfitting through dropout of nodes, which can be compared with the regularization that was mentioned earlier in the context of regularized linear models. Moreover, some of the most successful types of deep learning networks were constructed such that they are effective at encoding specific derived features. For example, convolutional neural networks, CNNs, were designed to represent image features regardless of where in an image they occur. For this purpose, they have convolutional layers that represent information on what happens within the vicinity of a point in an image. CNNs also have max-pooling layers that combine information regardless of where in the image it is found. A typical use of this capability would be plant identification (Grinblat et al., 2016).

Some deep learning networks have been designed to be particularly effective at encoding temporal relationships, in particular, recurrent neural networks (RNN) and extensions of RNNs that can capture long-term relationships and are called long-short term memory (LSTM) models. Jia et al. (2019) demonstrate the success of LSTM models in crop monitoring. They also show that when temporal aspects are to be combined with spatial machine learning, domain adaptation (DA) is a common and often successful approach. DA is a special case of transfer learning. In transfer learning, a machine-learning model is used on a data set that is different from the one that is used in prediction. DA refers to the special case that the data only differ in the time at which the data were collected.

6 Conclusion and future trends

There is much evidence that proximal and remote sensing technology can be used to improve crop management by farmers and their industry partners. There is also much evidence in the value of the use of multiple sensors over the use of one alone. Partnering sensors with different strengths to achieve a better prediction and management consequence will likely improve management decisions and promote more sustainable agricultural practices. However, this is dependent on the value obtained through investment in additional data sources (e.g. sensors). The challenge now is to provide the end-user with a package of sensor and analytical tools to make the use of these technologies simple to use, easy to maintain, and produce a minimal burden. The integration of relevant tools into a plug-and-play package, with sufficient developmental research support to support their value will be critical to the movement of these sciences into commercial adoption.

7 Where to look for further information

Several journals are particularly rich sources of information regarding the use of sensors in agriculture. These journals include *Agronomy Journal* (https://acsess.onlinelibrary.wiley.com/journal/14350645), *Sensors* (https://www.mdpi.com/journal/sensors), *Remote Sensing and Environment* (https://www.journals.elsevier.com/remote-sensing-of-environment), *Computers and Electronics in Agriculture* (https://www.sciencedirect.com/journal/computers-and-electronics-in-agriculture), *Remote Sensing* (https://www.mdpi.com/journal/remotesensing), and *Precision Agriculture* (https://www.springer.com/journal/11119).

A book was released in 2021 that provides a broad view of research into the use of sensors in agriculture: *Sensing Approaches for Precision Agriculture*, Ruth Kerry and Alexandre Escola, eds, Springer Cham (https://doi.org/10.1007/978-3-030-78431-7).

Published by Burleigh Dodds Science Publishing Limited, 2023.

Also, recent proceedings from the ISPA (International Society of Precision Agriculture), https://www.ispag.org/, and the European Society for Precision Agriculture, https://www.ecpa2021.hu/, may also be helpful, as well as attending their future conferences.

8 References

AMIS (2016). Crop yield forecasting: methodological and institutional aspects. Agricultural Market Information System, Food and Agriculture Organization of the United Nations. Available at: http://www.amis-outlook.org/fileadmin/user_upload/amis/docs/resources/AMIS_CYF-Methodological-and-Institutional-Aspects_web.pdf.

Aula, L., Omara, P. E., Nambi, E., Oyebiyi, F. B., Dhillon, J., Eickhoff, E., Carpenter, J. and Raun, W. R. (2021). Active optical sensor measurements and weather variables for predicting winter wheat yield. Agron. J. 113(3):2742–2751 doi:10.1002/agj2.20620.

Barzin, R., Pathak, R., Lofti, H., Varco, J. and Bora, G. C. (2020). Use of UAS multispectral imagery at different physiological stages for yield prediction and input resource optimization in corn. Remote Sens. 12(15):2392 doi:10.3390/rs12152392.

Basso, B. and Liu, L. (2019). Chapter four - seasonal crop yield forecast: methods, applications and accuracies. Adv. Agron. 154:201–255.

Beeri, O., Zhang, X., Newcomb, T., Carson, P. and Wagner, G. (2004). Using Landsat images to map quality and quantity sugar beet yield. 2004 *Sugarbeet Research and Extension Reports*. Vol. 35, p. 125. Sugarbeet Research and Education Board of Minnesota and North Dakota, Fargo, ND. Available at: https://www.sbreb.org/wp-content/uploads/2018/09/Production5.pdf.

Berglund, D. R. (2007). Sunflower production. Available at: https://www.ag.ndsu.edu/extensionentomology/recent-publications-main/publications/A-1331-sunflower-production-field-guide.

Boomsma, C. R., Vyn, T. J., Brewer, J. C., Santini, J. B. and West, T. D. (2009). Corn yield responses to plant height variability resulting from tillage and crop rotation systems in a long-term experiment. In: *17th Triennial Conference of the International Soil Tillage Research Conference Proceedings*.

Bu, H., Sharma, L. K., Denton, A. and Franzen, D. W. (2016). Sugar beet yield and quality prediction at multiple harvest dates using active-optical sensors. Agron. J. 108:273–284.

Cummings, C., Miao, Y., Paiao, G. D., Kang, S. and Fernandez, F. G. (2021). Corn nitrogen status diagnosis with an innovative multi-parameter Crop Circle PhenoM sensing system. Remote Sens. 13(3):401 doi:10.3390/rs13030401.

Daberkow, S., McBride, W. and Ali, M. (2003). Implications of remote sensing imagery and crop rotation for nitrogen management in sugar beet production. *Proceedings of the 2003 AAEA, Montreal, CA, 27-30 July, 2003*. file:///C:/Users/david.franzen/Downloads/sp03da01%20(1).pdf.

Donohue, R. J., Lawes, R. A., Mata, G., Gobbett, D. and Ouzman, J. (2018). Towards a national, remote-sensing-based model for predicting field-scale crop yield. Field Crops Res. 227:79–90. doi:10.1016/j.fcr.2018.08.005.

Doraiswamy, P.C., Akhmedov, B., Beard, L., Stern, A. and Mueller, R. (2007). Operational prediction of crop yields using MODIS data and products. p. 45-49. In: *ISPRS WG VIII/10 Workshop 2006 Remote Sensing Support to Crop Yield Forecast and Area Estimates* Barath, B., Royer, A. and Genovese, G. (Eds), Stesa, Italy. Available at: https://citeseerx.ist.psu.edu/viewdoc/download?doi=10.1.1.222.4442&rep=rep1&type=pdf#page=52.

Franzen, D. W. (2004). Delineating nitrogen management zones in a sugar beet rotation using remote sensing: a review. J. Sugar Beet Res. 41(1):47–60. Available at: https://www.researchgate.net/profile/Dave-Franzen/publication/237133722_Delineating_Nitrogen_Management_Zones_in_a_Sugarbeet_Rotation_Using_Remote_Sensing_-_A_Review/links/5473404d0cf216f8cfaecab0/Delineating-Nitrogen-Management-Zones-in-a-Sugarbeet-Rotation-Using-Remote-Sensing-A-Review.pdf.

Franzen, D. W., Kitchen, N., Holland, K., Schepers, J. and Raun, W. (2016). Algorithms for in-season nutrient management in cereals. Agron. J. 108(5):1775–1781.

Franzen, D. W., Long, D., Sims, A., Lamb, J., Casey, F., Staricka, J., Halvorson, M. and Hofman, V. (2011). Evaluation of methods to determine residual soil nitrate zones across the northern Great Plains of the USA. Prec Agri. 12:594–606. doi:10.1007/s11119-010-9207-0.

Franzen, D. W., Schultz, E. C., DeSutter, T. M., Sharma, L. K., Ashley, R. and Bu., H. (2019). Sunflower type influences yield prediction using active optical sensors. Agron. J. 111(2):881–888. doi:10.2134/agronj2018.07.0440.

Gabriel, J. L., Zarco-Tejada, P. J., Lopez-Herrera, P. J., Perez-Martin, E., Alonso-Ayuso, M. and Quemada, M. (2017). Airborne and ground level sensors for monitoring nitrogen status in a maize crop. Biosys. Eng. 160:124–133. doi:10.1016/j.biosystemseng.2017.06.003.

Gautam, R., Panigrahi, S., Franzen, D. and Sims, A. (2011). Residual soil nitrate prediction from imagery and non-imagery information using neural network technique. Biosys. Eng. 110(1):20–28.doi:10.1016/j.biosystemseng.2011.06.002.

Grinblat, G. L., Uzal, L. C., Larese, M. G. and Granitto, P. M. (2016). Deep learning for plant identification using vein morphological patterns. Comp. Elec. Agric. 127:418–424.

Hamann, A., Gylander, T. and Chen, P. Y. (2011). Developing seed zones and transfer guidelines with multivariate regression trees. Tree Genet. Genomes 7(2):399–408.

Hämmerle, M. and Höfle, B. (2016). Direct derivation of maize plant and crop height from low-cost time-of-flight camera measurements. Plant Methods 12:50 doi:10.1186/s13007-016-0150-6.

Hatfield, J. L., Prueger, J. H. and Meek, D. W. (1999). Spatial variation of rainfall over a large watershed in Central Iowa. Theor. Appl. Climatol. 64(1-2):49–60.

Herrero-Huerta, M., Rodriguez-Gonzalvez, P. and Rainey, K. M. (2020). Yield prediction by machine learning from UAS-based multi-sensor data fusion in soybean. Plant Methods 16:78 doi:10.1186/s13007-020-00620-6.

Jia, X., Khandelwal, A., Mulla, D. J., Pardey, P. G. and Kumar, V. (2019). Bringing automated, remote-sensed, machine learning methods to monitoring crop landscapes at scale. Agric. Econ. 50(S1):41–50.

Jones, H. G., Hutchinson, P. A., May, T., Jamali, H. and Deery, D. M. (2021). A practical method using a network of fixed infrared sensors for estimating crop canopy conductance and evaporation rate. Biosys. Eng. 165:59-69. doi:10.1016/j.biosystemseng.2017.09.012.

Kasampalis, D. A., Alexandridis, T. K., Devan, C., Challinor, A., Moshou, D. and Zalidis, G. (2018). Contribution of remote sensing on crop models: a review. J. Imaging 4(4):52, doi:10.3390/jimaging4040052.

Khosla, R., Fleming, K., Delgado, J. A., Shaver, T. M. and Westfall, D. G. (2002). Use of site-specific management zones to improve nitrogen management for precision agriculture. J. Soil Cons. 57:513–518.

LeCun, Y., Bengio, Y. and Hinton, G. (2015). Deep learning. Nature 521(7553):436–444.

Liakos, K. G., Busato, P., Moshou, D., Pearson, S. and Bochtis, D. (2018). Machine learning in agriculture: a review. Sensors (Basel) 18(8):2674. doi:10.3390/s18082674.

Licht, M. A., Wright, D. and Lenssen, A. W. (2013). Soybean response to drought. *Agriculture and Environment Extension Publications*. 190. Available at: https://lib.dr.iastate.edu/extension_ag_pubs/190.

Lowenberg-DeBoer, J. and Erickson, B. (2019). Setting the record straight on precision agriculture adoption. *Agron. J.*:1552–1569. doi:10.2134/agronj2018.12.0779.

Martin, K. L., Raun, W. R. and Solie, J. B. (2012). By-plant prediction of corn grain yield using optical sensor readings and measured plant height. J. Plant Nutr. 35(9):1429–1439.

Montesinos-Lopez, O. A., Montesinos-Lopez, A., Crossa, J., de los Campos, G., Alvarado, G., Suchismita, M., Rutkoski, J., Gonzalez-Perez, L. and Burueno, J. (2017). Predicting grain yield using canopy hyperspectral reflectance in wheat breeding data. Plant Methods 13(4):4 doi:10.1186/s13007-016-0154-2.

Moshou, D., Bravo, C., Oberti, R., West, J. S., Ramon, H., Vougioukas, S. and Bochtis, D. (2011). Intelligent multi-sensor system for the detection and treatment of fungal diseases in arable crops. Biosys. Eng. 108(4):311–321.

Nevavuori, P., Narra, N., Linna, P. and Lipping, T. (2020). Crop yield prediction using multitemporal UAV data and spatio-temporal deep learning models. Remote Sens. 12(23):4000 doi:10.3390/rs12234000.

Osco, L. P., Junior, J. M., Ramos, A. P. M., Furuya, D. E. G., Santana, D. C., Teodoro, L. P. R., Gonçalves, W. N., Baio, F. H. R., Pistori, H., Junior, C. AdS. and Teodoro, P. E. (2020). Leaf nitrogen concentration and plant height prediction for maize using UAV-based multispectral imagery and machine learning techniques. Remote Sens. 12(19):3237 doi:10.3390/rs12193237.

Patrick, N. A. and Stephenson, D. (1990). Spatial variation of rainfall intensities for short duration storms. Hydrol. Sci. J. 35(6):667–680 doi:10.1080/02626669009492471.

Paudel, D., Boogaard, H., de Wit, A., Janssen, S., Osinga, S., Pylianidis, C. and Athanasiadis, I. N. (2021). Machine learning for large-scale crop yield forecasting. Agric. Syst. 187:103016.

Peng, B., Guan, K., Pan, M. and Li, Y. (2018). Benefits of seasonal climate prediction and satellite data for forecasting U.S. maize yield. Geophys. Res. 45(18):9662–9671. doi:10.1029/2018GL079291.

Pineda, M., Baron, M. and Perez-Bueno, M.-L. (2021). Thermal imaging for plant stress detection and phenotyping. Remote Sens. 13(1):68 doi:10.3390/rs13010068.

Qin, Z., Myers, D. B., Ransom, C. J., Kitchen, N. R., Liang, S., Camberato, J. J., Carter, P. R., Ferguson, R. B., Fernandez, F. G., Franzen, D. W., Laboski, C. A. M., Malone, B. D., Nafziger, E. D., Sawyer, J. E. and Shanahan, J. F. (2018). Application of machine learning methodologies for predicting corn economic optimal nitrogen rate. Agron. J. 110(6):2596–2607. doi:10.2134/agronj2018.03.0222.

Ransom, C. J., Kitchen, N. R., Camberato, J. J., Carter, P. R., Ferguson, R. B., Fernandez, F. G., Franzen, D. W., Laboski, C. A. M., Myers, D. B., Nafziger, E. D., Sawyer, J. E.

and Shanahan, J. F. (2019). Statistical and machine learning methods evaluated for incorporating soil and weather into corn nitrogen recommendations. Comp. Elec. Agric. 164:104872 doi:10.1016/j.compag.2019.104872.

Ransom, C. J., Kitchen, N. R., Camberato, J. J., Carter, P. R., Ferguson, R. B., Fernandez, F. G., Franzen, D. W., Laboski, C. A. M., Nafziger, E. D., Sawyer, J. E., Scharf, P. C. and Shanahan, J. F. (2020). Corn nitrogen rate recommendation tools' performance across eight US Midwest corn belt states. Agron. J. 112(1):470–492. doi:10.1002/agj2.20035.

Ransom, C. J., Kitchen, N. R., Sawyer, J. E., Camberato, J. J., Carter, P. R., Ferguson, R. B., Fernandez, F. G., Franzen, D. W., Laboski, C. A. M., Myers, D. B., Nafziger, E. D. and Shanahan, J. F. (2021). Improving publicly available corn nitrogen rate recommendation tools with soil and weather measurements. Agron. J. 113(2):2068–2090 doi:10.1002/agj2.20627.

Ratjen, A. M. and Kage, H. (2015). Forecasting yield via reference-and scenario calculations. Comp. Elec. Agric. 114:212–220.

Sabini, M., Rusak, G. and Ross, B. (2017). Understanding satellite-imagery-based crop yield predictions. Available at: http://cs231n.stanford.edu/reports/2017/pdfs/555.pdf.

Salvador, P., Gomez, D., Sanz, J., and Casanova, J. L. (2020). Estimation of potato yield using satellite data at a municipal level: A machine learning approach. Int. J. Geo-Info. 9(6):343 doi:10.3390/ijgi9060343.

Schultz, E., DeSutter, T., Sharma, L., Endres, G., Ashley, R., Bu, H., Markell, S., Kraklau, A. and Franzen, D. (2018). Response of sunflower to nitrogen and phosphorus in North Dakota. Agron. J. 110: 1–11.

Schwalbert, R. A., Amado, T. J. C., Corassa, G. M., Pott, L. P., Prasad, P. V. V. and Ciamtitt, I. A. (2020). Satellite-based soybean yield forecast: integrating machine learning and weather data for improving crop yield prediction in southern Brazil. Agric. Forest Meteorol. 284 doi:10.1016/j.agrformet.2019.107886.

Setiyono, T. D., Nelson, A. and Holecz, F. (2014). Remote sensing based crop yield monitoring and forecasting. *FAO Expert Meeting Proceedings*. Available at: http://www.fao.org/fileadmin/templates/rap/files/Project/Expert_Meeting__17Feb2014_/P2-4_Setiyono_2014_Remote-Sensing_based_Crop_Yield_Monitoring.pdf.

Setiyono, T. D., Yang, H., Walters, D. T., Dobermann, A., Ferguson, R. B., Roberts, D. F., Lyon, D. J., Clay, D. E. and Cassman, K. G. (2011). Maize-N: a decision tool for nitrogen management in maize. Agron. J. 103(4):1276–1283.

Sharma, L. K. and Franzen, D. W. (2016). Comparison of two ground-based active-optical sensors for in-season estimateion of corn (Zea mays, L.) yield. J. Plant Nut. 39:957–966.

Sharma, L. K., Bali, S. K., Zaeen, A. A., Baldwin, P. and Franzen, D. W. (2018). Use of rainfall data to improve ground-based active optical sensors yield estimates. Agron. J. 110(4):1561–1571.

Sharma, L. K. and Franzen, D. W. (2014). Use of corn height to improve the relationship between active optical sensor readings and yield estimates. Precision Agric. 15(3):331–345.

Shrestha, D. S., Steward, B. L., Birrell, S. J. and Kaspar, T. C. (2002). Plant height estimation using two sensing systems. In: *Proceedings of the ASAE Annual International Meeting, 2002*. CD-ROM, St. Joseph, MO.

Sibley, A. M., Grassini, P., Thomas, N. E., Cassman, K. G. and Lobell, D. B. (2014). Testing remote sensing approaches for assessing yield variability among maize fields. Agron. J. 106(1):24–32.

Stepanov, A., Dubrovin, K., Sorokin, A. and Asseva, T. (2020). Predicting soybean yield at the regional scale using remote sensing and climatic data. Remote Sens. 12(12):1936 doi:10.3390/rs12121936.

Sui, R., Fisher, D. K. and Reddy, K. N. (2013). Cotton yield assessment using plant height mapping system. J. Agri. Sci. 5(1): 23–31. doi:10.5539/jas.v5n1p23.

Thompson, L. J., Ferguson, R. B., Kitchen, N., Franzen, D. W., Mamo, M., Yang, H. and Schepers, J. S. (2015). Model and sensor-based recommendation approaches for in-season nitrogen management in corn. Agron. J. 107(6):2020–2030. doi:10.2134/agronj15.0116.

Trepos, R., Champolivier, L., Dejoux, J.-F., Al Bitar, A., Casadebaig, P. and Debaeke, P. (2020). Forecasting sunflower grain yield by assimilating leaf area index into a crop model. Remote Sens. 12(22):3816 doi:10.3390/rs12223816.

Varela, S., Assefa, Y., Vara Prasad, P. V., Peralta, N. R., Griffin, T. W., Sharda, A., Ferguson, A. and Ciampitti, I. A. (2017). Spatio-temporal evaluation of plant height in corn via unmanned aerial systems. J. Appl. Remote Sens. 11:036013 doi:10.1117/1.JRS/11/036013.

Vasques, G. M., Rodrigues, H. M., Coelho, M. R., Baca, J. F. M., Dart, R. O., Oliveira, R. P., Teixeira, W. G. and Ceddia, M. B. (2020). Field proximal soil sensor fusion for improving high-resolution soil property maps. Soil Syst. 4(3):52. doi:10.3390/soilsystems4030052.

Viscarra Rossel, R. A. and Adamchuk, V. I. (2013). Chapter 6. Proximal soil sensing. In: Precision Agriculture for Sustainability and Environmental Protection Oliver, M., Bishop, T. and Marchant, B. (Eds). 99–118. doi:10.13140/2.1.2799.1847.

White, J., Berg, A. A., Champagne, C., Zhang, Y., Chipanshi, A. and Daneshfar, B. (2020). Improving crop yield forecasts with satellite-based soil moisture estimates: an example for township level canola yield forecasts over the Canadian Prairies. Int. J. Appl. Earth Obs. Geoinf. 89:102092 doi:10.1016/j.jag.2020.102092.

Xu, X., Fan, L., Li, Z., Meng, Y., Feng, H., Yang, H. and Xu, B. (2021). Estimating leaf nitrogen content in corn based on information fusion of multiple-sensor imagery from UAV. Remote Sens. 13(3):340 doi:10.3390/rs13030340.

Yin, X., Jaja, N., McClure, M. A. and Hayes, R. M. (2011a). Comparison of models in assessing relationship of corn yield with plant height measured during early- to mid-season. J. Agric. Sci. 3(3):14–24.

Yin, X., McClure, M. A., Jaja, N., Tyler, D. D. and Hayes, R. M. (2011b). In-season prediction of corn yield using plant height under major production systems. Agron. J. 103(3):923–929.

Yuan, W., Li, J., Bhatta, M., Shi, Y., Baenziger, P. S. and Ge, Y. (2018). Wheat height estimation using LiDAR in comparison to ultrasonic sensor and UAS. Sensors 18:3731. https://doi.org/10.3390/s18113731.

Zhao, Y., Potgieter, A. B., Zhang, M., Wu, B. and Hammer, G. L. (2020). Predicting wheat yield at the field scale by combining high-resolution Sentinel-2 satellite imagery and crop modelling. Remote Sens. 12(6):1024 doi:10.3390/rs12061024.

Chapter 11

Using remote and proximal sensor data in precision agriculture applications

Luciano S. Shiratsuchi and Franciele M. Carneiro, Louisiana State University, USA; Francielle M. Ferreira, São Paulo State University (UNESP), Brazil; Phillip Lanza and Fagner A. Rontani, Louisiana State University, USA; Armando L. Brito Filho, São Paulo State University (UNESP), Brazil; Getúlio F. Seben Junior, State University of Mato Grosso (UNEMAT), Brazil; Ziany N. Brandao, Brazilian Agricultural Research Corporation (EMBRAPA), Brazil; Carlos A. Silva Junior, State University of Mato Grosso (UNEMAT), Brazil; Paulo E. Teodoro, Federal University of Mato Grosso do Sul (UFMS), Brazil; and Syam Dodla, Louisiana State University, USA

1 Introduction

One of the biggest challenges in agriculture is to increase crop yields without expanding production into new areas. It has been suggested that agricultural production needs to double by 2050 to meet increasing demand (Foley et al., 2011; The Royal Society, 2016; Narvaez et al., 2017). Precision agriculture (PA) provides one potential way to improve crop yields on existing agricultural land and help ensure existing land is used more sustainably. The International Society of Precision Agriculture (ISPA) has defined PA as 'a management strategy that takes account of temporal and spatial variability to improve sustainability of agricultural production' (ISPA, 2022). By identifying varying soil and crop characteristics in a field, PA enables more precise (and potentially more sustainable) variable rate application of inputs such as fertilizers and pesticides. This is achieved by applying the most appropriate inputs at the

http://dx.doi.org/10.19103/AS.2022.0107.19

right location, rate, and time and in the right manner, often referred to as site-specific management (SSM). Due to factors such as the heterogeneity of soil and variability in crop response to changing environmental conditions, in-field variation can be significant both spatially and temporally.

Sensors are therefore a key component of PA. They allow patterns of in-field temporal and spatial variability to be identified cost effectively, which then allow more precise and targeted application of inputs (Queiroz et al., 2020). A good example is nitrogen (N) fertilization. Crops vary in their N requirements at differing stages of growth while individual plants and parts of fields may have differing N needs at differing times of the year. N fertilization management is always a challenge since the N cycle in the soil is highly dynamic and complex and varies significantly over short ranges. Management must also take account of multiple loss pathways (e.g. denitrification, volatilization, runoff, and leaching) which result in negative environmental impacts (Marchi, 2021). Variable rate N fertilization, based on measuring differences in plant nutrient status, and then establishing distinct crop management zones requiring different treatments, provides the opportunity to optimize N use efficiency and reduces both economically and environmentally costly losses (Elbl et al., 2021).

This chapter reviews key issues in using sensor data in PA and, in particular, their mode of deployment (proximal or remote). It assesses the relative strengths and weaknesses of proximal sensing techniques, compared with imaging data typically acquired from remote sensing (RS) platforms, before assessing trade-offs in sensor data resolution, as well as sources of error in the way data are processed. The chapter concludes by looking at ways of integrating remote and proximal sensor data, to utilize the beneficial characteristics of each type of data to improve the impact PA in improving efficiency and sustainability.

2 Remote and proximal sensing in agriculture

There are many different sensor principles and technologies that can be employed. However, the method of deployment has a significant impact on what sensors can measure and the value of the data obtained. Sensor deployment can be broadly classified as proximal (or terrestrial) and remote.

2.1 Proximal and terrestrial sensors

Proximal soil sensing (PSS) is a well-developed area of research and is defined as soil sensor measurements that are obtained either in contact with or within a short range (<2 m) of the soil. A key advantage of proximal sensors is their mode of operation, which can be either passive or active, non-contact or invasive. An example of the latter is soil water probes that can provide high temporal resolution and information through depth (i.e. the soil profile) but typically are limited in spatial coverage, i.e. the data obtained are typically representative of only a small volume of soil and are not broadly representative of soil across a field. As well as being deployed in situ, proximal soil sensors can also be deployed on the go, e.g. by mounting sensors on agricultural vehicles to provide better spatial coverage. However, spatial coverage may still be limited by how much of the field a vehicle covers while temporal coverage will be limited by how often a vehicle is deployed.

Crop or plant sensors can also be considered 'proximal' although terms such as 'ground-based' or 'terrestrial' are also used to distinguish these types of sensor from remote sensor systems (aerial or satellite-based - discussed later). Similarly to soil sensors, proximal plant sensor measurements can be active or passive, invasive (where plant samples are collected and processed for sensor measurement), in contact with the plant (e.g. the Soil Plant Analysis Development (SPAD) meter to measure leaf chlorophyll concentrations), or can operate within a short range of a crop (e.g. typically within a range of 2–3 m). Definitions of 'proximal' have been complicated by the development of sensors mounted on unmanned aerial vehicles (UAVs) capable of hovering a few feet above a crop, pointing to an increasing degree of overlap between proximal and remote sensor technologies in areas such as degree of spatial resolution (Ferguson, 2019).

As for PSS, plant or crop sensors can be deployed on the go by mounting sensors on agricultural vehicles such as tractors, with data collected during farming operations such as fertilizer/pesticide application or harvesting, as has shown in monitoring crops such as cotton (Schielack and Thomassen, 2016; Sui et al., 2012). Active canopy sensors (ACS), based on portable or vehicle-mounted active proximal sensors, provide real-time monitoring of crop characteristics such as nutrient (e.g. N) status. Types of active proximal sensors include:

- handheld chlorophyll meters, such as (SPAD) 502 plus (Konika Minolta® Inc., Tokyo, Japan) and LEAF (FT Green LLC®, Wilmington, DE, USA).
- handheld canopy reflectance sensors such as GreenSeeker (Trimble Inc., Sunnyvale, CA, USA), RapidSCAN CS-45 sensor (Holland Scientific® Inc., Lincoln, NE, USA) and Crop Circle Series (Holland Scientific®, Inc.).

ACS-based yield predictions conducted at the stem elongation stage of crop growth have been used for topdressing fertilizer recommendations. Active proximal sensors have now been combined with N fertilization algorithms to enable on-the-go variable N application (Colaço and Bramley, 2018). Active sensors such as GreenSeeker, OptRX, CropSpec, and N sensor can be installed on agriculture machines such as sprayers to monitor and map N status of plants and then target N application to match plant requirements. Combine harvesters equipped with yield monitors and global navigation satellite system sensors (to map position and allow navigation of vehicles) can be used to produce maps of spatial yield variability in a field while harvesting.

For both soil and plant sensing, deploying sensors proximally can provide a wide range of detailed information. It provides an opportunity to deploy a wide potential range of sensor technologies. Proximal crop sensor technologies include contact and in situ sensors (such as chlorophyll meters), ranging sensors (e.g. using acoustic or laser techniques), and electromagnetic (EM) sensors, as well as passive, active, and invasive systems (Ferguson, 2019). There is also a wide range of proximal soil monitoring techniques such as electrical, optical, temperature, electrochemical, radioactive radiation, and mechanical sensors (Gebbers, 2019). Proximal sensor technologies allow the rapid and inexpensive collection of precise, fine-resolution data on soil or crop characteristics (Viscarra Rossel et al., 2011). However, a key limitation is limited spatial information and the ability to obtain both high spatial and temporal coverage simultaneously.

2.2 Remote sensors

RS involves measuring objects from a distance without physical contact with the object being measured (e.g. aerial or orbital sensors on a drone, aircraft, or satellite). Sensors operating remotely can be deployed on aerial platforms (e.g. aircraft, unmanned aircraft systems (UASs) or UAVs (commonly known as drones) or orbital platforms (artificial satellites orbiting the Earth)) (Crepani, 1993; Jensen, 2009; Florenzano, 2011; Molin et al., 2015) (Fig. 1). As noted earlier, the development of UAVs capable of operating a few feet above a crop has blurred the distinction between remote and proximal sensors. To make it possible to map spatial variability on the scale needed for PA (e.g. mapping a

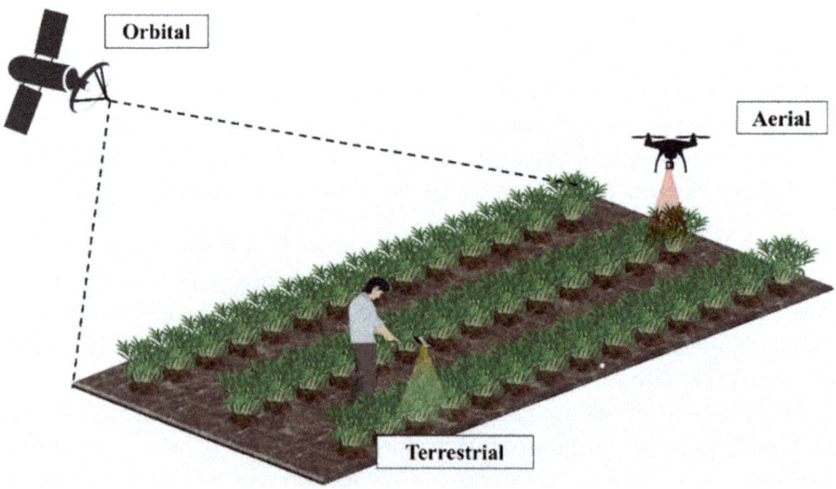

Figure 1 Collection of data using terrestrial, aerial, and orbital platforms.

whole field), UAVs can operate at heights ranging from 30 m to 100 m or more (Zhang and Kovacs, 2019). In identifying in-field variability, PA makes particular use of RS technologies.

The two main types of aerial sensors are aircraft designed to carry sensor systems and UAS or UAVs. They can carry high-definition sensors with high spatial resolutions (potentially in centimeters or even millimeters) (Zheng et al., 2020). They can also carry cameras able to capture different wavelengths (Fig. 2) (Guo et al., 2016; Adao et al., 2017; Yuan et al., 2021). Manned aircraft have advantages such as the ability to carry a large array of heavier sensors with high spatial resolution, to provide data in real-time/near-real-time, to cover large areas (which might otherwise be difficult to access), and to map over larger scales (Yang, 2019).

With their low cost, ease of operation, and ability to fly at low altitudes, UAS have become a popular RS platform, complementing ground-based and more traditional RS platforms such as aircraft and satellites (Hunt and Daughtry, 2018; Yang, 2019). UAS-mounted sensors are capable of ultra-high spatial imagery resolution (in cm or less), making them well-suited to field scouting (Zhang and Kovacs, 2019). Speed and quality in the processing of drone imagery are developing rapidly.

A key limit to commercial UAS usage is battery life, but other constraints include the weight of instrumentation drones can carry, the distance drones can fly away from the operator (e.g. in terms of maintaining contact for control), altitude limits (typically 120 m/400 ft), and regulations restricting use (e.g. proximity to potential hazards).

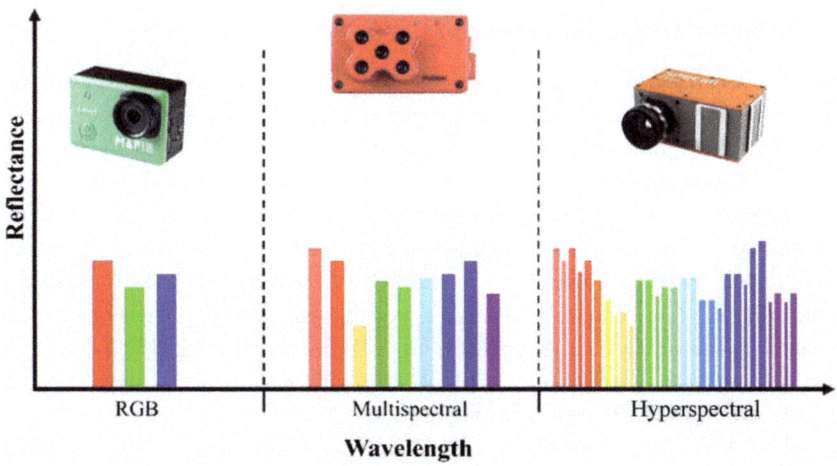

Figure 2 Different spectral resolutions as RGB (visible rays), multispectral, and hyperspectral.

The use of orbiting satellites to gather data about the Earth itself had military origins but has now been applied in many other areas such as weather and environmental monitoring (Ge et al., 2020; Singh et al., 2020), monitoring patterns of urban development (Huang et al., 2020), and agricultural management (Tedesco et al., 2021). A key advantage is the ability to image very large areas. Agricultural applications of satellite imaging have been limited by problems such as relatively coarse spatial resolution, interruptions in coverage (depending on the speed of orbit), cloud cover, and delays in communicating data back to Earth (Mulla, 2013), though recent developments in high-resolution satellite sensors have significantly narrowed the gap in spatial resolution between traditional satellite and airborne imagery (Yang, 2019). Satellites can now carry a range of high-quality sensors able to capture images in hundreds of spectral bands and spatial resolutions that vary from kilometers to centimeters (Zanotta et al., 2019). The development of sensor technology has been accompanied by advances in mathematical modeling to analyze this data rapidly in identifying crop characteristics and the in-field variability required for PA applications (Dash et al., 2010; Santos et al., 2021).

When comparing the unique characteristics of proximal and RS, key differences are the mode of operation (e.g. active or passive), the spatial and temporal resolution of information obtained, and also the specificity of this information, i.e. does it capture variation in crop response due to some factor or does it specifically measure and quantify this factor (e.g. N status). Generally, RS can be considered most suited to addressing variability within a field, to apply treatments more precisely across the field, while proximal sensing can better attribute the cause of this variability, allowing the selection of correct treatments and rates.

3 Active and passive sensors

Most sensors used in agricultural operations are imaging sensors which analyze objects by measuring EM radiation (EMR) generated from the sun and reflected or emitted from targets on the Earth's surface (Mobley, 1994; Moreira, 2011). EMR is composed of an EM spectrum split into bands called wavelengths, each of which has its own frequency (Lorenzzetti, 2015). While some waves (gamma rays to microwave rays) are blocked by the atmosphere, others (visible light to microwave rays) reach the Earth's surface (Novo, 2008). The EMR bands most used in agriculture research are visible waves, near-infrared (NIR) waves, mid-infrared waves, and microwaves (Moreira, 2011; Molin et al., 2015). When EMR is in contact with a target (e.g. a plant), it is either absorbed, transmitted, or reflected. RS sensors typically collect reflectance data.

Plant leaf reflectance over the visible and NIR range contains useful information about crop growth and health (Fig. 3) (Jensen, 1996; Thenkabail et al., 2000; Hansen and Schjoerring, 2003; Brandão, 2009; Shiratsuchi, 2014).

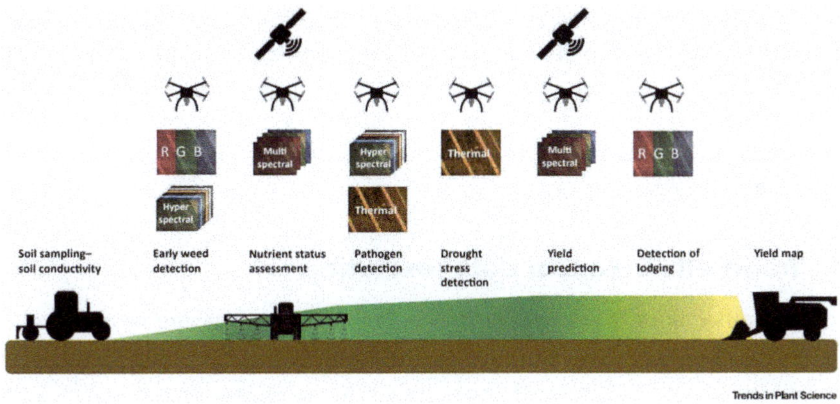

Figure 3 Application of remote sensing in agriculture.

Within the visible range, leaf reflectance is lower due to light absorption by leaf pigments involved in photosynthesis, in particular chlorophylls which absorb light in the blue and red region of the spectrum (Lichtenthaler and Buschmann, 2001; Gitelson et al., 2006). In the NIR region, however, these leaf pigments do not absorb light and leaf reflectance is many times higher in magnitude (Kumar and Silva, 1973). Given this pattern of light scattering, a sharp increase between red and NIR regions can be observed in the reflectance of green parts of plants, named the red edge. Red edge reflectance has been correlated with many different crop parameters.

Using sensors to measure differences in reflectance can be used to identify spatio-temporal variation within a field which may be linked to phenomena such as differences in chlorophyll content. These, in turn, may reflect crop nutritional status or availability of water, corresponding to differences in crop growth or biomass, or may indicate the presence of pests or diseases. Active sensors measuring light transmittance by red (650 nm) and NIR (940 nm) spectral bands correlate strongly with leaf chlorophyll content (N is a component of chlorophyll) and are used in targeted, variable rate N application, e.g. in applying fertilizer in response to plant deficiency during the vegetative stage (Solari et al., 2008).

One use of such reflectance data is vegetation indices (VIs) (Thenkabail et al., 2000; Hansen and Schjoerring, 2003). A VI is a single value based on comparing observations from multiple spectral bands. An example is the Simple Ratio (SR), which is the ratio between the reflectance recorded in the NIR and red bands: by distinguishing leaf greenness, it can be used to assess plant health or biomass. VIs frequently used to analyze crop characteristics include SR, Normalized Difference Vegetation Index (NDVI), Normalized Difference Red Edge (NDRE), Red Green Blue Vegetation Index, Soil Adjusted Vegetation Index, Infrared Percentage Vegetation Index, and Enhanced Vegetation Index (EVI and EVI2). The NDVI is a spectral index that relates NIR with red wavelength reflectance in a normalized form and has been shown to correlate well with plant biomass. In contrast to the NDVI, other indices relate to narrow wavelength ranges. One such index is the NDRE with the wavelengths 790 nm and 720 nm, which shows a strong correlation with N content in plants. VIs can be used to interpret sensor data in assessing plant growth and health.

4 Trade-offs in sensor data resolution

Spectral reflectance is one dimension of resolution, a measure of the level of detail captured in an image. There are four basic types of resolution:

- Spatial resolution: images have matrix structures composed of pixels: their smallest elements (Schrader and Pouncey, 1997). The spatial resolution of an image refers to the size that the pixel represents in the image. The smaller the pixel size, the higher the spatial resolution and the more detailed the image.
- Spectral resolution: refers to the number of bands (range of the EM spectrum) that a sensor can capture.
- Radiometric resolution: the level of gray that each pixel has, ranging from black to white. The gray level represents the intensity of the emitted/reflected EM energy measured by the sensor, corresponding to the pixel size.
- Temporal resolution: corresponding to when a satellite is able to return to the same area, i.e. the time required to collect a new image at the same point on the Earth's surface.

These are discussed in more detail later.

4.1 Spatial resolution

The spatial resolution of an image refers to the ground dimension represented by a pixel (representing a square area within an image, typically measured in square meters) (Fig. 4). The smaller the pixel size (and thus the area covered per pixel), the higher the image quality and resolution (Shiratsuchi, 2014; Paranhos Filho, 2021; Bullock, 2021). For example, the spatial resolution of wavelength bands 2, 3, 4 (visible), and 8 (infrared) of the Sentinel-2 satellite is 10 m, while bands 1, 9, and 10 (infrared) have a resolution of 60 m. Bands 5, 6, and 7 have a resolution of 20 m (Fletcher, 2012). Figure 5 shows different bands on the Sentinel-2 satellite for spatial resolution comparison. With the right-hand image shown in Fig. 5, we can observe a greater level of detail due to the smaller pixel size, including variations between and within fields.

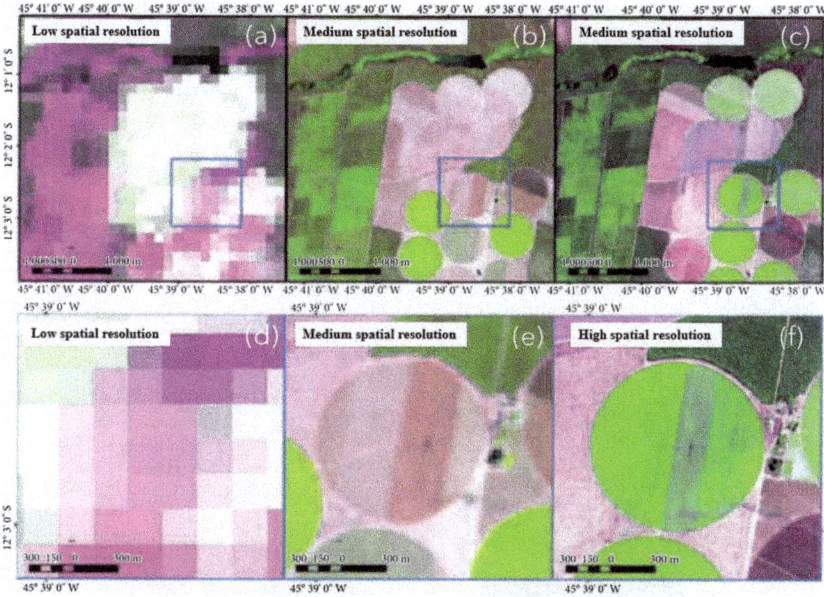

Figure 4 Comparison of spatial resolutions. Source: Shiratsuchi (2014).

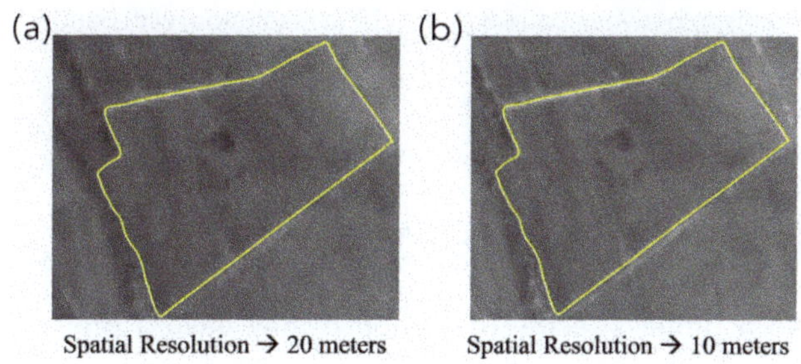

Figure 5 Comparison of Sentinel-2 bands resolutions between the red edge (a) and the near-infrared (b) waves.

4.2 Spectral resolution

Spectral resolution measures the sensitivity of a sensor in distinguishing differences in EM spectra (Paranhos Filho, 2021). For example, a sensor working in the range of 0.4–0.5 μm (micrometers) has a higher resolution compared to a sensor working in the range of 0.4–0.6 μm, meaning the first sensor will be able to register smaller variations in the EM spectrum (Novo, 2010). Figure 6 shows sensors with different spectral resolutions.

Ideally, remote imaging sensors need to capture images with multiple bands. Spectral resolution comprises three measurement parameters (Meneses and Almeida, 2012):

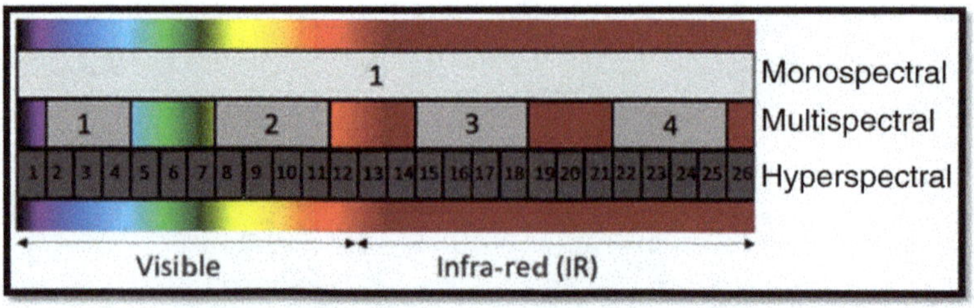

Figure 6 Different spectral resolutions. Source: Lira (2016).

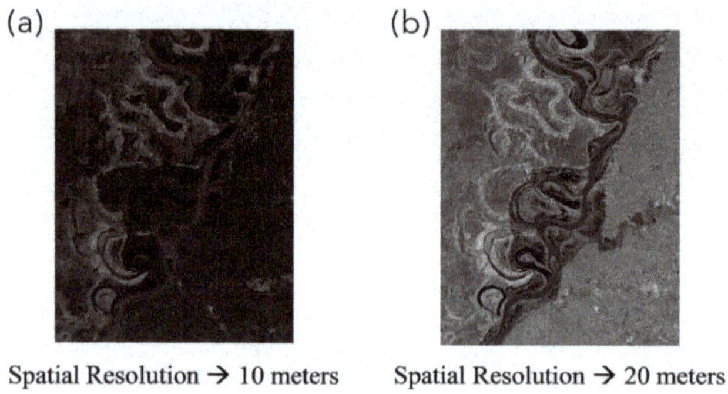

Figure 7 Comparison between band 3 green (10 m resolution) with band 6 red edge with (20 m resolution), Sentinel-2.

- Number of bands;
- The wavelength width of bands; and
- The positions of bands in the EM spectrum.

Sensors have a better spectral resolution if they have a greater number of bands located in different spectral regions and with a narrow wavelength. This greater number of spectral bands is necessary to differentiate a greater number of targets on the ground surface, e.g. targets that have a reflectance at a certain wavelength with a small difference from their surroundings.

Figure 7 shows the effect of spectral resolution in detecting different objects in a forest area. Even with a higher spatial resolution, Fig. 7a does not show the river despite its size. This is because the reflectance of water and vegetation is practically the same at this wavelength, and thus, there is no contrast between the targets (Meneses and Almeida, 2012). Figure 7b uses another band where the river can be easily identified: even with low reflectance the water can be detected and differentiated from the rest of the forest, establishing a contrast ratio between the targets observed.

Hyperspectral sensors have particularly high resolutions with bands measured in nanometers. An example is the Airborne Visible/Infrared

Imaging Spectrometer developed by the National Aeronautics and Space Administration, which is capable of acquiring continuous or practically continuous spectra along the reflected solar spectrum (0.4 μm to 2.5 μm) reaching 224 bands (Carvalho Júnior et al., 2003). Another example is the Moderate Resolution Imaging Spectroradiometer sensor that has 36 bands in the spectral ranges of near and visible infrared (400 nm to 2500 nm) and thermal infrared (8000 nm to 15 000 nm). In 2000, the Earth Observing-1 satellite was launched using the Hyperion sensor with 220 bands (Pizarro et al., 2001).

4.3 Radiometric resolution

Radiometric resolution refers to the number of levels of gray (ranging from black to white) the pixel can use to represent the information acquired by the sensor. The gray level represents the intensity of EM energy reflected or emitted by the sensor for the observed terrestrial surface area. The number of gray levels is measured in bits (Lira, 2016). The higher the gray scale, the greater the radiometric resolution and the higher the sensor's ability to differentiate small variations in the intensity of the reflected signal (Barbosa et al., 2019; Jensen, 2009). For example, the Landsat-5 satellite carries the Thematic Mapper sensor with an 8-bit radiometric resolution that corresponded to 256 gray levels (28 = 256 gray levels), while the Landsat-8 satellite equipped with the Operational Land Imager (OLI) sensor has a 12-bit resolution (128 = 4096 gray levels) (Barbosa et al., 2019). Figure 8 shows that an 8-bit image (28 = 256 gray levels) has better visual details compared to the lower radiometric resolution images of 6, 4, and 2 bits. The 2-bit image has only 4 levels of gray.

4.4 Temporal resolution

Sensors mounted on fixed structures can capture video images every few seconds. In RS, it takes time for a satellite to complete an orbit around the Earth (Lira, 2016) (Fig. 9). Satellite temporal resolution indicates the frequency with which the satellite revisits images in a particular area of interest; it is also known as the periodicity or repetitiveness with which the satellite records images

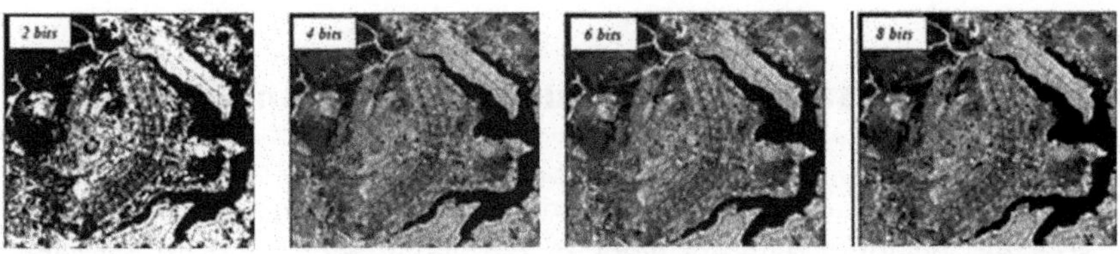

Figure 8 Band radiometric resolution in bits. Source: Meneses and Almeida (2012).

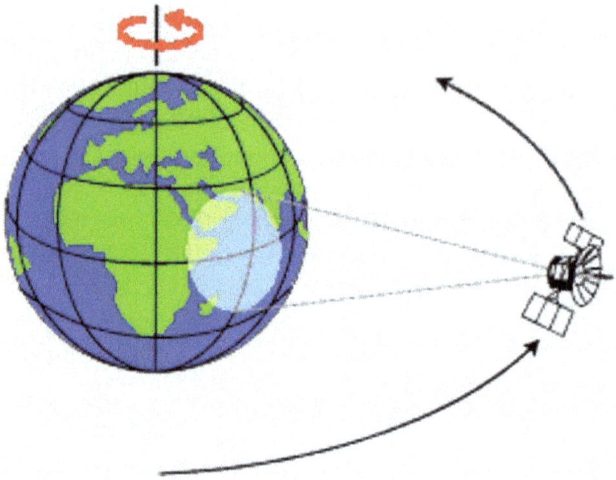

Figure 9 Path of a satellite around the Earth.

(CCRS, 2019). The CBERS 4 satellite, e.g. has a temporal resolution of 26 days which limits its value for agricultural applications. In contrast, the Landsat-8 satellite has a temporal resolution of 16 days (USGS, 2018). The Sentinel series of satellites (some of which consist of two polar-orbiting satellites) has a temporal resolution of 10 days or 5 days where two satellites are paired (Van Der Meer et al., 2014; ESA, 2012). The Suomi-NPP platform provides daily imaging due to its wide imaging range (Martins, 2019).

4.5 Trade-offs in sensor data resolution

All sensors involve some degree of compromise between spatial and temporal resolution, and specificity. It is challenging in RS, e.g. to combine many spectral bands, high spectral resolution, and high spatial resolution, given the amount of data that needs to be collected. Normally high spatial resolution sensors have limited visible and NIR bands in order to reduce the large volume of data transmitted back to Earth. Some sensors with high spatial resolution and low spectral resolution (Meneses and Almeida, 2012). A sensor like the ASTER has a low spatial resolution that is compensated by a greater number of bands (14 in total).

5 Processing sensor data: sources of error and their resolution

Data to support PA and other applications is derived from a growing number of sources, including different types of terrestrial, aerial, and orbital sensor combinations. If correctly interpreted, this data can be used to better understand variations in crop response to fertilizer application (Basso et al., 2016) and to develop models to better target crop fertilization (Alesso et al., 2019). Sensors can also be used to detect patterns of crop biotic and abiotic stress (Bhandari et al., 2018) or to predict potential crop yields (Elavarasan et al., 2018; Peerlinck et al., 2018; El-Hendawy et al., 2019). However, this technology has several inherent sources of error which can cause incorrect spatial application rates. Sources of error include the accuracy of application maps, as well as errors in navigation and variable rate application (Chan et al., 2004).

The increasing volume and variety of data from a range of sensors and sensor platforms also bring with it challenges such as corrupting or 'noisy' data (data that contain a large amount of meaningless information) (Fan et al., 2014). Noisy data amass in proportion to the volume and variety of data collected. Working with spatial data presents additional challenges. A particular problem is spatial autocorrelation which describes the degree to which observations at different spatial locations resemble one another. Models based on data rely on observations being independent of one another. Spatial autocorrelation related to the geographic proximity of observations can undermine model reliability, e.g. through underestimation of degrees of model error. Additional data processing is required to account for how the location of the sample point influences its value and to avoid spatial autocorrelation effects when using spatial data in modeling (Ruß and Brenning, 2010). Effective methods to process 'big data' in this and other ways are critical to ensuring it can be translated into accurate and meaningful information for decision-making (McMillen, 2010; Chi et al., 2016; Kamilaris et al., 2017).

5.1 Sensor calibration

Calibration is the process of determining the relationship between the sensor signal and the characteristic(s) being measured in a given target. Calibration methods include empirical and mechanistic techniques (Weiss et al., 2020). A common empirical technique to calibrate sensors measuring VIs, such as NDVI, is the use of an N-rich strip area. This area has an adequate or slightly excess supply of a given input and is then used to generate a sensor signal value corresponding to 100% sufficiency for the variable being evaluated. Subsequent measurements taken of areas having different, and possibly insufficient, input amounts are then compared to the value of the reference area to calculate the sufficiency index of each sample area. This technique has been used to calibrate sensors used to measure N sufficiency in various crops using both active and passive sensors (Blackmer and Schepers, 1995; Holland and Schepers, 2013).

Calibration is particularly important in using passive sensors to create N variable-rate prescription maps. Small variations in sensor N readings can be masked by factors such as time of day, soil moisture, crop physiology, water stress, cloud coverage, shadow, etc. Figure 10 shows how UAS-collected reflectance data can be skewed compared to data from an ACS (Teixeira et al., 2020). If a passive sensor is used, it should be calibrated against data from active sensing technologies (Fig. 11). The use of active sensors in calibration is important, e.g. in estimating leaf area index (Viña et al., 2011) (Fig. 12).

5.2 Data filtering

Effective data filtering can reduce erroneous values in output datasets. Filtering techniques are generally performed to smooth an overall dataset by removing extreme and spatially nonstandard values (Blackmore, 1999). Common statistical data filtering methods can be used, such as removing values falling outside of a given range. This can, however, result in the removal of valid observations, so other methods of handling outlier values should be considered, such as Winsorization or a robust estimation method (Kwak and Kim, 2017).

It is important to account for the spatial nature of the data when considering filtering techniques. This is largely due to Tobler's first law of geography, which states, 'everything is related to everything else, but near things are more related than distant things.' One popular technique is to use

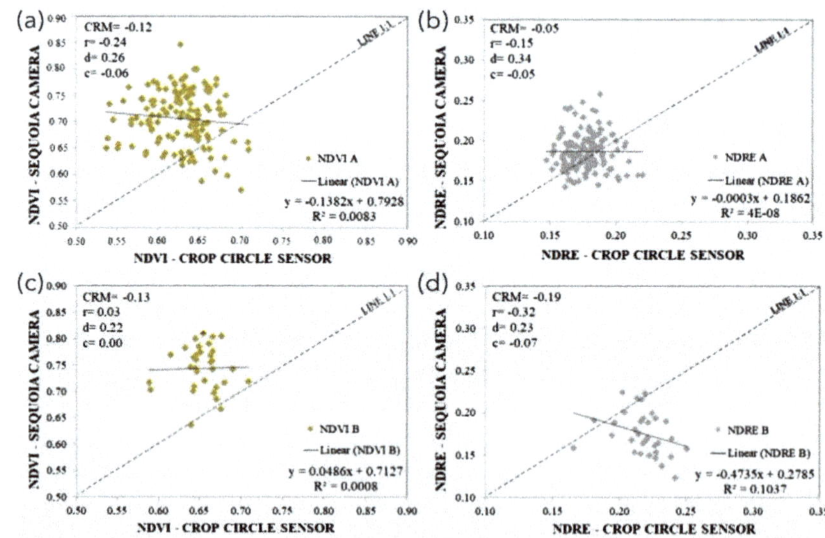

Figure 10 Comparison of active and passive sensors. Source: Adapted from Teixeira et al. (2020).

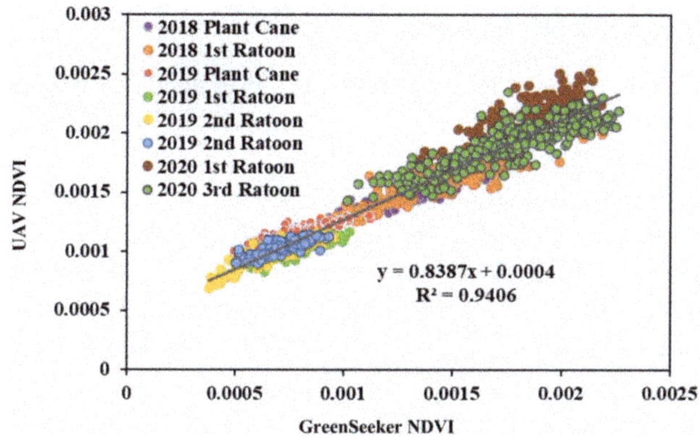

Figure 11 Correlation between active canopy sensors and unmanned aircraft systems. Source: Adapted from Forestieri (2021).

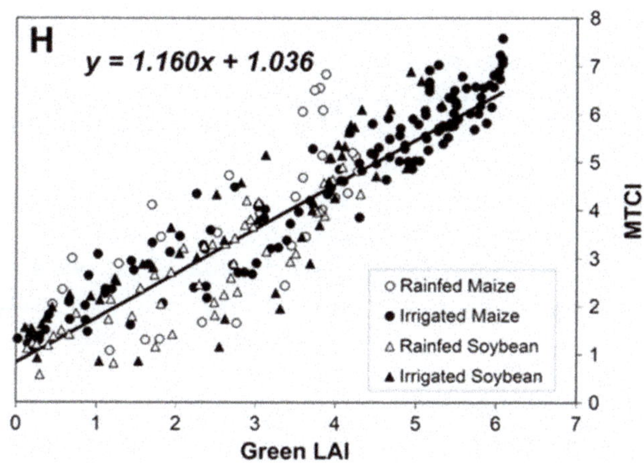

Figure 12 Leaf area index correlation with different crop systems and phenological stages. Source: Adapted from Viña et al. (2011).

a moving window-type calculation of the coefficient of variation (CV). Points falling within a given radius of the data point are used to calculate CV, which is then stored. This is done for every data point in the data set. Data points having a CV exceeding a given threshold are then discarded. This has the effect of smoothing the data by eliminating areas of abnormally high and low values (Spekken, 2013). Another technique is to remove data points occurring within a given distance of the field border since field borders are often subject to nonstandard growing conditions. In developing yield maps, semi-automated data filtering software has also been developed to facilitate the removal of erroneous data points, such as the USDA's Yield Editor software (Sudduth and Drummond, 2007).

5.3 Data analysis and evaluation

There have been significant developments in artificial intelligence and its use to analyze very large and complex datasets. Ruan et al. (2022) have used machine learning (ML) algorithms to analyze proximal sensing data from field experiments from multiple locations and years, combined with publicly available weather data, in order to predict winter wheat yields. ML algorithms have proved very promising for analyzing large volumes of sensor data. Some examples of ML algorithms are K-Nearest Neighbor (KNN), Multiple Linear Regression, Artificial Neural Networks – Multilayer Perceptron, and Random Forest (Morelli-Ferreira et al., 2021).

Researchers such as Fue et al. (2018), Xu et al. (2018), and Yeom et al. (2018) have used deep learning to identify regions of interest in RS images of cotton fields. Xu et al. (2018) developed a convolutional neural network to detect and count the number of newly opened cotton flowers in aerial photos as well as an algorithm to identify and calculate bolls in order to predict cotton yield. Tedesco-Oliveira et al. (2020) have used similar methods to rapidly estimate cotton yield via the identification and counting of bolls obtained from smartphone cameras taking pictures in fields. Morelli-Ferreira et al. (2021) have evaluated Sentinel satellite images for cotton yield prediction using eight VIs and four ML algorithms and concluded that the SR with the KNN algorithm improved the prediction of the timing of harvest and yield. Based on data from proximal active crop canopy sensors, Marchi (2021) used two VIs to identify variations in N requirements in cotton plants. They developed an algorithm for variable rate top dressing of cotton plants with N to optimize N use efficiency (NUE) and minimize N losses to the environment.

Habyarimana and Baloch (2021) have used Bayesian-based ML algorithms to correlate different VIs: fraction of absorbed photosynthetically active radiation (fAPAR) derived from Sentinel-2, chlorophyll content, and NDVI derived from handheld optical reflectance sensors. These techniques and data were used to develop and compare models to more accurately predict sorghum biomass yields. Modeling based on optical reflectance meter data measuring chlorophyll concentration index in the red edge at 700 nm was found to outperform NDVI meter-based models and was comparable to a model based on Sentinel-2-derived fAPAR.

6 Integrating remote and proximal sensor data for precision agriculture

Given issues such as the inherent complexity of the crop systems being studied, the differing strengths and weaknesses of different sensor systems, trade-offs in data resolution and quality, as well as potential sources of error from any data source, it makes sense to combine data from remote and proximal sensors. Multi-sensor fusion thus seeks to offset potential omissions and errors from one type of sensor data as well as minimize the uncertainty resulting from any estimated variable used (Barbedo, 2022). Figure 13 shows the integration of remote and proximal sensor data into a precision crop management system (Schellberg et al., 2008; Atzberger, 2013). An example is 'ground truthing.' The complexity of natural surfaces, the effects of atmosphere and the complex of relationship of spectral signatures to crop characteristics can limit the reliability of RS data which then needs to be validated by comparison with data from aerial or ground-based sensors (Steven, 2003). As an example, Hegarty-Craver et al. (2020) used images from UAVs to create a ground-truthing dataset which was then used to train a model capable of analyzing data from the Sentinel-1 and Sentinel-2 satellites in more accurately mapping maize crops. Techniques to combine data (data fusion) from the ground, aerial, and satellite sensors to identify site-specific variables such as N availability are discussed by Ahmad et al. (2022). The following discusses some examples of data fusion initiatives.

Pantazi et al. (2015) have combined data from satellite images, combine-harvest-mounted yield sensors and mobile proximal NIR soil sensors, using self-organizing clustering statistical techniques to create more accurate management zones for SSM. Benedetto et al. (2013) have used proximal EMI sensors, vegetative indices, and radiance data derived from satellite images, processed

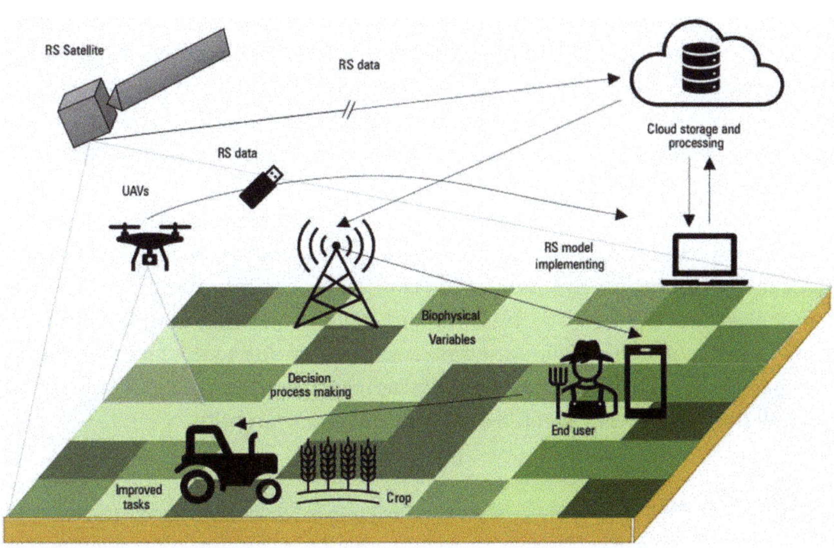

Figure 13 Collection of remote and proximal sensors data in agriculture.

using multivariate geostatistics and clustering, to improve site-specific irrigation. More accurate yield maps have been developed using combinations of remote and proximal sensing data, crop modeling, and ML (Arab et al., 2021; Canata et al., 2021; Khanal et al., 2021) and by developing machine-vision-based yield monitor technologies (Jacques et al., 2021). Drones and other platforms carrying a range of sensors (hyperspectral, multispectral, and red-green-blue) have been used in areas such as weed control (Roslim et al., 2021). Tefera et al. (2022) used a combination of ground-based active sensors (Crop Circle) and aerial-based passive sensors (carried by UAS) to establish an optimal time for field pea establishment to outcompete the emergence of weeds in the field.

Dado et al. (2020) have compiled a ground-truth dataset of high-resolution (5 m) data from yield maps (derived from combine harvester data) from the US Midwest over the period 2008–18, representing over one million observations. This dataset was used to test the reliability of models derived from Landsat and Sentinel-2 sensor data and suggest ways these might be refined to better predict soybean yields. In better exploiting existing datasets both to develop and validate models, Woodard (2018) has developed the AgAnalytics platform for researchers to share publicly available data more efficiently by improving the ways it is documented, disseminated, and processed.

Maestrini and Basso (2018) analyzed yield monitor data (involving 1625 yield maps) collected by proximal combine-mounted sensors for 338 fields growing maize, soybean, cotton, and wheat across 8 states in the US Midwest. These data were used to identify patterns of in-field spatial and temporal variation. These data were correlated with publicly available datasets relating to soil type, topography, and rainfall patterns to explain yield fluctuations in the same locations in fields (characterized as unstable) which the authors were able to relate to the interaction of topography and rainfall. A Landsat-derived NDVI (red band at 0.3 m resolution and temperature at 2 m resolution) was found to be helpful in the timing of N application in these unstable field zones to improve NUE and yields.

Current data fusion techniques are reviewed by Barbedo (2022). While they have improved accuracy in some areas (e.g. improving image identification), applications to more complex problems such as improving the accuracy of management zones have been mixed. This is, in part, due to continuing challenges in correlating different data sources and due to the fact that, though large amounts of data are being generated, they are often still not sufficient to cover all the variables in complex agricultural environments.

7 Conclusion

The use of data from sensors has been critical in developing effective PA applications. Research suggests that site-specific variable rate N application, e.g. can increase NUE and reduce N leaching losses without compromising yield, with some research suggesting potential savings of 2.25–4.5 kg N/hectare (Long, 2019; Pedersen et al., 2019). A recent study by the US Association of Machinery Manufacturers in alliance with the American Soybean Association, the National Corn Growers Association, and CropLife America suggested that PA techniques were responsible for a 7% improvement in fertilizer placement efficiency, a 9% reduction in herbicide, and 4% reduction in water use with the potential for far greater improvements (https://www.aem.org/news/the-environmental-benefits-of-precision-agriculture-quantified).

However, the process of collecting and processing data to deliver actionable outcomes remains complex, with a significant potential for error in the interpretation of sensor data to achieve more sustainable crop management. Understanding and avoiding such errors requires an understanding of the key elements in this process, the limitation of sensor information, and what different sensors are able to detect or measure, as well as parameters such as spatial, spectral, and temporal resolution that meet the spatial and temporal needs of management, i.e. match the spatial and temporal variation of the target property. It includes an understanding of different types of sensor as well as the different platforms on which sensors can be mounted (terrestrial/proximal, aerial, and orbital), their relative strengths and weaknesses, and the ways they can complement each other. In general terms, RS can be seen to be best suited to identifying where variability occurs and where and when an intervention might be targeted, while proximal sensing can provide more precise information on why variability has occurred and what treatment and rate of application might be needed. When combined, the two can achieve the goal of PA in applying the right product or treatment at the right location, rate, and time.

8 References

Adao, T., Hruška, J., Pádua, L., Bessa, J., Peres, E., Morais, R. and Sousa, J. (2017). Hyperspectral imaging: A review on UAV-based sensors, data processing and applications for agriculture and forestry. *Remote Sens.* 9(11), 1110. doi: 10.3390/rs9111110.

Ahmad, U., Nasirahmadi, A., Hensel, O. and Marino, S. (2022). Technology and data fusion models to enhance site-specific crop monitoring. *Agronomy* 12(3), 555.

Alesso, C. A., Cipriotti, P. A., Bollero, G. A. and Martin, N. F. (2019). Experimental designs and estimation methods for on-farm research: A simulation study of corn yields at field scale. *Agron. J.* 111(6), 2724–2735. doi: 10.2134/agronj2019.03.0142.

Arab, S. T., Noguchi, R., Matsushita, S. and Ahamed, T. (2021). Prediction of grape yields from time-series vegetation indices using satellite remote sensing and a machine-learning approach. *Remote Sens. Appl. Soc. Environ.* 22, 100485. doi: 10.1016/j.rsase.2021.100485.

Atzberger, C. (2013). Advances in remote sensing of agriculture: Context description, existing operational monitoring systems and major information needs. *Remote Sens.* 5(2), 949–981. doi: 10.3390/rs5020949.

Barbedo, J. G. A. (2022). Data fusion in agriculture: Resolving ambiguities and closing data gaps. *Sensors (Basel)* 22(6), 2285.

Barbosa, C. C. F., Novo, E. M. L. M. and Martins, V. S. (2019). *Introdução ao sensoriamento remoto de sitemas aquáticos: Princípios e aplicações.* São José dos Campos, Brazil: INPE. Available at: http://www.dpi.inpe.br/labisa/livro/res/conteudo.pdf (accessed 28 February 2022).

Basso, B., Fiorentino, C., Cammarano, D. and Schulthess, U. (2016). Variable-rate nitrogen fertilizer response in wheat using remote sensing. *Precis. Agric.* 17(2), 168–182. doi: 10.1007/s11119-015-9414-9.

Benedetto, D., Castrignano, A., Diacono, M., Rinaldi, M., Ruggieri, S. and Tamborrino, R. (2013). Field partition by proximal and remote sensing data fusion. *Biosyst. Eng.* 114(4), 372–383. doi: 10.1016/j.biosystemseng.2012.12.001.

Bhandari, S., Raheja, A., Chaichi, M. R., Green, R. L., Do, D., Pham, F. H., Ansari, M., Wolf, J. G., Sherman, T. M. and Espinas, A. (2018). Lessons learned from UAV-based remote sensing for precision agriculture. In: *International conference on unmanned aircraft systems (ICUAS)*, USA.

Blackmer, T. M. and Schepers, J. S. (1995). Use of a chlorophyll meter to monitor nitrogen status and schedule fertigation for corn. *J. Prod. Agric.* 8(1), 56–60. doi: 10.2134/jpa1995.0056.

Blackmore, S. (1999). Remedial correction of yield map data. *Precis. Agric.* 1(1), 53–66.

Brandão, Z. N. (2009). *Estimativa da produtividade e estado nutricional da cultura do algodão irrigado via técnicas de sensoriamento remoto. 152 f. Tese* (PhD student in Natural Resources), Universidade Federal de Campina Grande, Campina Grande, Brazil.

Bullock, D. S. (2021). Using satellite imagery to estimate optimal site-specific N side-dressing strategies [abstract]. In: *SSSA International Annual Meeting*, CSSA, ASA, Salt Lake City, UT. Available at: https://scisoc.confex.com/scisoc/2021am/meetingapp.cgi/Paper/132907.

Canata, T. F., Wei, M. C. F., Maldaner, L. F. and Molin, J. P. (2021). Sugarcane yield mapping using high-resolution imagery data and machine learning technique. *Remote Sens.* 13(2), 232. doi: 10.3390/rs13020232.

Carvalho Júnior, O. A., Guimarães, P. H., Lopes, R. A. S., Guimarães, R. F., Martins, É. D. S. and Pedreño, J. N. (2003). *Estratificação dos Ambientes de Mistura em Imagens Hiperespectrais.* Simpósio Brasileiro de Sensoriamento Remoto, Brazil, pp. 1045–1051.

CCRS: Canada Centre for Remote Sensing. (2019). Fundamentals of remote sensing. Government of Canada. Available at: https://www.nrcan.gc.ca/sites/www.nrcan.gc.ca/files/earthsciences/pdf/resource/tutor/fundam/pdf/fundamentals_e.pdf (accessed 19 February 2022).

Chan, C. W., Schueller, J. K., Miller, W. M., Whitney, J. D. and Cornell, J. A. (2004). Error sources affecting variable rate application of nitrogen fertilizer. *Precis. Agric.* 5(6), 601–616.

Chi, M., Plaza, A., Benediktsson, J. A., Sun, Z., Shen, J. and Zhu, Y. (2016). Big data for remote sensing: Challenges and opportunities. *Proc. IEEE* 104(11), 2207–2219. doi: 10.1109/JPROC.2016.2598228.

Colaço, A. F. and Bramley, R. G. V. (2018). Do crop sensors promote improved nitrogen management in grain crops? *Field Crops Res.* 218, 126–140. doi: 10.1016/j.fcr.2018.01.007.

Crepani, E. (1993). *Princípios Básicos de Sensoriamento Remote*. CNPQ/INPE, São José dos Campos, Brazil, p. 45.

Dado, W. T., Deines, J. M., Patel, R., Liang, S. and Lobell, D. B. (2020). High-resolution soybean yield mapping across the US midwest using subfield harvester data. *Remote Sens.* 12(21), 3471. doi: 10.3390/rs12213471.

Dash, J., Jeganathan, C. and Atkinson, P. M. (2010). The use of MERIS terrestrial chlorophyll index to study spatio-temporal variation in vegetation phenology over India. *Remote Sens. Environ.* 114(7), 1388–1402. doi: 10.1016/j.rse.2010.01.021.

Elavarasan, D., Vincent, D. R., Sharma, V., Zomaya, A. Y. and Srinivasan, K. (2018). Forecasting yield by integrating agrarian factors and machine learning models: A survey. *Comput. Electron. Agric.* 155, 257–282. doi: 10.1016/j.compag.2018.10.024.

Elbl, J., Mezera, J., Kintl, A., Širůček, P. and Lukas, V. (2021). Comparisons of uniform and variable rate nitrogen fertilizer applications in real conditions–Evaluation of potential impact on the yield of wheat available for use in animal feed. *Acta Univ. Agric. Silvic Mendelianae Brun* 69(1), 33–43.

El-Hendawy, S., Al-Suhaibani, N., Elsayed, S., Refay, Y., Alotaibi, M., Dewir, Y. H., Hassan, W. and Schmidhalter, U. (2019). Combining biophysical parameters, spectral indices and multivariate hyperspectral models for estimating yield and water productivity of spring wheat across different agronomic practices. *PLoS ONE* 14(3), e0212294. doi: 10.1371/journal.pone.0212294.

ESA: European Space Agency. (2012). Sentinel-2: ESA's optical high-resolution mission for GMES operational services. European Space Agency. Available at: https://sentinel.esa.int/documents/247904/349490/s2_sp-1322_2.pdf (accessed 12 February 2022).

Fan, J., Han, F. and Liu, H. (2014). Challenges of Big Data analysis. *Natl. Sci. Rev.* 1(2), 293–314. doi: 10.1093/nsr/nwt032.

Ferguson, R. B. (2019). Proximal crop sensing. In: Stafford, J. (Ed.), *Precision Agriculture for Sustainability*. Burleigh Dodds Science Publishing, Cambridge, UK, pp. 3–28.

Fletcher, K. (2012). Sentinel-2: ESA's optical high-resolution mission for GMES operational services. *Esa SP-1322*. Available at: https://sentinel.esa.int/documents/247904/349490/s2_sp-1322_2.pdf (accessed 2 March 2012; accessed 15 February 2022).

Florenzano, T. G. (2011). *Iniciação em Sensoriamento Remoto* (3rd edn.). ampl. e atual, Oficina de Textos, Sao Paulo, Brazil, p. 128.

Foley, J. A., Ramankutty, N., Brauman, K. A., Cassidy, E. S., Gerber, J. S., Johnston, M., Mueller, N. D., O'Connell, C., Ray, D. K., West, P. C., Balzer, C., Bennett, E. M., Carpenter, S. R., Hill, J., Monfreda, C., Polasky, S., Rockström, J., Sheehan, J., Siebert, S., Tilman, D. and Zaks, D. P. (2011). Solutions for a cultivated planet. *Nature* 478(7369), 337–342. doi: 10.1038/nature10452.

Forestieri, Daniel Erneso, *Integrated Nutrient Management and Cover Cropping Practices in Louisiana Sugarcane Production Systems* (2021). LSU Doctoral Dissertations. 5665.

Fue, K. G., Porter, W. M. and Rains, G. C. (2018). *Deep Learning-Based Real-Time GPU-Accelerated Tracking and Counting of Cotton Bolls Under Field Conditions Using a Moving Camera.* American Society of Agricultural and Biological Engineers. doi: 10.13031/aim.201800831.

Ge, X., Schaap, M., Kranenburg, R., Segers, A., Reinds, G. J., Kros, H. and de Vries, W. (2020). Modeling atmospheric ammonia using agricultural emissions with improved spatial variability and temporal dynamics. *Atmos. Chem. Phys.* 20(24), 16055–16087. doi: 10.5194/acp-20-16055-2020.

Gebbers, R. (2019). Proximal soil surveying and monitoring techniques. In: Stafford, J. (Ed.), *Precision Agriculture for Sustainability*. Burleigh Dodds Science Publishing, Cambridge, UK, pp. 29–78.

Gitelson, A. A., Viña, A., Verma, S. B., Rundquist, D. C., Arkebauer, T. J., Keydan, G., Leavitt, B., Ciganda, V., Burba, G. G. and Suyker, A. E. (2006). Relationship between gross primary production and chlorophyll content in crops: Implications for the synoptic monitoring of vegetation productivity. *J. Geophys. Res.* 111(D8). doi: 10.1029/2005JD006017.

Guo, Q. H., Wu, F., Hu, T., Chen, L., Liu, J., Zhao, X., Gao, S. and Pang, S. (2016). Perspectives and prospects of unmanned aerial vehicle in remote sensing monitoring of biodiversity. *Biodivers. Sci.* 24(11), 1267–1278. doi: 10.17520/biods.2016105.

Habyarimana, E. and Baloch, F. S. (2021). Machine learning models based on remote and proximal sensing as potential methods for in-season biomass yields in commercial sorghum fields. *PLoS ONE* 16(3), e0249136. doi: 10.1371/journal.pone.0249136.

Hansen, P. M. and Schjoerring, J. K. (2003). Reflectance measurement of canopy biomass and nitrogen status in wheat crops using normalized difference vegetation indices and partial least squares regression. *Remote Sens. Environ.* 86(4), 542–553. doi: 10.1016/S0034-4257(03)00131-7.

Hegarty-Carver, M., Polly, J., O'Neil, M., Ujeneza, N., Rineer, J., Beach, R. H., Lapidus, D. and Temple, D. S. (2020). Remote crop mapping at scale: Using satellite images and unmanned aerial vehicle-acquired data as ground truth. *Remote Sens.* 12: 1984.

Holland, K. H. and Schepers, J. S. (2013). Use of a virtual-reference concept to interpret active crop canopy sensor data. *Precis. Agric.* 14(1), 71–85. doi: 10.1007/s11119-012-9301-6.

Huang, X., Cao, Y. and Li, J. (2020). An automatic change detection method for monitoring newly constructed building areas using time-series multi-view high-resolution optical satellite images. *Remote Sens. Environ.* 244, 111802. doi: 10.1016/j.rse.2020.111802.

Hunt, L. and Daughtry, C. (2018). What good are unmanned aircraft systems for agricultural remote sensing in precision agriculture? *J. Remote Sens.* 39(15–16), 5345–5376.

ISPA. (2022). International society of precision agriculture. *Precision AG Definition*. Available at: https://www.ispag.org/about/definition (accessed 20 May 2022).

Jacques, A. A. B., Adamchuk, V. I., Park, J., Cloutier, G., Clark, J. J. and Miller, C. (2021). Towards a machine vision-based yield monitor for the counting and quality mapping of shallots. *Front. Robot. AI* 8, 41. doi: 10.3389/frobt.2021.627067.

Jensen, J. R. (1996). *Introductory Digital Image Processing: A Remote Sensing Perspective*. Prentice Hall, Upper Saddle River, p. 316.

Jensen, J. R. (2009). *Sensoriamento Remoto do Ambiente: Uma perspectiva Em Recursos Terrestres*. Parêntese, São José dos Campos, Brazil, p. 598.

Kamilaris, A., Kartakoullis, A. and Prenafeta-Boldú, F. X. (2017). A review on the practice of big data analysis in agriculture. *Comput. Electron. Agric.* 143, 23–37. doi: 10.1016/j.compag.2017.09.037.

Khanal, S., Klopfenstein, A., Kc, K., Ramarao, V., Fulton, J., Douridas, N. and Shearer, S. A. (2021). Assessing the impact of agricultural field traffic on corn grain yield using remote sensing and machine learning. *Soil Till. Res.* 208, 104880. doi: 10.1016/j.still.2020.104880.

Kumar, R. and Silva, L. (1973). Light ray tracing through a leaf cross section. *Appl. Opt.* 12(12), 2950–2954. doi: 10.1364/AO.12.002950.

Kwak, S. K. and Kim, J. H. (2017). Statistical data preparation: Management of missing values and outliers. *Korean J. Anesthesiol.* 70(4), 407–411. doi: 10.4097/kjae.2017.70.4.407.

Lichtenthaler, H. K. and Buschmann, C. (2001). *Chlorophylls and Carotenoids: Measurement and Characterization by UV-Vis Spectroscopy, Current Protocols in Food Analytical Chemistry*. John Wiley and Sons, New York, NY, pp. F4.3.1–F4.3.8.

Lira, C. (2016). Sistemas de informação geográfica: análise de dados de satélite, Lisboa, Portugal. Available at: https://www.researchgate.net/publication/312383752_Sistemas_de_Informacao_Geografica_Analise_de_Dados_de_Satelite (accessed 26 March 2022).

Long, D. S. (2019). Site-specific nutrient management systems. In: Stafford, J. (Ed.), *Precision Agriculture for Sustainability*. Burleigh Dodds Science Publishing, Cambridge, UK, pp. 299–322.

Lorenzzetti, J. A. (2015). *Princípios Físicos de Sensoriamento Remote*. Blucher, São Paulo, Brazil, p. 293.

Maestrini, B. and Basso, B. (2018). Predicting spatial patterns of within-field crop yield variability. *Field Crops Res.* 219(1), 106–112. doi: 10.1016/j.fcr.2018.01.028.

Marchi, J. de (2021). Active Sensor Algorithm Approach to Optimize Nitrogen Rate Fertilization in Cotton Production. *Preprints*. doi: 10.20944/preprints202110.0183.v1.

Martins, V. S. (2019). *Introdução Ao Sensoriamento Remoto de Sistemas Aquáticos: Principios e Aplicações* (1st edn.). INPE, São José dos Campos, Brazil. Available at: http://www.dpi.inpe.br/labisa/livro/res/conteudo.pdf (accessed 22 February 2022).

McMillen, D. P. (2010). Issues in spatial data analysis. *J. Reg. Sci.* 50(1), 119–141. doi: 10.1111/j.1467-9787.2009.00656.x.

Meneses, P. R. and Alemida, T. de (2012). Introdução ao Processamento de Imagens em Sensoriamento Remoto, Brasília, Brazil. Available at: https://docero.com.br/doc/ss1x1x5 (accessed 23 March 2022).

Mobley, C. D. (1994). *Light and Water: Radiative Transfer in Natural Waters*. Academic Press, San Diego, p. 592.

Molin, J. P., Amaral, L. R. and Colaço, A. F. (2015). *Agricultura de Precisão*. Oficina de Textos, São Paulo, pp. 119–153.

Moreira, M. A. (2011). *Fundamentos de Sensoriamento Remoto e Metodologias de Aplicação*. 4 Universidade Federal de Viçosa (UFV), Viçosa, Brazil, p. 422.

Morelli-Ferreira, F., Maia, N. J. C., Tedesco, D., Kazama, E. H., Carneiro, F. M., Santos, L. B., Junior, G. F. S., Rolim, G. S., Shiratsuchi, L. S. and Silva, R. P. (2021). Comparison of machine learning techniques in cotton yield prediction using satellite remote sensing. *Preprints* 1, 1–17. doi: 10.20944/preprints202112.0138.v2.

Mulla, D. J. (2013). Twenty-five years of remote sensing in precision agriculture: Key advances and remaining knowledge gaps. *Biosystems Engineering* 114(4), 358–371.

Narvaez, F. Y., Reina, G., Torres-Torriti, M., Kantor, G. and Cheein, F. A. (2017). A survey of ranging and imaging techniques for precision agriculture phenotyping. *IEEE ASME Trans. Mechatron.* 22(6), 2428–2439. doi: 10.1109/TMECH.2017.2760866.

Novo, E. M. L. M. (2008). *Sensoriamento Remoto: Princípios e Aplicações*. Blucher, São Paulo, Brazil, p. 363.

Novo, E. M. L. de M. (2010). Sensoriamento remoto. In: *Princípios e aplicações* (Vol. 4). Blucher, São Paulo, Brazil. Available at: https://www.atenaeditora.com.br/wp-content/uploads/2018/10/E-book-Aplica%C3%A7%C3%B5es-e-Princ%C3%ADpios-do-Sensoriamento-Remoto-1.pdf (accessed 10 March 2022).

Pantazi, X., Moshou, D., Mouazen, A. M., Alexandridis, T. and Kuang, B. (2015). Data fusion of proximal sensing and remote sensing for the delineation of management zones in arable precision farming. In: Anon (Ed.), *Proceedings of the Seventh International Conference of Information and Communication Technologies in Agriculture, Food and the Environment*, Kavala, Greece.

Paranhos Filho, A. C. (2021). Chapter 3: Resolução de imagens de satelite. In: Filho, P. (Ed.), *Geotecnologia para Aplicaçao Ambiental*. PR Uniedusul, Maringá, Brazil, pp. 25–59.

Pedersen, S. M., Medici, M., Anken, T., Tohidloo, G., Pedersen, M. F., Carli, G., Canavari, M., Tsiropoulos, Z. and Fountas, S. (2019). Financial and environmental performance of integrated precision farming systems. In: Stafford, J. (Ed.), *Agriculture*. Wageningen Academic Publishers, Wageningen, The Netherlands, pp. 833–839.

Peerlinck, A., Sheppard, J. and Maxwell, B. (2018). Using deep learning in yield and protein prediction of winter wheat based on fertilization prescriptions in precision agriculture. In: *International Conference on Precision Agriculture (ICPA)*, Canada.

Pizarro, M. A., Epiphanio, J. C. N. and Galvão, L. S. (2001). Caracterização mineralógica de solos tropicais por sensoriamento remoto hiperespectral. *Pesq. Agropec. Bras.* 36(10), 1277–1286. Available at: https://www.scielo.br/j/pab/a/Wqsh4D8XyKN5JGyfwJBC7mc/?lang=pt (accessed 14 February 2022).

Queiroz, D. Md, Coelho, A. LdF., Valente, D. S. M. and Schueller, J. K. (2020). Sensors applied to digital agriculture: A review. *Rev. Cienc. Agron.* 51(5), e20207751. doi: 10.5935/1806-6690.20200086.

Roslim, M. H. M., Juraimi, A. S., Che'Ya, N. N., Sulaiman, N., Manaf, M. N. H. A., Ramli, Z. and Motmainna, M. (2021). Using remote sensing and an unmanned aerial system for weed management in agricultural crops: A review. *Agronomy* 11(9), 1809. doi: 10.3390/agronomy11091809.

Ruan, G., Li, X., Yuan, F., Cammarano, D., Ata-Ul-Karim, S. T., Liu, X., Tian, Y., Zhu, Y., Cao, W. and Cao, Q. (2022). Improving wheat yield prediction integrating proximal sensing and weather data with machine learning. *Comput. Electron. Agric.* 195(1), 106852. doi: 10.1016/j.compag.2022.106852.

Ruß, G. and Brenning, A. (2010). Data mining in precision agriculture: Management of spatial information. In: *International Conference on Information Processing and Management of Uncertainty in Knowledge-Based Systems*. Springer, pp. 350–359. Available at: https://link.springer.com/chapter/10.1007/978-3-642-14049-5_36 (accessed 28 May 2022).

Santos, A. F., Corrêa, L. N., Lacerda, L. N., Tedesco-Oliveira, D., Pilon, C., Vellidis, G. and da Silva, R. P. (2021). High-resolution satellite image to predict peanut maturity variability in commercial fields. *Precis. Agric.* 22(5), 1464–1478. doi: 10.1007/s11119-021-09791-1.

Schellberg, J., Hill, M. J., Gerhards, R., Rothmund, M. and Braun, M. (2008). Precision agriculture on grassland: Applications, perspectives and constraints. *Eur. J. Agron.* 29(2–3), 59–71. doi: 10.1016/j.eja.2008.05.005.

Schielack, V. and Thomassen, J. (2016). Harvester-based sensing system for cotton fiber quality. *J. Cotton Sci.* 20, 393.

Schrader, S. and Pouncey, R. (1997). *Erdas Field Guide. 4*. Erdas Inc., Atlanta, USA, p. 656.

Shiratsuchi, L. S. (2014). Sensoriamento Remoto: conceitos básicos e aplicações na Agricultura de Precisão. In: Bernardi, A. C. C. (Ed.), *Agricultura de Precisão: Resultados de um Novo Olhar* (Vol. 4). EMBRAPA, Brasília, Brazil, pp. 58–73.

Singh, G., Reynolds, C., Byrne, M. and Rosman, B. (2020). A remote sensing method to monitor water, aquatic vegetation, and invasive water hyacinth at national extents. *Remote Sens.* 12(24), 4021. doi: 10.3390/rs12244021.

Solari, F., Shanahan, J., Ferguson, R., Schepers, J. and Gitelson, A. (2008). Active sensor reflectance measurements of corn nitrogen status and yield potential. *Agron. J.* 100(3), 571–579. doi: 10.2134/agronj2007.0244.

Spekken, M. (2013). A simple method for filtering spatial data. *Precision Agriculture* 13, 259–266. Springer. Available at: https://www.agriculturadeprecisao.org.br/wp-content/uploads/2019/11/cgr-2013_07.pdf (accessed 28 May 2022).

Steven, M. D. (2003). Ground truth: An underview. *Int. J. Remote Sens.* 8(7), 1033–1038.

Sudduth, K. A. and Drummond, S. T. (2007). Yield editor: Software for removing errors from crop yield maps. *Agron. J.* 99(6), 1471–1482. doi: 10.2134/agronj2006.0326.

Sui, R., Thomasson, J. A. and Ge, Y. (2012). Development of sensor systems for precision agriculture in cotton. *Int. J. Agric. Biol. Eng.* 5(4), 1–14.

Tedesco, D., de Oliveira, M. F., dos Santos, A. F., Silva, E. H. C., de Souza Rolim, G. and da Silva, R. P. (2021). Use of remote sensing to characterize the phenological development and to predict sweet potato yield in two growing seasons. *Eur. J. Agron.* 129, 126337. doi: 10.1016/j.eja.2021.126337.

Tedesco-Oliveira, D., Pereira da Silva, R., Maldonado, W. and Zerbato, C. (2020). Convolutional neural networks in predicting cotton yield from images of commercial fields. *Comput. Electron. Agric.* 171(1), 105307. doi: 10.1016/j.compag.2020.105307.

Tefera, A. T., Banerjee, B. P., Pandey, B. R., James, L., Puri, R. R., Cooray, O., Marsh, J., Richards, M., Kant, S., Fitzgerald, G. J. and Rosewarne, G. M. (2022). Estimating early season growth and biomass of field pea for selection of divergent ideotypes using proximal sensing. *Field Crops Res.* 277, 108407. doi: 10.1016/j.fcr.2021.108407.

Teixeira, A. A. D., Mendes Júnior, C. W., Bredemeier, C., Negreiros, M. and Aquino, RdS. (2020). Evaluation of the radiometric accuracy of images obtained by a sequoia multispectral camera. *Eng. Agric.* 40(6), 759-768. doi: 10.1590/1809-4430-eng.agric.v40n6p759-768/2020.

The Royal Society. (2016). *Reaping the Benefits: Science and the Sustainable Intensification of Global Agriculture*. Available at: http://royalsociety.org/Reapingthebenefits (accessed 12 December 2016).

Thenkabail, P. S., Smith, R. B. and De Pauw, E. (2000). Hyperspectral vegetation indices and their relationships with agricultural crop characteristics. *Remote Sens. Environ.* 71(2), 158–182. doi: 10.1016/S0034-4257(99)00067-X.

USGS: United States Geological Survey. (2018). Landsat 8 OLI (operational land Imager) and TIRS (thermal infrared sensor). Available at: https://lta.cr.usgs.gov/L8 (accessed 28 March 2022).

Van Der Meer, F. D., Van Der Werff, H. M. A. and Van Ruitenbeek, F. J. A. (2014). Potential of ESA's Sentinel-2 for geological applications. *Remote Sens. Environ.* 148, 124–133. doi: 10.1016/j.rse.2014.03.022.

Viña, A., Gitelson, A. A., Nguy-Robertson, A. L. and Peng, Y. (2011). Comparison of different vegetation indices for the remote assessment of green leaf area index of crops. *Remote Sens. Environ.* 115(12), 3468–3478. doi: 10.1016/j.rse.2011.08.010.

Viscarra Rossel, R. A., McKenzie, N. J., Adamchuk, V. I., Sudduth, K. A. and Lobsey, C. (2011). Proximal soil sensing: An effective approach for soil measurements in space and time. *Adv. Agron.* 113, 237–282.

Weiss, M., Jacob, F. and Duveiller, G. (2020). Remote sensing for agricultural applications: A meta-review. *Remote Sens. Environ.* 236, 111402, doi: 10.1016/j.rse.2019.111402.

Woodard, J. (2018). *Aganalytics: Webplatform*. Personal Communication. Available at: http://analytics.ag.

Xu, R., Li, C., Paterson, A. H., Jiang, Y., Sun, S. and Robertson, J. S. (2018). Aerial images and convolutional neural network for cotton bloom detection. *Front. Plant Sci.* 8(1), 17. doi: 10.3389/fpls.2017.02235.

Yang, C. (2019). Airborne and satellite remote sensors for precision agriculture. In: Stafford, J. (Ed.), *Precision Agriculture for Sustainability*. Burleigh Dodds Science Publishing, Cambridge, UK.

Yeom, J., Jung, J., Chang, A., Maeda, M. and Landivar, J. (2018). Automated open cotton bol detection for yield estimation using unmanned aircraft vehicle (UAV) data. *Remote Sens.* 10(12), 1895. doi: 10.3390/rs10121895.

Yuan, G., Wang, Y., Zhao, F., Wang, T., Zhang, L., Hao, M., Yan, S., Dang, L. and Peng, B. (2021). Accuracy assessment and scale effect investigation of UAV thermography for underground coal fire surface temperature monitoring. *Int. J. Appl. Earth Obs. Geoinf.* 102, 102426. doi: 10.1016/j.jag.2021.102426.

Zanotta, D. C., Ferreira, M. P. and Zortea, M. (2019). Chapter 1: Iniciação aos dados de sensoriamento remoto. In: Zanotta, D. C., Ferreira, M. P. and Zortea, M. (Eds.), *Processamento de Imagens de Satélite*. Oficina de Texto, São Paulo, Brazil, pp. 11–63.

Zhang, C. and Kovacs, J. M. (2019). The use of unmanned aerial systems (UASs) in precision agriculture. In: Stafford, J. (Ed.), *Precision Agriculture for Sustainability*. Burleigh Dodds Science Publishing, Cambridge, UK.

Zheng, H., Zhou, X., He, J., Yao, X., Cheng, T., Zhu, Y., Cao, W. and Tian, Y. (2020). Early season detection of rice plants using RGB, NIR-G-B and multispectral images from unmanned aerial vehicle (UAV). *Comput. Electron. Agric.* 169, 105223. doi: 10.1016/j.compag.2020.105223.

Index

www.ingramcontent.com/pod-product-compliance
Ingram Content Group UK Ltd.
Pitfield, Milton Keynes, MK11 3LW, UK
UKHW021826100926
8556IPUK00003B/170

9 781786 769770